The Best of Time
ROLEX
WRISTWATCHES
An Unauthorized History

James M. Dowling & Jeffrey P. Hess

77 Lower Valley Road, Atglen, PA 19310

Fly leaf sculpture copyrighted by John Read and used with his permission.

Library of Congress Cataloging-in-Publication Data

Dowling, James M.
    Rolex wristwatches: The best of times: an unauthorized history/ James M. Dowling & Jeffrey P. Hess.
      p. cm.
    Includes index.
    ISBN 0-7643-0011-3
      1. Montres Rolex S.A. 2. Wrist watches–Switzerland–History. 3. Wrist watches–Switzerland–Collectors and collecting. I. Hess, Jeffrey P. II. Title.
TS543.S9068 1996
681.1'14'09492–dc20   96-9370
                                                CIP

Copyright © 1996 by Write Time Partners II

All rights reserved. No part of this work may be reproduced or used in any forms or by any means–graphic, electronic or mechanical, including photocopying or information storage and retrieval systems– without written permission from the copyright holder.

None of the companies referred to in this book either authorized or furnished or approved of any of the information contained therein. This book is derived from the authors' independent research.

Printed in Hong Kong.
ISBN: 0-7643-0011-3

Published by Schiffer Publishing, Ltd.
77 Lower Valley Road
Atglen, PA 19310
Phone: (610) 593-1777
Fax: (610) 593-2002
Please write for a free catalog.
This book may be purchased from the publisher.
Please include $2.95 for shipping.
Try your bookstore first.

We are interested in hearing from authors with book ideas on related subjects.

AUTHORS' NOTES

This book attempts to tell the story of the Rolex watch and the men who brought it to its pre-eminent position in the world of wristwatches. It is, however, a completely unauthorized book. The authors have received no help in any way from Rolex or any of its associated companies or any serving employee. Although many ex-employees have been of inestimable help.

The following are trade marks of the Rolex Watch Company, Limited, Geneva, Switzerland: Rolex, the coronet logo, Oyster, Oysterquartz, Cosmograph, Perpetual, Datejust, Day-Date, GMT-Master, Sea-Dweller, Explorer, Submariner, Tridor, Jubilee, Milgauss, Air-King, Cellini, Tudor, Tudor Prince, Tudor Princess, Tudor Day-Date, Mini-Sub, Lady-Sub, Tudor Monarch, and the Tudor shield logo and we acknowledge the intellectual and proprietary rights of the company and all their subsidiary companies and distributors.

CONTENTS

	Foreword	5
	Acknowledgments	6
	Introduction	7
Chapter 1	Forty Years On	8
Chapter 2	Oyster	40
Chapter 3	The Perpetual	76
Chapter 4	Someday My Prince Will Come	110
Chapter 5	They Also Serve...Other Models	134
Chapter 6	Chronometers	150
Chapter 7	The 1940s: Out of the Empire into the Fire	158
Chapter 8	The 1950s: 250,000 Chronometers Can't Be Wrong	190
Chapter 9	The Chronograph: Stopping Time	214
Chapter 10	A Date(just) with destiny	232
Chapter 11	The Explorer	240
Chapter 12	The Submariner and the Sea-Dweller: The Stock Car, The Diver, and The Gas Escape Valve	254
Chapter 13	The Kew "A": The Rarest Rolex	270
Chapter 14	The GMT Master: The Correct Time Times Two	276
Chapter 15	The President or Day-Date: Attaining a New Summit	280
Chapter 16	The Quartz Revolution	284
Chapter 17	The Lady's Rolex	288
Chapter 18	Bracelets: Around the wrists and Around the World	302
Chapter 19	Movements: Under the Covers	314
Chapter 20	From Merchant to Market Leader	326
Chapter 21	Oddities: WSH	330
Chapter 22	Patents: Patently Superior	340
Chapter 23	Collecting Rolex Watches	350
Chapter 24	Fakes: Imitation is the Nastiest Form of Flattery	364
Appendix 1	A 1932 visit to the Rolex factory	374
Appendix 2	List of names & trademarks	379
Appendix 3	Oyster case numbers	381
Appendix 4	Notes & Queries	382
	Bibliography	383
	Price Guide	385
	Subject Index	390
	Index to Model Numbers in Photographs	391
	Index to Model Numbers in the Text	392

This book, like everything, is for Minda.
—James Dowling

To Mr. Paul Graehling, a true collector, a true friend, and in many ways my mentor. I am most grateful for his guidance, kindness, and friendship over the years.
—Jeffrey P. Hess

FOREWORD

It is an honor to be associated with the writing of this book on the history of the Rolex Watch Company. Its contents have been carefully researched and I believe strongly that such a famous and successful company should have its history recorded. Personally, having been part of Rolex for 30 years in all its different departments, from working at the bench to managing an international division. Having had the privilege of knowing and working for its founder, Hans Wilsdorf and for his successor, André J. Heiniger, I can only congratulate the writers and publishers for undertaking such a task. It is a tribute to all who have made such an unusual international giant in the horological world. They are definitely "A Landmark in the History of Time Measurement."

Every successful company starts with a vision, a dream - but a dream is not enough. Talent, tenacity and strength of character are required to triumph. The start of this success was on March 22, 1881 when Hans Wilsdorf was born in Kulmbach, Bavaria. This "gentle giant" was the son of a merchant and had no technical background, but he was destined for a brilliant future. Orphaned at the age of 12, he left Bavaria for Switzerland seven years later and, while serving as an interpreter and correspondent for Cuno, Korten (a watch and clock exporter), he became fascinated with the horological business. Hans Wilsdorf foresaw the trend in watches. He was so sure that wristwatches, rather than pocket watches were the obvious future of the industry that he invested every effort and all his money in developing, promoting and selling them. He realized the potential of a small, accurate and reliable watch strapped to the wrist. During his endeavors his greatest enemies were the Great Depression, two World Wars, and currency devaluations. However his inbred resilience turned desperation into victory. Where others failed, he succeeded.

Mr. Wilsdorf was very fond of recounting the past while sitting around the tea table on a Sunday afternoon. Many were the happy occasions that I joined him. He had a wonderful and forceful way of conveying his thoughts to others, but was also a good listener. He had deep, melancholy eyes which spoke of both sadness and happiness. He loved to travel and visit his agencies and friends around the world, to the extent that he was extremely well versed in the international market. He was a kind, fatherly man who would always give his employees a second chance if necessary.

During the 1950s a great change had to take place as a new leader was needed to modernize and prepare for the future. Mr. Wilsdorf was aging and his health was becoming a cause for concern. He had prepared, during his lifetime, a solid platform from which could be launched an even greater company. On July 6, 1960 Hans Wilsdorf passed away and his successor André J. Heiniger took the reins and lost no time in reorganizing and preparing for the future.

Rolex grew from being a mysterious, little-known company to a world leader in its field. Personally I would be tempted to suggest that a history of Rolex should comprise two volumes - the pioneering years under the founder Mr. Wilsdorf and the later success story under Mr. Heiniger.

This book gives an honest representation of the history of this great company, with its failings as well as its successes. Almost my entire working life was with this company and I believe that it will go on ever successfully so long as it adheres to the tenets laid down by its only two managing directors, an asset which any company today would envy.

John David Read OBE, FBHI, Dip. Fed. Horo. Geneva
Former Managing Director, Rolex (Far East)

ACKNOWLEDGEMENTS

This book would not have been possible without the help, guidance, advice and encouragement of countless people; we would like to list as many of them as possible but if space and our memories run out please excuse us.

Firstly we would like to thank John Read OBE, FBHI, former Managing Director of Rolex (Far East) whose encyclopedic knowledge of the company was vital; Richard Good, John Leopold and David Thompson, the three most recent assistant keepers of the Horology Department at the British Museum for their help and encouragement; Gregory A. Brigham, Nikki Ahlefeld, Paul Denicolo, Ed Hess, and Shery Lundy of Hess Fine Art; Daryn Schnipper and Barbara Feldspaugh of Sotheby's New York; Tina Millar and Johnathan Durracott of Sotheby's London; Roger Lister and Tim Bourne of Christie's London; Stuart Pollitt and Sue Osborne of the NPL; Uwe Kroeger of ZDF German Television; Werner Eschenbacher, Oberstudiendirektor, Hans Wilsdorf Schule, Kulmbach, Germany; Michael Wood, Merseyside; Stewart Foreman, San Francisco; Yuki Shirakawa; Paul Grzeszczak; Andy Lunt and Peter Cooper at Ripping Image, all from London; and last but not least Peter Schiffer and Douglas Congdon-Martin at Schiffer Publishing Ltd. for leaning on us just the correct amount.

We would also like to thank all the collectors and dealers who graciously provided watches for photography;

Keith Haltom, Dallas; Gianni Paolini, Firenze; Steve Dubinsky, Hartsdale; Pete Davis, Kansas City; Bruce Ellison, Knoxville; Giuseppe and Dan Pizzigoni, London; Graham Simmonds, London; Joa Baptista, London; John Cole, London; Malcolm Hammond, London; Nick Urul, London; Nino Santi, London; Pascall Isbell, London; Martyn Ralph, London; Trevor Waite, London; Giacomo Rapacioli, London and Lucca; Malcolm and Kelley McDowell, Los Angeles; Richard Glassman, Memphis; Isami Shiroma, New York City; Jay Cocks, New York City; Jeff Morris and Peter Fosner, New York City; Richard Paradies, Atlanta; Norifumi Ôsawa, Gifu; Laup and Fej Armah, Phoenix; Tom Rumpf, Ohio; Paul Graehling, Polo, Illinois; Maurizio and Loretta, Rome; Jeffrey Mattson, San Francisco; Doug Giard, St. Clair Shores; Kevin O'Keefe, St. Hellier; Carl Schway, St. Paul; James Watters, St. Petersburg, Florida; Kesaharu Imai, Tokyo; Akira Ishino, Tokyo; Steve Oltuski, Toronto; Kenichi Kurisaki, Yokosuka.

The front cover photograph is by Ray Smart, London, and the back cover photograph is by Ed Parinello, New York. The main body of photography is by Ray Smart and the authors. Flyleaf photograph provided by John Read.

INTRODUCTION

In 1990 the respected international branding consultants, Interbrand Group plc, produced a survey of the world's most well known and successful brands. Nestled among the giant multi-nationals such as Marlboro, Coca-Cola, Mercedes-Benz, and Kodak one found Rolex, the only watch company in the top 50 brands.

It is, however, doubtful if more than a handful of the people who read the survey knew that Rolex was founded in turn-of-the-century London, by an exiled German pearl dealer and his English partner. The story of where Rolex came from and how it attained its pre-eminent position is the story this book will attempt to tell.

Gentleman's 13''' sterling silver Rebberg watch with metal sunburst finish dial, sunken seconds, and stylized Arabic numerals, hinged bezel and rear. Notice the drum shaped bezel on the case and hinged lugs.
Circa 1924

Chapter 1
FORTY YEARS ON

It was November 24, 1945. The war in Europe had been over for less than six months, and Hans Wilsdorf was looking to the future. Tonight he would celebrate the launch of two important new models and a whole new brand, but not before paying a little respect to the past. The heads of Rolex from countries all over Europe would join him at the Hotel des Bergues in Geneva to celebrate 40 years, or the ruby jubilee of Rolex.

In fact it was not the birthday of Rolex, but the fortieth anniversary of the founding of a little known company which had lasted only ten years and one whose only lasting legacy was a small brass nameplate discreetly displayed in the foyer of the Geneva head office of Rolex. The plaque had been removed from the door of 83, Hatton Garden, London. It had announced to the world that this was the headquarters of a company called "Wilsdorf & Davis Ltd." Now, forty years after the plaque had been screwed to that door in London, one of those two men was celebrating not only forty years in business but also his firm's recent entry into the United States and, most importantly, his company's emergence from the shadows of a world war into the promise of peace.

The stories of Hans Wilsdorf and Rolex are inextricably bound. Wilsdorf had dominated the company he founded. Every patent up until 1930 was in his name and every Rolex advertisement[1] from the very first also bore his name and continued to do so until over a year after his death. So it would be futile to look at the history of the company that he founded without examining the man himself.

Hans Eberhard Wilhelm Wilsdorf was born in Kulmbach, a tiny village 50 kilometers from Nuremberg in Bavaria on March 22, 1881 at 11:30 pm. Today Kulmbach, connected by fast roads, is close enough to be within commuting distance of Nuremberg. But when Hans was born, the second son to Ferdinand and Anna (nee Maisel) Wilsdorf, the village was a day's horseback ride away. Hans grew up thinking of himself as Bavarian, not German. It had been only ten years before, in 1871, that Chancellor Bismarck finally convinced the Bavarian royal family to join Prussia in forming the greater German Reich (or empire).

Wilsdorf came from a family of ironmongers, a business initially founded by his grandfather Carl Traugger Wilsdorf in 1842 when he moved from his native Saxony to Bavaria. It was natural that the young Wilsdorf would look forward to going into the family business with its large shop standing proudly in the market square of Kulmbach[2]. But when he was only twelve years old, tragedy struck; his parents died within months of each other. Wilsdorf and his siblings were left in the trust of other family members, the other partners in the family business. For reasons that can only be guessed at, the family decided, then and there, to sell the business. The proceeds were put in a trust for the Wilsdorf heirs until their majority.

Young Hans was sent to a boarding school in Coburg, which was a slightly larger town only 25 miles away. We would venture to suggest that this did not go down too well with young Hans, for as soon as he was old enough and before he had achieved his majority (and received his inheritance) he left Kulmbach, Bavaria and Germany to seek a new life abroad.

It is difficult for us now to imagine the pain Hans went through during these formative years. We think of him as Mr. Rolex but forget that he once was a 12 year-old boy who had been brought up in a small provincial town where his family were well-respected merchants, the second generation to run the store. He would have been what we now call upper middle class with a strong likelihood of having either taking over the family business or entering a profession. Everything in his life had been planned out, then it all fell apart.

While at the boarding school in Coburg he put all his energies into schoolwork and excelled at math and languages (particularly English). He also developed an interest in Switzerland through a friendship with a Swiss fellow pupil. When he left school his first move was to Geneva, Switzerland, where he found work in the export department of a major pearl merchant with business throughout the world. Through its agents this company would buy pearls from fishermen all over the world. At its headquarters in Geneva, it would grade and sort them before selling matched pearls to an international market of jewelers. The company manufactured nothing, it simply used its connections and knowledge to profit wisely from the demands of the world's women for this most fashionable jewelry item.

After learning as much as he could, Wilsdorf left and joined the watch exporters Cuno, Korten. Hans was employed there as a correspondent at an annual salary of 80 Swiss Francs, his job being, simply, to answer letters. But he was much more than a secretary or a clerk. In a time when every letter was hand-written, Hans communicated in English to the company's clients in the British Empire and the U.S., who at that time were the richest in the world. In this position he would have had knowledge of price negotiations conducted by the agents and who the customers were. Most importantly for his future development, he would have gained the knowledge that you didn't have to make something to make money.

It is worth taking a moment off to understand why a boy from a small German village should be employed by a Swiss company as a correspondent in English. Despite the following unpleasantries over the next few years, Germany and Britain were, in the late nineteenth and early twentieth centuries, the closest that the two countries have ever been. The German Prince Albert had married Queen Victoria and then half of her children had gone on to marry into other German princely families. It was only 50 years since the English and German armies had fought together at Waterloo to finally defeat the scourge of Europe, Napoleon. Britain and Germany were bound together in 1875 as closely as Britain and the U.S. were in 1945 when the two countries were fighting a war together and the British leader, Churchill, had an American mother.

Wilsdorf took time away from his responsibilities to learn more about the construction and timing of watches and somewhere during the three years he spent at Cuno Korten he made up his mind to make watches his profession. By coming to work before everyone else and leaving after they had gone home, Hans was able to spend time learning about watches. He liked to experiment by winding and setting all the watches in a batch, then comparing them for accuracy the following morning. He would wind watches so often, that from this point until his death, he had a large callus on his right thumb. One day he took three of the most accurate watches and submitted them to the Neuchatel Observatory; he had exceeded his authority but he was sure the watches would pass the strict observatory tests. When they did, his superiors at Cuno Korten were delighted and used the results in their advertising and promotion; Hans watched and learned.

It is difficult not to draw the conclusion that young Hans felt that he had something to prove to the world. Once again he chose to change countries; this time to England. The decision to move to London in 1903, when he was still only 22, was influenced, no doubt, by the amount of business he had seen being transacted through London during his years at Cuno, Korten.

While in Geneva Hans had seen the increasing industrialization of the Swiss watch industry at the end of the nineteenth century and begun to realize that perhaps there was an international market for these small, accurate (for their time) and inexpensive watches.

When he arrived in London, Wilsdorf once again worked for a watchmaking company, the name of which is unknown, but stayed with it for less than two years. At the age of 24 Hans was ready to go into business for himself, with a friend as his partner. Wilsdorf met Alfred James Davis through the ministrations of Wilsdorf's solicitor (lawyer) who knew that Davis had money to invest and that Wilsdorf was looking for a partner. The meeting took place in the solicitor's London club and the two seem to have hit it off almost immediately. The relationship was further strengthened when Davis married Wilsdorf's younger sister.

The company called Wilsdorf and Davis was founded and commenced trading in 1905. Hans invested his inheritance from the family trust and also used some money from his brother and sister, allowing him to match Davis's investment. They began as equal partners. They traded from offices at 83, Hatton Gardens; the heart of London's jewelry world, but within 18 months they had moved to 44 Holborn Viaduct, less than 500 yards from Hatton Gardens and less than a mile from Clerkenwell, then as now the center of the English watchmaking trade.

The business was, at the time, one of simply importing movements from Jean Aegler's ebauche factory in the Rebberg, Bienne, and cases and dials from other Swiss suppliers. The watches would then be tested for performance by English watchmakers (the reason for the company's decision to operate so close to Clerkenwell) before being cased and shipped to their customers, many of whom were just down the road in Hatton Gardens. The watches were unsigned on the dials, allowing the retailer to place his own name there. On the movement was the simple mark W&D, which was also repeated on the inside of the case back, standing, of course, for Wilsdorf and Davis.

When the company began, only two watches were traded, pocket and purse (or portfolio) models. Reading an account of the recent Boer War, Wilsdorf came across a mention of the advantages of wristwatches in combat.[3] At once, Hans saw a small niche in the market and decided to exploit it. He would specialize in wristwatches. Most of the major watch companies already made some but they were regarded as a passing fancy, not something serious or lasting. Hans reckoned that there was room for a company specializing just in wristwatches and if they were to be just a passing phase, what was wrong with exploiting that phase for as long as it lasted?

While living in Switzerland and working at Cuno, Korten, Wilsdorf had learned of the Jean Aegler company based in Bienne. They had been producing small sized watch movements for over 20 years and had succeeded in selling them all over the world. Hans went to them and gave them an initial order for movements to fit into ladies' and men's wristwatches.

These new products proved very successful, particularly in the far flung dominions of Australia and India. The success convinced Hans and Alfred that there really was a demand for the wristwatch and so in 1912 he returned to Bienne and negotiated with Jean Aegler's son Hermann for a contract in which Aegler would supply the young company with movements to a value of £125,000 ($500,000 at the time). This sum was five times the value of the capital of Wilsdorf and Davis and the contract was by far the largest for wristwatch movements ever signed. Wilsdorf and Davis were now totally committed to the production of wristwatches; the die was cast.

In 1884 George Eastman introduced a new camera which promised to simplify the whole procedure of photography, it was pre-loaded at the factory with a 100 exposure roll and after use the whole thing was then returned to the factory for processing and then returned to the customer with fresh film along with the finished photographs. He did something that may be much more important than just inventing that camera: he decided against calling it the Eastman Model 25 and invented a new word for it. He called it the Kodak and he later explained the name this way. "I knew a trade name must be short, vigorous, incapable of being misspelled to an extent that will destroy its identity, and, in order to satisfy the trademark laws, it must mean nothing." It was the first invented word, it meant nothing in any language and sounded the same in all of them. It set the company and its new product apart.

Before long everybody was at the game, trademarks, logo styles and invented words became the rage. In 1908, Hans and Alfred decided to join in. The fruit of their efforts was the word "Rolex," first registered in Chaux de Fonds, Switzerland on July 2, 1908 and four years later in London on July 6, 1912. We do not now know the origins of the word; the partners left no notes. In fact we do not even know if it was either or both of them who chose it. The oft repeated explanation that it was a contraction of the French phrase "Horlogerie Exquisite" can be fairly quickly dismissed, as neither of the partners were native French speakers and it was to be another ten years before the partners ever even entered the French market. What can be said is that, in English, it has the sound of movement; and that of course is what you want to imply with a watch. In fact, at around the same time, Isaac Dittisheim was also inventing a watch trade mark name and he also used the sound of movement for his: Movado.

The invention of a brand name for the product allowed the partners to differentiate their product from all others. They began to mark the watches with the new trademark, first on the movement only and then on the inside of the case back. They had just commenced this program when the sky fell in on them and their new company.

A Serbian nationalist named Gavrillo Prinçip murdered the Austrian Archduke Franz Ferdinand on a summer's day in the Bosnian capital of Sarajevo. Six weeks later the world was at war, with Britain on one side and Germany on the other. Hans was a native German (with an incredibly German sounding name, Hans Otto Wilhelm Wilsdorf) living in and doing business in a country which was at war with his own. This was not good, but things got worse. After the German invasion of Belgium, news of German atrocities (real and imagined) spread like wildfire throughout Britain. Soon the anti-German riots began. Throughout the United Kingdom there were many Germans, most of whom were employed as waiters, but many who owned and ran small butchers shops. They were the first to suffer. Within days of the breaking news the waiters were expelled from their jobs and the merchants' windows were smashed. Soon the more bizarre facets of the anti-German feelings were exposed. Dachshunds were kicked in the street and the breeders of German Shepherd dogs then chose to rename the breed Alsatians (after the first part of France invaded by the Germans) a name they keep to this day in Britain. The First

Lord of the Admiralty (a position similar to that of the Secretary of the Navy in the U. S.) was forced from office because his name was Battenberg.[4] Most interestingly the British Royal Family changed their name from Saxe-Coburg-Gotha to Windsor.

It is difficult now, for us who are looking back, to imagine the depths of anti-German feeling in Britain at that time. The two advertisements shown below from the British magazine "The Horological Journal," dated December, 1914 may give some idea.

THE LONGINES FACTORY

The LONGINES WATCH FACTORY is situated in the Valley of St. Imier, in the French part of Switzerland.

It gives employment to 1,200 hands, in the small town of St. Imier.

The Factory is owned and directed by a private Company, all the members of which are French Swiss, viz. :—

The family of the late Ernest Francillon. *Founder.*
The family of the late Jaques David. *Director.*
E. Gagnebin. *Managing Director.*
B. Savoye (Member of Swiss Parliament). *Managing Director.*
Arthur Baume, London.

Telegraphic Address: "BAUME WATCH, LONDON."   Telephone: HOLBORN 486.

 NOTICE TO THE TRADE.

MESSRS. STAUFFER, SON & CO.

beg to remind their customers that all the **Watch Bracelets** supplied by them are **British Made** as Messrs. S., S. & Co. have **never** Stocked German Bracelets

Also all <u>Leather</u> Goods : Leather Folding Watches, Wristlets, Straps, etc., are <u>British Made</u>.

STAUFFER, SON & CO.,
Established in Chaux-de-fonds, Switzerland, and England, 1830,
Swiss Watch Manufacturers,
13, Charterhouse St., Holborn, London, E.C., & Chaux-de-fonds, Switzerland.
Telephone, Central 1386.   Telegraphic Address, Stauffers, London.

Advertisements emphasizing the Swiss and British manufacturer, and making sure that the public did not associate the company with German products.

Nobody wanted any trait of German about them or their company, so it was obvious that Wilsdorf & Davis would have to do something. Fortunately Hans had already become a British Citizen a few years earlier when he had married his English wife, May Frances. They quickly renamed the company "Rolex", consigning the "Wilsdorf & Davis" name to history. It had served them well for ten years, but on November 15, 1915, The Rolex Watch Company Ltd. was registered with the company number 142138, by their lawyers White and Leonard.

The documents surrounding the registration of the company are interesting and give a vivid impression of the national psyche at the time. The first document dated November 12, 1915 is a declaration that the company was not formed for the purpose of trading with the enemy. The second was a letter from Mr. Wilsdorf's lawyer, also dated November 12, 1915, affirming that Mr. Wilsdorf was a British citizen by naturalization. The renaming of the company was not something they undertook lightly; it was a very expensive proceeding, there seems to have been a very large number of lawyers involved and the fees to the government alone were £74.00.00 (or $296 in 1915 dollars), a considerable sum.

It must have seemed to Alfred and Hans that the world was conspiring against them, apart from the problems related above a much more important one now reared its head: geography. Simply put, Britain was physically off the west coast of Europe, Switzerland (where their movements came from) was in the center of Europe, and Germany was smack in between them, ideally placed to intercept shipments from one to the other. Fortunately their geographical fears did not materialize. In fact, the Swiss were ideally positioned, both commercially and mentally, to exploit the problems of the rest of Europe and throughout the war shipments got through without too many problems, usually through neutral countries.

Ironically, the war, which initially had seemed a major problem to the fledgling company, provided a significant boost to the wristwatch. Prior to the outbreak of war the wristwatch had been almost exclusively a feminine adornment. Because of this all wristwatches were quite small. 10-1/2''' was the largest available movement, and many were much smaller. This gave rise to the another fact about wristwatches of the time: they were fairly inaccurate. In that era (and may the Political Correctness Police forgive us) accurate time was not very important to the few ladies able to afford a wristwatch.

Because of the combination of its feminine identification and its lack of accuracy, it was inconceivable that any self-respecting gentleman would even consider a wristwatch. Real men used pocket watches, usually kept in the pocket of a waistcoat draped with a watch chain liberally festooned with fobs and keys. Even cowboys used pocket watches; the small pocket inset just above the right front pocket on your Levi's 501 jeans is called a watch pocket. With that kind of background getting any turn-of-the-century man to wear a skirt would have been easier than to get him to wear a wristwatch (after all most of the photographs of Queen Victoria's bailiff and alleged lover, John Brown, show him wearing a skirt/kilt, but never a wristwatch).

Utility was the key to change. The first stirrings were in Paris in 1904 when the Brazilian born balloonist, Albert Santos-Dumont, asked his friend, Louis Cartier the jeweler, to design a watch for him that would allow him use of both his hands while he was aloft. Although wristwatches for ladies had existed since the sixteenth century, Cartier put a wrist strap on a pocket watch and announced that he had invented the wristwatch. He promptly named it the "Santos," in honor of his friend. This watch had some impact on the popular consciousness, but did not sway the British males. After all the man who wore it was Brazilian and he lived in France; that was enough for most Britons to consign the wristwatch to the scrap heap of Frenchified aestheticism.

The first world war changed all of that. For the first time, it was necessary for very large groups of men to act simultaneously, so every officer needed a watch. But in the mud, cold, and rain of Flanders' trenches, pocket watches proved to be a major encumbrance when under the layers of Jerkin, Greatcoat and Uniform. So the wristwatch became a necessity, and of course, no-one would dare accuse the men at the front of being effeminate just because they wore wristwatches.

The first officers who used wristwatches were most probably regarded strangely by their compatriots, but one of the things about warfare is that people care a great deal about function and not very much about form. Soon Rolex was working hard to produce enough man's size wristwatches to keep up with the demand. It is interesting to note that almost all of the Rolex wristwatches seen with hallmarks dated prior to 1914, the start of the war in Europe, are gold and in ladies sizes, while those with hallmarks after 1915 are mostly silver and man's 13''' size.[5]

Sterling silver was used at this time as the least expensive case material and was regarded as utilitarian, as steel is now. The simple watches of this period represent a transitional stage in the development of the wristwatch. Almost all have porcelain dials with sunk subsidiary seconds dials at the "6" position and hinged cases with domed backs. Some of them are even more reminiscent of pocket watches, being either full or half-hunters. These were a popular option during the war. As the officers began to depend upon their wristwatches, they realized the risks of damage to the (very breakable) glass and dial and so adopted these more protected models. Many companies offered perforated protectors to fit over the glass to shield them, while still affording some dial visibility. Rolex designed their own version of this in 1917 under Swiss patent 77129, but it was too close to the end of the war to see production.

This Swiss patent is the first in a line of over 500 patents taken out by Rolex in the period between 1912 to 1990 added to the 25 or so Wilsdorf took out in his own name. There were also a significant number of patents taken out in Britain, even after the company relocated to Switzerland. Wilsdorf's first recorded patent in Britain was for a new way of luminescing watch dials and was obtained on 19th September 1917.

The end of the war saw Rolex in a much improved trading position, but in a much worse financial condition. Wilsdorf had founded the company in London to take advantage of being at the heart of the British Empire. But the costs of fighting a World War had forced Britain, in 1915, to impose an import duty of 33.33% on watches. The duty of 33.33% on all imported watches, clocks and parts was imposed on the September 21, 1915 and caused chaos in the British Jewelry trade. But this was as nothing compared to the problems which arose a year later, when, on November 16, 1916 the government banned completely the importation of all gold and silver, with the exception of watch cases. The respite given to the watch trade was short lived, however, when on the December 5, 1916 a further proclamation was issued banning the import of even these items. This now meant that all watches brought into Britain, even if they were due to be re-exported to the colonies would now be subject to this additional charge.

To Wilsdorf the answer to this new challenge was obvious. Rolex had had a presence in La Chaux de Fonds for over three years for the purpose of working in close contact with Aegler. Rolex moved this operation to the factory at Bienne, where it was expanded and became the new center of operations for Rolex. The head office remained in London, but the teams of watchmakers who checked and adjusted the watches became surplus to requirements now that this process was carried out in Switzerland. Operating from rooms in Aegler's factory Rolex employees now checked all the watches whether they were destined for London or some far flung outpost of Empire. This of course had the effect of reinforcing the relationship between Rolex and Aegler.

They became even closer when Wilsdorf decided to open offices in Geneva and to relocate there himself. This was accomplished in early 1919. The Geneva offices were to handle exports, design and administration, while the technical side of the business was continued in Bienne. Seventy-five years later the company is still divided between these two cities along the guidelines Wilsdorf originally established.

In 1919, the relationship between Aegler and Rolex was still mainly that of supplier and customer, but the following year it was decided to convert the relationship into a formal alliance. Wilsdorf and Davis gave 6,960 shares in their company to Hermann Aegler. These shares, representing almost 15% of their equity in the company were not of course handed over for nothing, because in return we can see that, from this time on, Aegler's company begins to call itself Aegler S.A. Rolex Watch Co.

We leave the first decade of Rolex with their foundations firmly established, but still subject to the vicissitudes of international trade and economies. The next decade was to see them begin to triumph even over these adversities.

---

[1] Apart from the advertisements in the US, for legal reasons.

[2] The store is still there, but now is owned by the Heinlein family.

[3] As early as 1901, a testimonial in the catalog of the Goldsmiths & Silversmiths Company read, "...I wore it continually in South Africa on my wrist for 3 1/2 months...Faithfully yours, Capt. North Staffs Regt," obviously a soldier in the Boer War.

[4] The family name was changed to Mountbatten and his son, Lord Louis, became the First Lord of the Admiralty during World War II.

[5] As late as 1921 Rolex was still offering mostly lady's watches, although some gent's models were available. They were also offering eight-day clocks with metal stands in enamel, as well as a good range of clocks designed to be fixed to a car dashboard and wound via the bezel.

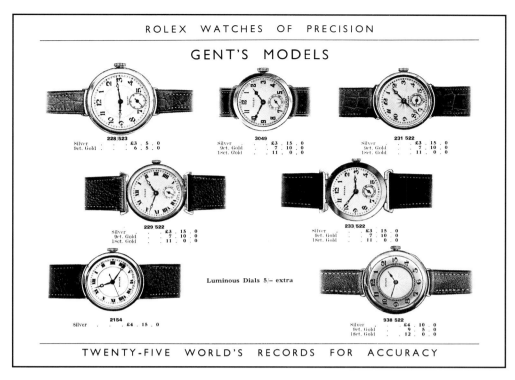

This and the following page show Rolex pages from an early catalog.

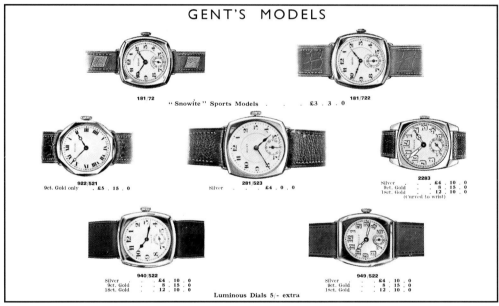

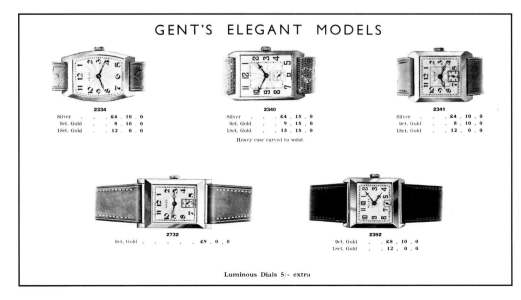

GENT'S ELEGANT MODELS

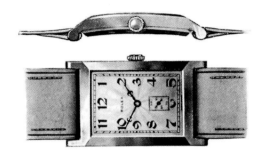

2607

Curved to wrist
Extra Heavy Cases
Silver with Chrome Finish   £6 . 0 . 0
9ct. Gold   .   .   .   12 . 0 . 0
18ct. Gold   .   .   .   16 . 0 . 0

2709

Luminous Dials
5/- extra

2618

Latest Flat Cases curved to wrist
in Red and White Gold
9ct. Gold   .   .   £14 . 0 . 0
18ct. Gold   .   .   18 . 0 . 0

2608

GENT'S FANCY MODELS
For the Small Wrist

2448

2613

2622

2437

9ct. Gold   .   .   .   .   .   .   £9 . 10 . 0
18ct. Gold   .   .   .   .   .   .   13 . 10 . 0

Luminous Dials 5 - extra

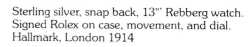

Sterling silver, snap back, 13''' Rebberg watch.
Signed Rolex on case, movement, and dial.
Hallmark, London 1914

Tonneau shaped sterling silver, metal dial man's watch. Rolex is signed on the movement rather than the wheel, as you can see on page 332.
Hallmark, London 1913
Case no. 576083

16 Forty Years On

Lady's 10-1/2''' Rebberg, 9kt gold with flexible hinged lugs, metal dial with Breguet hands. Hallmark, London 1913

Lady's 9kt gold, porcelain dial 10-1/2''' Rebberg with expanding bracelet. An unusual dial with blue numbers and red twelve.
Circa 1914

Early sterling silver, cut corner rectangle with sunken metal dial and exaggerated numerals.
Circa 1914

Lady's 9kt gold, faux Demi-Hunter with enameled bezel with engraved numbers, porcelain dial 9-3/4''' Rebberg Movement, 9kt gold expanding bracelet.
Case no. 669615
Hallmark, London 1915

18 Forty Years On

9kt gold, cushion porcelain dial Rolex, 13‴ Rebberg.
Hallmark, Glasgow 1926
Case no. 104546

Lady's faux Demi-Hunter 18kt gold with blue enameled bezel and bracelet.
18kt gold.
Hallmark, London 1912
Case no. 483362

Gentleman's 9kt gold Hunter, snap back.
Porcelain dial.
Hallmark, London 1919
Case no. 972252

Forty Years On 19

Gentleman's sterling silver flexible lug wrist-watch, 13''' Rebberg movement.
Hallmark, London 1917
Case no. 720538

Lady's 10-1/2''' Hunter porcelain dial wristwatch, 9kt pink gold with expanding bracelet, porcelain dial with gold minute markers and red "12".
Circa 1925

Sterling silver faux Demi-Hunter Rebberg with engraved enameled bezel, black enamel numerals, and solid lugs. Movement 11''' Rebberg
Hallmark, Glasgow 1924
Model no. 1
Case no. 51528

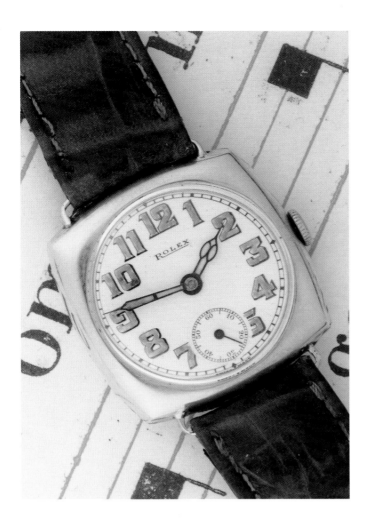

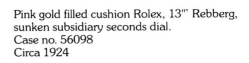

Pink gold filled cushion Rolex, 13''' Rebberg, sunken subsidiary seconds dial.
Case no. 56098
Circa 1924

Gentleman's 13''', sterling silver, screw back and front wristwatch with sunken subsidiary seconds, porcelain dial, full Arabic numerals, Rebberg Movement.
Hallmark, Glasgow 1914. Notice the inscription on the rear of the case showing that it was presented to someone going into World War I.

Movement of the watch on page 8.

Forty Years On 23

Sterling silver octagonal Unicorn, 13''' Movement
Case no. 104105
Model no. 73
Hallmark, Glasgow 1925

Gentleman's sterling silver Demi-Hunter, full Arabic numerals on porcelain dial with sunken subsidiary seconds, lid with black enameled numerals, lid release at six, screw back, 13''' Rebberg.
Circa 1917

Gentleman's 13''' 9kt gold, Rebberg wristwatch with porcelain dial, snap bezel, screw on back, round dial with red twelve and sunken subsidiary seconds, 13''' movement.
Circa 1919

9kt red gold, 11''' Rebberg, porcelain Roman numeral dial with sunken center and red twelve, hinged back and bezel.
Case no. 630952
Circa 1919

Sterling silver cushion wristwatch with black and white porcelain dial, sunken subsidiary seconds and luminous hands. 13''' Rebberg movement.
Case no. 586415
Hallmark, Birmingham 1929

Nurse's watch. 13''' Rebberg, designed to be worn suspended from a button on the uniform. Notice the unusual configuration of the double sunken porcelain dial. The winder is at twelve and the subsidiary seconds is at three. This is because the whole dial has been turned 90 degrees. Notice also the very unusual case with only one lug. This watch was the subject of one of Wilsdorf's very first patents.
Circa 1925

Nurse's watch. 9-3/4''' Rebberg. The earliest known sweep seconds Rolex watch. Notice, on the movement, the additional pinion and wheel to drive the center seconds.
Circa 1925

A comparison of the sizes of two previous nurse's watches.

This is the earliest known Rebberg wristwatch. It features an 8-3/4''' movement and it is extremely unusual in that the case and movement are both signed Rebberg. Bezel shown has blue translucent enamel on top of the 14kt gold.
Case no. 517535
Circa 1910

Gentleman's long rectangular wristwatch, gold filled, marked Rolex Geneva Swiss, rolled gold to wear 20 years, 10-1/2''' Rebberg Movement, metal dial with Arabic numerals and Breguet hands.
Circa 1922

9kt pink gold cushion shaped watch with Roman numeral double sunken enamel dial, cathedral hands. The 13''' Rebberg movement from this watch shows later finish on bridges, which made them look more like the recently introduced Hunter movement.
Hallmark, Glasgow 1927

18kt gold gentleman's tonneau case, 10-1/2'" Hunter, metal sunburst finished dial with "Fleur d'Lys" hands and sunken subsidiary seconds bit. Note again the solid hallmarked gold bars.
Circa 1927

Rolex rectangular wristwatch, 9kt gold solid bars, Roman Numerals 10-1/2'" Hunter, no seconds.
Circa 1930

9kt gold gentleman's two-piece wristwatch. Slightly curved back. Metal dial signed with the retailer's name, Meader of Boscombe, under the twelve and the name Rolex just visible above the sunk subsidiary seconds.
Model no. 554
Case no. 41592
Hallmark, Glasgow 1927

Forty Years On 31

Pink gold, square wristwatch with unusual landscape format dial, metal sunburst finish, sunken subsidiary seconds square. Signed above the seconds square with the retailer's name - The Alex Clark Company, London.
Model no. 535
Case no. 33610
Circa 1931

Sterling silver, black and white porcelain dial wristwatch with hinged back and bezel. Hallmarked London 1915, making this one of the earliest 13''' Rebberg sweep seconds watches.
Hallmark, London 1915

32 Forty Years On

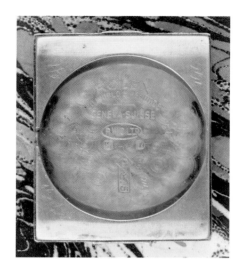

Gentleman's rectangular wristwatch, 9kt gold, metal dial, cathedral shaped hands, sunken subsidiary seconds, 10-1/2''' Hunter movement signed "Prima."
Model no. 34
Case no. 57362
Circa 1931

Forty Years On 33

Gentleman's white metal Marconi wristwatch, rectangular movement signed "Rolex Watch Company," case signed "Rolex Watch Company Snowite." Silver metal dial with applied Arabic numerals
Model no. 35
Outside of movement stamped with Case no.
1060 515989
Circa 1932

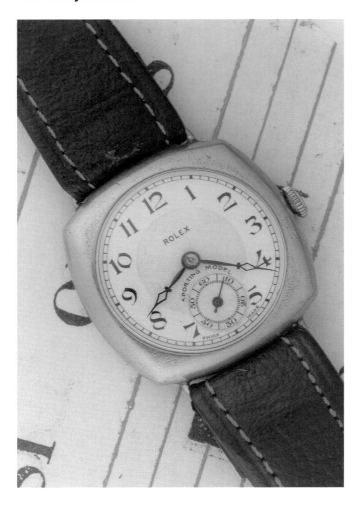

Rolex sporting model. Porcelain dial, cathedral hands, sunken subsidiary seconds, signed "Sporting Model" above the seconds.
Model no. 20
Case no. 61738
Circa 1932

A very unusual steel hooded wristwatch. Case signed "Rolex," "Seven World's Records," and "Staybrite." Movement 10-1/2''' Hunter extra prima rare 17 jewel version. Please note this dial has been refinished and this watch is not an Oyster. This is one of the first styled Rolexes ever made. *See drawings on page 160.*
Model no. 2361
Case no. 12539
Circa 1933

Sterling silver two-piece cushion wristwatch with curved back and solid sterling silver case. Sunburst finish, metal dial, Roman numerals signed with the retailer's name, "Emmanuel Southhampton," below the twelve and "Rolex" above the subsidiary seconds.
Late 10-1/2''' Rebberg, 15 jewels.
Model 554
Case no. 57419
Case signed inside with "15 World's Records"
Hallmarked Glasgow 1927

Unusual two-tone tonneau shaped wristwatch. 9kt gold. Movement 9-3/4''' Hunter signed "Rolex Prima," 15 jewels. This is the earliest known two-tone Rolex wristwatch. Note the hallmarked case bars. Case stamped inside "7 World's Records".
Model 870
Case no. 55966
Hallmarked Glasgow 1928

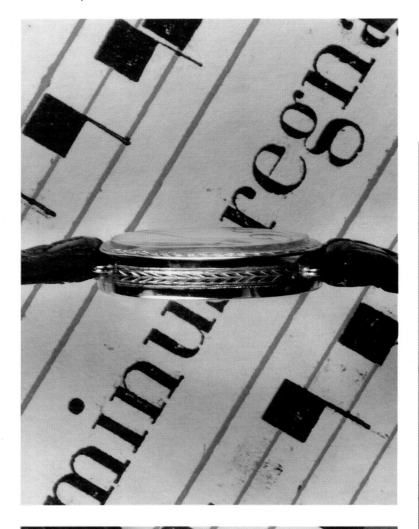

Lady's 9kt gold wristwatch. Unusual oval shape with square dial. Bezel and side of case completely chased. Movement fits exactly into case and fills it totally
Case no. 53346
Circa 1924

Forty Years On 37

18kt gold Gentleman's rectangle. Silvered sunburst finished dial with subsidiary seconds rectangle and solid bars. Note the solid case bars are also hallmarked. Note also the early transitional movement which uses the winding wheels of the Hunter but the earlier center bridge of the Rebberg.
Model no. 34
Case no. 0330798
Circa 1927

Sterling silver screw back and front, "Dennison" transitional wristwatch with porcelain dial and skeleton hands. Very rare 18 jewel "Extra Prima" version of the 13''' Rebberg. Note that the movement is jeweled to the center, has a different finish and has all of the engraving on the bridges rather than on the wheels.
Case no. 257947
Hallmark Birmingham 1915

9kt pink gold cushion with two-piece case and wire lugs. "Sunburst" finish metal dial with stylized arabic numerals and subsidiary seconds dial. Note that the dial is signed "Dunkling's Rolex," Dunkling being the name of the Australian retailer. 10-1/2''' Hunter, 15 jewel Extra Prima movement.
Model no. 2494
No case number
Circa 1924

Ultra-rare lady's sterling silver watch with original faceted expanding bracelet. 11''' Rebberg movement. Exotic hexagon-shaped porcelain dial with four color inlays in the enamel and hexagonal mask cut into the bezel to frame the dial.
Model no. 4
Case no. 7859066
Circa 1913

Chapter 2
OYSTER

Oyster, a. A special delicacy. b. Something from which benefits may be extracted.[6]

It is the generally accepted view that Hans Wilsdorf, the famous Swiss watchmaker, sitting alone in his Geneva workshop and working late into the night, came up with the idea of the Oyster case and thereby set Rolex off in a new direction. It is a great story except for the fact that none of it is true. Wilsdorf was never either Swiss or a watchmaker and the Oyster was the result of at least four attempts by the company to produce a waterproof case prior to the introduction of the Oyster.

It is dangerous to ascribe certain characteristics to a whole nation, but we would venture to say that the prime German characteristic is thoroughness, not blinding flashes of inspiration or intuitive leaps of logic. German industry succeeds on the precise and thorough development of an existing product. For examples look at everything from Leica cameras to Mercedes Benz. Despite his British Nationality and Swiss domicile Hans was very much a German and his greatest strengths were his persistence and his dedication to slow progressive development of his existing product. It is in closely examining the route that led Wilsdorf and Rolex to the Oyster case that we see these strengths at their finest.

Like most watch companies in the early part of this century, Wilsdorf & Davis bought their cases from specialist case making firms, of which there were dozens. The very first cases that they used were simply just scaled down versions of the traditional pocket watch cases of the day. They consisted of three distinct parts, the case body (which held the movement and dial) and the back and bezel, which were both hinged to the body. The back and the bezel were slightly domed, a design that allowed them to be snapped shut.

This style of construction, which had worked perfectly for pocket watches for over two centuries, was not really suited to a wristwatch. Apart from the hinge area, there was a visible gap around the rest of the circumference of the watch. Unlike pocket watches, the size of the early wristwatches precluded the fitting of an inner dust cover, so the ingress of dust, dirt and detritus was a major problem for the early wristwatch. The resulting problems for the first owners of such watches mitigated against wristwatches gaining a wider acceptance. Any snap style back will have this problem unless a pressure tight gasket is used. At the time there were no such gaskets.

Two attempts at constructing a dustproof case are notable, prior to the First World War. In Paris, Jacques Cartier was making his first wristwatches after his success with the one for his friend Santos-Dumont, the famous Brazilian balloonist. Cartier's solution was to reduce the number of parts of the case to two and to hold them almost hermetically together by means of screws passing through one half of the case into the other. In Switzerland the major case makers produced a range of solutions all based on the fact that threaded surfaces can be closed tighter and with a higher degree of precision than pressure fitted surfaces can. Wilsdorf & Davis introduced their first watch with a threaded bezel and back in 1914. This man's watch used their normal "Rebberg" 13''' movement in a silver case made by one of their regular casemakers. Because of the outbreak of war in 1914, the case was produced in very small quantities.

With the English tariffs imposed in 1915 and 1916, Rolex turned to a British company, Dennison, to make the threaded cases. The Dennison family was one of the greatest families in the history of watchmaking, having owned or operated, Howard and Waltham. However even their greatest admirers could hardly call them business geniuses and a string of bankruptcies and plant closures had driven them from the U.S. to settle in the English Midlands town of Birmingham, then, as now, a great jewelry center. They started the firm of A.L. Dennison and Company as case makers there in 1874.

Dennison was already making a pocket watch case with threaded front and back for many other companies and so it was a simple matter to fit the 13''' Rebberg into it. Made in gunmetal, sterling silver, and 9kt and 18kt gold, it proved to be a very strong and very reliable case. In the dangerous and muddy conditions of World War I trench warfare it proved its worth. The Dennison case was important in one particular way; on this case, unlike the earlier Swiss threaded cases, both the bezel and the back had milling on their edges like a coin edge. Though there was no matching key, this milling enabled even fingers to tighten the case to a degree hitherto impossible. From this moment on, all of the watches produced by Rolex on the route to the Oyster, would use this milling. Look at the back of the 1915 Dennison case and any Oyster from 1926 to today and you will see these same milled edges.

Meanwhile in Switzerland progress was also being made, Francis Baumgartner is one of the greatest of all the Swiss case making concerns, having made cases for every major concern. The original Patek Philippe model 96 "Calatrava" cases were made by them, as were the famous rectangular clip-back waterproof cases of the 1930s, supplied to everyone from Movado to Jaeger-le-Coultre.

In the early 1920s Baumgartner began to construct cases made to the patent of Borgel. This was the first real attempt at a wristwatch case specifically designed to be dustproof. The secret of the Borgel case was that it was a two-piece case in which the entire movement and dial were held in a ring secured to a milled bezel. This whole assembly then screwed into the other part of the case, which was a single spun form that took the place of the case center and back.

The only problem with this system was the winding. The movement was attached to the bezel, but the winding stem had to come through the side of the case. The problem was solved by fitting a clutch on the winding button and leaving the button permanently attached to the case. The clutch was disengaged by pulling on the winding button and the bezel and movement could then be unscrewed and removed. This left another problem: if the pulling out of the winding button served to release the stem clutch, rather than setting the hands as it did on all other watches, how do you set the hands? Baumgartner reverted to a pin-setting system using a separate pin just below the winding crown.

The case proved to be very popular and many makers, including Longines and Omega, adopted it. Rolex was not one of the major users, but in 1922 did produce a small series. The total number of Rolex watches produced using this case style is not known, but as of 1995 only three had surfaced. Despite the small number of watches produced in this case style, it is a very important development in Rolex watch design. It was the first model produced by Rolex in which the case was specifically designed to give protection against some of the elements.

When the First World War ended, Rolex was larger than ever before, safe and secure in the capital city of the British Empire. It was obviously time to expand and that was exactly what Rolex did, to the farther reaches of the Empire. Strange as it may seem to us now, Rolex opened offices in India and Burma many years before they entered the U.S.A. or France. Their watches proved very popular in the Indian sub-continent and by the late 1920s there were only two major companies dominating the market there: West End Watch company, who were the agents for Longines as well as selling some pieces under their own name; and Rolex.

Wristwatch sales took off in the tropical markets of India and East Asia long before they did in the more temperate areas of Europe because of the climate. In the hot and humid conditions there it was natural for the men to wear fewer layers of clothes, and the first thing to be discarded was the waistcoat (or vest) which obviously made it much more difficult to carry a pocket watch.

While creating a strong demand, the climate also created new challenges. As watches were taken from the temperate areas around Delhi and Bombay to the more extreme climates of Madras and Calcutta, they encountered a humidity that was overwhelming and gave serious problems to the watch mechanisms, even with the newly introduced Borgel cases. Movements would seize up and metal dials would corrode within months.

There was a need for a new solution and once again Baumgartner came to the rescue. The new case style, known as the Hermetic, was a simple solution, but one that worked very well. The movement and porcelain dial were contained in a simple snap bezel and back, three-piece case. The novel twist was to enclose the whole of this watch inside a second case that was made from only two parts, a spun back and a bezel. The milled bezel then screwed firmly on to the case back securing the complete watch inside. This watch case style was patented by Wilsdorf in London on May 10, 1923 (*see the patent drawings on page 344*).

How the watches patented by Wilsdorf differ from those made by Baumgartner is difficult to tell, but they seem to have been made concurrently. Some of the watches are proudly stamped with Baumgartner's trade mark of FB and "Double Boitier Hermetique" inside (meaning double hermetic case), while many others bear just the classic W&D mark.

Wilsdorf was so proud of this watch that even before submitting his patent he applied for three new model names, "Aqua", "The Submarine" and "Diver". In an attempt to demonstrate the impermeability of this new case he also registered a new style of window display, involving the suspension of a working watch in an aquarium. The display was registered in November 1922 and, as far as we know, was not produced at that time, but five years later it was put into production as the main window display for the new Oyster[7].

This style of watch proved to be remarkably successful. The only major problem was that the bezel had to be unscrewed and removed each day to allow the watch to be wound. This also led to a subsequent problem which only occurred after a few years use. On the silver cases, particularly, the bezel milling would wear almost completely away, making it very difficult to then open the case.

This case style was important for it marks the first time Rolex attempted to match the needs of an individual market with a particular product. All examples of this watch that have so far surfaced with the retailer's names still on the dial appear to have been sold in either India or Burma[8]. To further strengthen the hypothesis that none of these watches were ever sold in Britain (which was of course Rolex's home and main market), none of the watches yet seen have been stamped with British hallmarks, without which it would have been illegal to sell them in the U. K. This is the conclusive proof that these watches were for the "tropical" markets only.

Despite the success of the "tropical" watch, Baumgartner proceeded with their search for a more functional replacement and in 1924 introduced a case which incorporated the best features of the Borgel and the tropical. Known usually as the "semi-tropical," it is again a two-piece case with a threaded bezel. However, instead of the movement and dial being fixed to the bezel with a ring, the movement and dial, both held in a movement ring, simply lay in the case back, the movement protected by a simple dust cover. The winder projected through a cut out in the case and therefore the bezel need no longer be removed for winding and hand setting. The case was protected from dampness entering through the winding stem by a double gasket.

This case, much simpler than the "tropical," was almost as humidity proof and much more user-friendly. More importantly, it brought together, for the first time, the main components of the Oyster: a threaded closing system, a movement and dial contained in a movement ring, and a protected winder. All of these items had been in use previously, but this watch case was the first to use all three. It is not surprising that within a year of the semi-tropical being introduced to the market that someone would try and cure the only weak link in the chain: the winding button.

Now comes the great mystery. On October 30, 1925 in La-Chaux-de-Fonds, Paul Perregaux and Georges Peret, two prototype makers who had never previously obtained a patent or ever filed one again, filed a patent that was later given the number 114,948. In this patent they describe an invention for a moisture proof winding stem and button. The button utilized springs and double helical screws to provide the first real solution to water proofing a watch stem. Wilsdorf was able to see how important this patent was and moved rapidly to negotiate with the two partners, the result was that the two assigned their rights to the patent to Wilsdorf. In this context the word "assigned" means sold. Less than a year after the filing of the original Swiss patent a British patent was then issued, this time under Wilsdorf's name; and this one, bearing the number 260,554, is always seen as the Original Oyster Patent but an examination of the two patents will reveal their identical content.

The question as to who actually invented the winder is interesting, but essentially irrelevant. The facts are that Wilsdorf saw the importance of the design and, more importantly, was in a position to exploit the design. He had the company, the resources, and, of course, the vision. Also, as we have seen above, he had been working on other routes to this solution himself[9].

The name "Oyster" itself was Wilsdorf's own contribution, he said that he was inspired by the difficulty he experienced in opening an oyster while preparing a dinner party. He registered the name in Switzerland on July 29, 1926 and two months later in London. On February 28, 1927 Wilsdorf followed it up by registering the name in four more languages French, German, Italian and Spanish as respectively: Huitre, Auster, Ostica and Ostra.

It is worth noting that the British patent was filed on September 1, 1926.[10] There were only 121 days left in the year (less than 100 of them working days), yet we have seen more than a dozen Oysters bearing the Glasgow hallmark for 1926. What this means is that Wilsdorf knew he had something special with the Oyster and got right on with the manufacture and sale of the product. The first promotion of the Oyster was in the trade press, specifically the English publication "Practical Watch & Clock Maker," but Wilsdorf saved his retail advertising for almost a year. On November 24, 1927 (one month to Christmas) he took the front page of Britain's "Daily Mail" with a whole page advertisement saying "Make it a Rolexmas." The often repeated story is that the whole page was to publicize the Oyster, when in fact the Oyster portion covered around 15 to 20%, the balance of the space being given over to lady's and gent's dress watches. It was not, by the way, something unusual to take over the whole of the front page of the Daily Mail; the newspaper never printed news on its front page until well into the 1930s.

The Oyster portion of the Daily Mail advertisement publicized the swimming of the English Channel by a young lady named Mercedes Gleitz, who despite her name, was English. The 26 year-old London typist became famous in October, 1927, by being the first English woman to swim the channel and, of course, the first person of any nationality or gender to do so while wearing a wristwatch. Her swim from Cap Gris Nez, in France, to Dover was all the more remarkable because it was made in very heavy fog and miserable weather. When her feet hit the shore at 6:10 in the evening, after more than 15 hours, she was exhausted and immediately collapsed. She was carried into her accompanying boat and then ferried unconscious to Folkestone.

This triumphal crossing was, in fact, her eighth attempt at the Channel swim. This feat gave Wilsdorf something to hang his advertising campaign around. The campaign in the Daily Mail and in Practical Watch & Clock Maker was run by Garland Advertising (now incidentally part of the world-wide Saatchi advertising group) and was one of the first national advertising campaigns ever run in the press by a wristwatch manufacturer. The huge sums Wilsdorf was spending in advertising and promotion later came to the surface in a law suit brought by Wilsdorf against Schmitz Frères, whom he had sued in the Swiss Federal Court for infringement of the Oyster patent. Enquiries made by Fidhor (the semi-official watch industry institute) showed that Rolex had spent over 1,200,000 Swiss francs on their Oyster advertising campaign alone.

But Rolex did not use just print advertising. It was one of the first ever companies (in any business) to identify its chosen market and aim sales promotion events specifically at this group. As early as the summer of 1930, Rolex gave an Oyster to each member of the Swiss water polo team to wear while playing. It publicized the fact during the Lake Geneva Water Carnival by having the four members swim while pushing a 3 foot model of the Oyster which bore the sign "Nous portons la fameuse montre etanche Rolex Oyster" (We wear the famous waterproof watch Rolex Oyster). It is worth remembering that during the late twenties and early thirties, as the world plunged into a recession which soon became a depression, most people did not play sports because of a lack of interest or time, but through a lack of money (usually through the lack of a job).

An article in The Watch & Clock Maker of October 15, 1930 discusses how Rolex would verify the claim of water resistance by immersing every Oyster in a bath of water, though only at a depth of 6 inches or so, for 24 hours. The testing rig in which more than 60 Rolex Oysters would be laid was known, of course, as the "Oyster Bed."

Although the first Oysters were produced in late 1926, Wilsdorf was patenting improvements to the crown within the year. His first was patent no. 120,848. This was followed with 130,195 on December 28, 1927, and patent 140,165 on July 11, 1929. Less than six months later, on January 31, 1930, he applied for patent 151,574, and on August 22, 1933, it was time for patent 170,143. Patent 170,500 was applied for on July 20, 1933, and, finally, patent 171,802 was applied for on August 21, 1933. This ferocious rate of inventing and submitting patents, seven in less than ten years, was the first example of Rolex (and Wilsdorf) modifying and improving their product constantly even though they were far in the lead of all their competitors and had the market to themselves.

They had the market to themselves because they protected their franchise with vigor. We may think that copying a Rolex is a comparatively recent phenomenon, but less than seven years after the introduction of the Oyster someone was making copies. The oyster crown was copied by Schmitz Frères, a case-making company from Granges, Switzerland, who even had the nerve to inform Rolex that they were about to copy the crown. The wheels of Swiss justice may grind fine, but they do grind exceedingly slow. It took well over two years, during which Schmitz continued production, before Rolex were granted the injunction they had sought so long before.

After the 1929 change from an onion-shaped winder to a flat-faced steel winder, the next major change was 1930 adoption of the two-piece case which had been patented the year previously under patent no 304,291. The two-piece case was not only simpler to manufacture, it was much more reliable as it obviously had half the number of seals to fail (1 instead of 2). Parenthetically, the early Oyster literature makes much of the fact that the oyster employed no seals made of leather or rubber, using only metal to metal contact. This led most people to believe that the early oysters had no seals, when in fact they used deforming lead alloy ones[11]. The two-piece case was initially cushion shaped when it appeared as model no. 2416 in 1930. It was the introduction, two years later, of model no. 2280 (now universally known as the "boy's size") that finally brought Rolex to the case design we still recognize. A year later Rolex introduced the Oyster Royal in gold and in steel. First introduced with a sweep seconds hand, it was described as being ideal for the medical profession. It was this model that became, with the addition of a deeper back, the case of the first Oyster Perpetual.

---

[6] *The American Heritage Dictionary of the English Language*, Third Edition, licensed from Houghton Mifflin Company. Copyright © 1992 by Houghton Mifflin Company. All rights reserved.

[7] The aquarium displays proved very popular with Rolex dealers; however these dealers were not experts in the care and feeding of tropical fish and many fish died. It was not really a major help to the company to have their watches displayed in a tank of dead fish and so Rolex had to publish books in the 1930's on the care and feeding of tropical fish.

[8] The name SAB seen on the dial of many of these watches means "Societe Anonyme Belge", or Belgian Corporation, which was the Rolex importers for Siam (as Thailand was then known) and still are, although today they are part of the giant Inchcape group.

[9] Wilsdorf undoubtedly learned of the original patent via his habit of reading every new patent applied for in Switzerland, whether horologically related or not.

[10] It is also worth noting that the name "Oyster" was not registered as a trademark until September 28, 1926.

[11] Which, of course, enabled Rolex to correctly use the phrase "metal to metal contact."

44 Oyster

British advertisement from 1930, listing only Indian jewelers.

1932 Oyster brochure

1930 Rolex catalog showing "special models with centre seconds for doctor's use."

Daily Mail

FOR KING AND COUNTRY

LONDON. MANCHESTER. PARIS. NO. 9,856. THURSDAY, NOVEMBER 24, 1927. ONE PENNY. **20 PAGES** — BROADCASTING P. 16

The Marvellous ROLEX WRIST WATCH
The World's Best by Every Test

GIVE lifetime gifts this Christmastime—gifts worthy of a lifetime's use. Give 'Rolex' watches! They are the world's best by every test. Rare, beautiful productions, gem set and in precious metals, are never shown to fairer and more practical advantage than when they encase the Rolex Watch movement. Twenty awards in the Observatories of Kew, Geneva and Neuchatel have won for this incomparable little 'keeper' of time a record for accuracy eclipsing that of any other make of wrist watches in the wide world. That such precision can be contained within so tiny a compass is a constant source of wonder to many thousands of persons who daily regulate their lives by the faithful 'Rolex.' The 'Rolex' is a movement for punctuality, that lasts a lifetime — a timely gift for *all* time!

MAKE IT A ROLEXMAS

When diamonds are brought into requisition further to enhance 'ROLEX' beauty, surely the last word in watch luxury is attained!

ALL GOOD JEWELLERS throughout the British Empire stock ROLEX Watches. Genuine ROLEX Watches have the name on the dial, or on the movement, or inside the case.

A LIFETIME GIFT FOR CHRISTMASTIME.

ROLEX PRICES RANGE FROM £3 15 0 to £100
Men's Silver Wrist Watches from £3 15 0
Men's Gold Wrist Watches from £7 10 0
Ladies' Gold Wrist Watches from £4 0 0

BECAUSE IT IS THE BEST.
The 'Rolex' has emerged from every recognised test as the best timekeeper in the world, a fact that is vouched for by Twenty World's Records.

BECAUSE PUNCTUALITY IS WORTH WHILE.
The possession of a 'Rolex' Wrist Watch brings punctuality by its unfailing precision and accuracy. It turns punctuality by its very promptitude into a daily habit. It takes the guesswork out of timekeeping by its reliability.

BECAUSE IT IS BEAUTIFUL.
Care has been taken to encase the priceless 'Rolex' movement in Gold and Silver and in gems which make it, as a gift, all the more acceptable.

BECAUSE IT IS EXCLUSIVE.
In every country where English is spoken the 'Rolex' Watch sells to discriminating people who desire the most exclusive production in watches, and in both design and movement the 'Rolex' Wrist Watch is supreme.

THE ROLEX TEST
Every Rolex Wrist Watch is subjected to prolonged tests which reveal deviations as slight as does a chemical balance. Each must show accurate results in six different positions, over three weeks period—conditions infinitely more exacting than those of actual wear—before being approved.

Rolex introduces for the first time the greatest Triumph in Watch-making
ROLEX 'OYSTER'
The Wonder Watch that Defies the Elements.

MOISTURE PROOF
WATER PROOF
HEAT PROOF
VIBRATION PROOF
COLD PROOF
DUST PROOF

BEING hermetically sealed the Rolex 'Oyster' is proof against changes of climate, dust, water, damp, heat, moisture, cold, sand or grease; it can, in consequence, be worn in the sea or bath without injury, nor would arctic or tropical conditions affect the wonderful precision of its beautifully poised movement. The introduction of the Rolex 'Oyster' model marks an unique development in the forward stride of the chronometric science, and perfect timekeeping under all conditions is at last a possibility.

A HANDSOMELY printed, fully informative Brochure illustrating every variety of OYSTER Models available, together with the list of nearest jeweller stocking ROLEX Watches, sent post free to any reader who makes application by postcard or letter, direct and addressed to our London Office.

Send for this coloured Brochure it's FREE!

Miss Mercedes Gleitze carried an 'Oyster' throughout her recent Channel Swim. More than ten hours of submersion under the most trying conditions failed to have its perfect timekeeping. No moisture had penetrated and not the slightest corrosion or condensation was revealed in the subsequent examination of the Watch.

For MEN or WOMEN.

ROLEX OYSTER PRICES
The Silver £5 . 15 . 0
9-ct. Gold: £10 . 10 . 0
18-ct. Gold: £15 . 15 . 0

Fitted with good quality silver leather straps for Men, or Moiré Silk bands for Women. If fitted with the unbreakable "TRAYBUF" (tested Woven Flexible Wire Milanese Gold-Filled Bands (White or Yellow) 20/- extra to above.

THE ROLEX WATCH CO. Ltd.
40/44, Holborn Viaduct, London, E.C.4.

Patent Nos. 260554 and 8136.

Resolutely refuse substitutes — none of which have earned so many records for accuracy.

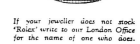

If your jeweller does not stock 'Rolex' write to our London Office for the name of one who does.

THE ROLEX WATCH CO. LTD. H. Wilsdorf, *Managing Director*. **GENEVA SWITZERLAND**

Prior to the introduction of the Oyster, Rolex produced a few of these hermetic watches. The color photographs show a 9kt pink gold version with a porcelain dial, while the black and white photographs show the interior of a sterling silver version.
Circa 1924

Another attempt at a hermetic case was Rolex's adaptation of the Borgel case. In this case, the entire movement and the dial come out with the bezel. The hands are set by the small pusher near the four. This watch has a 10-1/2''' Rebberg movement, porcelain dial with sunken subsidiary seconds, Roman numerals with a red twelve and quite unusual flexible lugs.
Circa 1921

The last attempt at a moisture-proof watch prior to the introduction of the Oyster was the semi-hermetic. In this model the bezel came off completely and the whole movement then came out. This watch has a 10-1/2''' Rebberg, 9kt gold case and a sunken porcelain dial with Arabic numerals and a double sunken center.
Circa 1925

48 Oyster

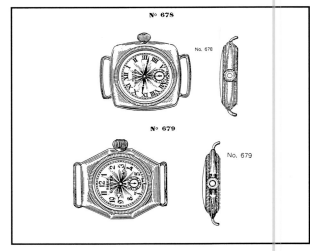

Original Oyster design drawing

White gold cushion Oyster, 14kt gold with black metal dial, sunken seconds, luminous dial markers and hands. Note the original Oyster winding crown which is always onion shaped. The dial on this watch is signed Bucherer after a retailer in Lucerne.
Model no. 698 Case no. 20005
Circa 1925 or 1926

Small sized octagon, 18kt gold. Porcelain dial with subsidiary seconds, skeleton markers and hands. Signed Rolex Oyster in a curve around the subsidiary seconds.
Case no. 20680
Hallmark, Glasgow 1927

Oyster 49

9kt gold porcelain dial cushion Oyster with unusual textured black dial. Dial signed Ultra Prima. Movement also signed Ultra Prima. In original oyster shaped box with papers. Original case number erased and restamped at factory.
Hallmarked Glasgow 1927.

9kt gold porcelain dial cushion Oyster.
Movement signed Rolex Prima.
Case no. 23186
Hallmarked, Glasgow 1930

9kt gold cushion Oyster, metal dial, sunken subsidiary seconds, retailer's name under the twelve, Rolex Oyster, above the subsidiary seconds and triple pierced hands.
Circa 1930

1939 advertisement.

9kt gold cushion, Oyster, porcelain dial, double sunken dial with Rolex Oyster around the subsidiary seconds.
Hallmarked in both Glasgow and Dublin in 1931 and 1933

18kt pink gold, boy's size cushion Oyster, sunburst finish dial, leaf hands and signed around the subsidiary seconds dial. Note the solid, rather than wire, lugs.
Circa 1931

18kt gold, boy's size, octagon Oyster, sunburst finished dial, Fleur d'Lys hands.
Hallmark 1931

Sterling silver cushion Oyster, black metal dial with painted numerals and luminous hands. Movement signed around subsidiary seconds Ultra Prima. Note the original onion shaped winder. This dial seems to be the earliest precursor of the Explorer.
Hallmark, Glasgow 1929

14kt gold cushion Oyster, porcelain dial, signed Rolex Oyster around the subsidiary seconds. Movement signed Extra Prima.
Case no. 26786
Notice the flat-faced winder introduced in 1929.
Circa 1931

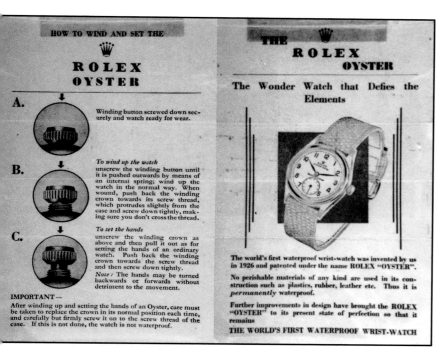

Instructions for winding and setting the Rolex Oyster.

18kt gold cushion Oyster, metal dial with luminous hands and dial. Signed Ultra Prima around subsidiary seconds. Rolex Oyster below twelve and indistinctly signed SAB below Rolex Oyster. SAB standing for Societie Anonyme Belge, the Rolex agents for Bangkok then and now.
Circa 1932

18kt gold cushion Oyster signed Rolex Oyster around subsidiary seconds and Lund & Blockley Bombay under twelve.
Case no. 24258
Circa 1932

18kt gold octagon Oyster, double sunken porcelain dial, signed Rolex Oyster below the twelve, Ultra Prima above the subsidiary seconds.
Circa 1932

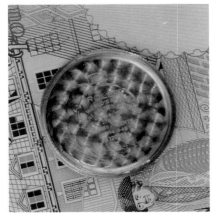

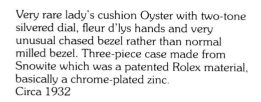

Very rare lady's cushion Oyster with two-tone silvered dial, fleur d'lys hands and very unusual chased bezel rather than normal milled bezel. Three-piece case made from Snowite which was a patented Rolex material, basically a chrome-plated zinc.
Circa 1932

Chromed octagon Oyster. Luminous dial and hands, subsidiary seconds. Notice smooth bezel rather than conventional milled one.
Circa 1933

Sweep seconds cushion Oyster. Note that the outer track for seconds is printed in red with the 5 seconds divisions, as is the word chronometer. Notice the large skeletonized hands which are originally designed to be filled with luminous paint.
Model no. 678
Case no. 34336
Hallmark, Glasgow 1935

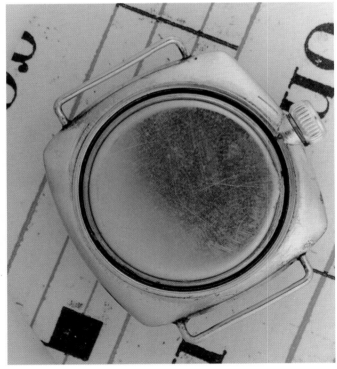

9kt gold cushion Oyster, porcelain dial, luminous hands and dial, sunken subsidiary seconds. Note that this is one of the first watches with an inner protective cap, which protects the movements from undue knocks.
Case no. 30823
Hallmark, Glasgow 1934

9kt gold cushion Oyster. Plain bezel, metal dial with applied minute track and applied raised Arabic numerals.
Circa 1935

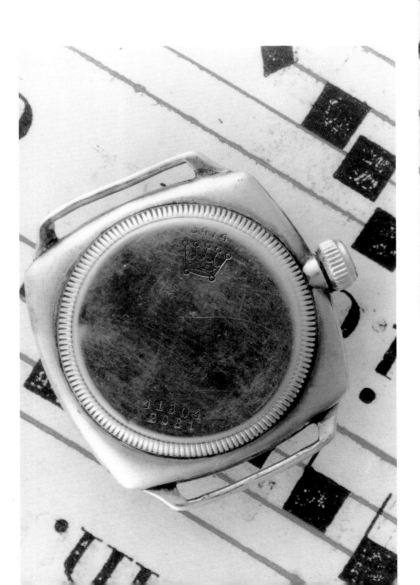

9kt gold cushion Oyster. This is one of the first Rolex Oysters using a super balance signed movement rather than earlier Prima, Extra Prima or Ultra Prima. Sweep seconds with three-tone metal finished dial. Notice the solid lugs rather than wire lugs. The top lug is also hallmarked. There is an additional number 3474 stamped above the Rolex crown on the back, but its origin is unknown.
Model no. 2081
Case no. 41804
Circa 1939

58 Oyster

Cushion Oyster made from Snowite metal.
Luminous dial and hands. Movement, case
and dial all signed Oyster Watch Company.
Notice the solid lugs rather than wire lugs.
Model no. 1069
Case no. 27433
Circa 1929

Stainless steel Rolex Oyster Imperial signed
"Chronometer" and "Swiss Made." Applied
minute track and applied Arabic numerals.
Two-tone criss-cross dial.
Model no. 2595
Case no. 117109
Circa 1942

9kt gold hooded cushion Oyster. Movement signed "Rolex Ultra Prima," "15 jewels" and "chronometer." Metal dial with luminous hands, numbers and sweep seconds. This was the final Rolex cushion Oyster model and it was the first hooded Oyster made by Rolex. These watches are very rare and have their own case model number. They are not converted from early cushions. Notice the lug end of the hood has a Swiss hallmark showing that it is obviously a factory model.
Model no. 3096
Case no. 41168
Circa 1939

Boy's size stainless steel Oyster. This is one of the very first of the boy's size Oysters. Movement is signed "Super Balance. 17 jewels" and it has a subsidiary seconds luminous dial. Notice the name of the retailer, H.G. Bell, Salisbury under the Oyster Rolex Royal signed on the dial. Salisbury is now known as Harare in the country of Zimbabwe.
Model no. 4220
Case no. 230878
Circa 1943

Stainless steel two-piece cushion-shaped Rolex Oyster. Super Balance, 17 jewels. Painted metal dial with luminous leaf hands, sweep seconds and luminous Arabic dial.
Model no. 4647
Case no. 556148
Circa 1944

Rolex Oyster Royal with dial signed "Observatory" above the subsidiary seconds. Three-tone luminous dial. Movement signed "Super Balance, 17 jewels." One of the very first boy's sized Oysters with subsidiary seconds and inner 13 to 24 track.
Model no. 2280
Case no. 142948
Circa 1942

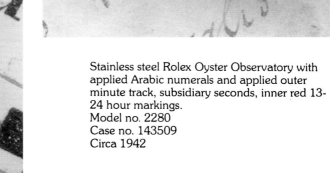

Stainless steel Rolex Oyster Observatory with applied Arabic numerals and applied outer minute track, subsidiary seconds, inner red 13-24 hour markings.
Model no. 2280
Case no. 143509
Circa 1942

62 Oyster

Stainless steel Oyster signed "Rolex Oyster Royal" and with the retailer's name, Asprey, above the subsidiary seconds, luminous dial and hands. 10-1/2''' Hunter Super Balance
Model no. 2280
Case no. 99775
Circa 1940

Lady's 9kt gold Oyster. Metal three color dial printed with subsidiary seconds marked "Extra Precision" around subsidiary seconds. 9¾ Hunter signed "Rolex Extra Ultra Prima." Notice solid bars which are also hallmarked.
Model no. 3019
Case no. 40920
Hallmark, Glasgow 1937

9kt gold two-piece cushion Oyster. Metal dial, applied Arabic numerals and applied outer minute track with subsidiary seconds. 10-1/2''' Hunter Extra Prima
Model no. 2416
Case no. 36856
Glasgow Hallmark 1936

Oyster 63

9kt gold two-piece cushion Oyster, metal dial, sweep seconds with three-tone dial with outer minute track and split into 1/5ths of a minute. Signed "Rolex Oyster" below the twelve and "Swiss Made" at the bottom. Solid bars. Notice hallmark on bars. 10-1/2''' Hunter movement, signed "Super Balance."
Model no. 2420
Case no. 37479
Glasgow Hallmark 1936

Two-piece stainless steel Oyster with applied Arabic numerals, applied outer minute track and painted 13-24 hour markers, subsidiary seconds.
Model no. 3121
Case no. 141248
Circa 1942

Rolex two-piece cushion Oyster, dial signed
"Army." Mercedes hands, luminous Arabic
numerals.
Model no. 3139
Case no. 214015
Circa 1943

Stainless steel and gold Viceroy. Machined
bezel with markers for each of the hours and
two-tone luminous sweep seconds dial.
Movement signed "Rolex Super Balance, 17
jewels." It is extremely rare to find one of these
watches in such perfect condition.
Model no. 3359
Case no. 278994
Circa 1945

Two-piece cushion Oyster, stainless steel,
signed "Rolex Oyster Imperial Chronometer."
Subsidiary seconds, luminous hands and metal
dial.
Model no. 3139
Case no. 106047
Circa 1941

Rolex stainless steel Viceroy. Two-piece case with very rare silver Roman and Arabic dial. Dial and movement both signed "Chronometer."
Model no. 3116
Case no. 111796
Circa 1942

Stainless steel Rolex Oyster Viceroy. Luminous hands and metal dial, subsidiary seconds. Movement signed "Rolex Chronometer" and "18 Jewels." Notice capped endstones on movement.
Circa 1942

Stainless steel two-piece Rolex Oyster Imperial. Signed "Chronometer" above subsidiary seconds. Note the luminous leaf hands and the luminous metal dial. Movement signed patented "Super Balance" and Chronometer. This movement is extremely unusual in that it has both a micrometer regulator and capped sapphire endstones rather than the usual rubies. This grade of movement is the highest Rolex production quality movement prior to the Kew A Certificates in the late 1940s.
Model no. 3116
Case no. 186719
Circa 1943

9kt gold, two-piece Rolex Oyster Imperial. Two-tone metal dial with painted Arabic numerals and outer seconds track. Dial also signed "Extra Precision." Movement, signed "Rolex Oyster Prima," has capped endstones. Note that this watch uses the very rare "open face" movement configuration.
Circa 1941

Stainless steel two-piece cushion Oyster. Metal dial with luminous Arabic numerals and gilt leaf luminous hands, sweep seconds. Movement signed "Rolex Patented Super Balance."
Circa 1943

Rolex Oyster stainless steel Athlete. Sunken subsidiary seconds dial and rare silvered Roman/Arabic dial with Mercedes hands.
Circa 1943

Gold filled Rolex Oyster Pall Mall Observatory. Black dial with luminous numbers and printed gilt outer seconds track and printed 13 to 24 inner hour track. Gilt luminous hands and gilt sweep seconds hand.
Model no. 3121
Case no. 187532
Circa 1943

Rolex Oyster Observatory. Two-piece stainless steel cushion case with double stamped bezel. Three-tone metal dial with printed luminous numerals. Outer minutes and seconds track. The seconds track divided into 5ths. Red sweep seconds hand. 10-1/2''' Hunter Patented Super Balance, 17 jewel movement. This was the first Rolex watch to be signed "Everest," probably in honor of the RAF pilots who flew over the mountain in the mid-1930s wearing Rolex watches.
Model no. 3139
Case no. 276000
Circa 1945

Gold filled Rolex Oyster Centregraph with sweep seconds and painted quarter numerals, outer minute and seconds track, two-tone dial.
Model no. 3474
Case no. 185641
Circa 1943

Stainless steel Rolex Oyster Royalite Observatory. Sweep seconds, luminous hands and dial, outer seconds marked into fifths. Note the inner red 13-24 hour markers.
Model no. 2280
Case no. 184606
Circa 1943

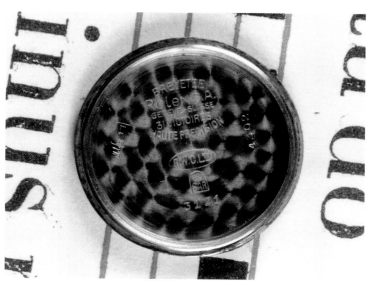

Rolex Oyster Royalite. Sweep seconds, Mercedes hands, luminous two-tone dial with outer seconds track and red 13-24 between seconds track and hours.
Model no. 4220
Case no. 224451
Circa 1943

Rolex Oyster Royal, stainless steel with gold bezel, winder and lug area. Original three-tone metal dial with luminous quarter numbers, luminous hands and red sweep seconds hand.

18kt gold Rolex Oyster Scientific three-piece case, metal dial with separate hours, minutes and seconds area. Sweep seconds. 10-1/2''' Hunter Ultra Prima, signed "Chronometer."
Case no. 32783
Circa 1937

Stainless steel two-piece Rolex Oyster with exaggerated, wide stepped bezel. Movement 8-3/4''' Hunter. Signed "Observatory Quality." Movement also numbered 8286. This style of watch, sometimes known as the "lifesaver" was the first exaggerated style Rolex Oyster. 18 jewels.
Model no. 2849
Case no. 36597
Circa 1936

Stainless steel three-piece Oyster Royal. 10-1/2''' Hunter Extra Prima Chronometer movement. This and the watch opposite are the antecedents of the bubbleback case.
Circa 1934

Stainless steel Rolex Oyster Egyptian. Metal dial with luminous Arabic numerals and hands, subsidiary seconds. 10-1/2''' Hunter transitional movement. This model of the Egyptian was the first hooded stainless steel Oyster.
Model no. 2518
Case no. 65726
Circa 1939

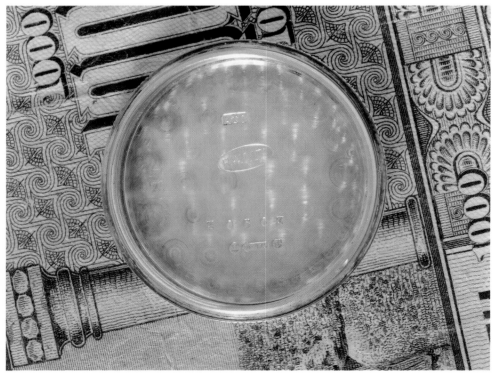

Rolex 9kt gold Oyster Royal. Three-piece case, metal two-tone dial with sunken subsidiary seconds. This watch is extremely unusual in that it is one of only two known with an English-made Oyster case. Case is signed by the famous English casemaker, Alan T. Oliver, and was hallmarked in London, 1934. The presentation engraving on the back shows that it was given as a gift in 1935. Like most gold Rolexes sold in England during that period it has solid strap bars on the lugs and these are also hallmarked.
Hallmark London 1934

Two stainless steel and gold Oyster "Egyptian" watches, both with two-tone silvered and gilt subsidiary seconds dials. Steel bodies with stepped sides and gold double stepped hoods. Both watches have 10-1/2''' Hunter extra prima 17 jewel movements. In the catalogs of the period, there were three different models called the "Egyptian," all with double stepped hooded lugs.
Circa 1934

A tonneau form Viceroy in 14kt retailed by Birks (who were the agents for Rolex in Canada) and signed Observatory above the subsidiary seconds.
Circa 1935 © Sotheby's Inc 1990.

The two watches above, showing that they have consecutive case serial numbers 32020 and 32021.

9kt yellow gold Oyster pocket watch, essentially an Oyster wristwatch magnified four times. The case construction is identical down to the three-piece case construction and the screw down winder. The two-tone dial is signed around the subsidiary seconds bit, leaving room for the retailer to sign at the top. It is also interesting that the watch bears only the original Perregaux/Perret patent number, rather than the later UK patent. This is only the second watch to surface bearing solely this patent number. For the other see Chapter 21. Glasgow import hallmark for 1926, making this one of the earliest known datable oysters.
Movement no. 776526
Case no. 2000237
Hallmark Glasgow 1926

18kt yellow gold Oyster "Imperial." Three-piece case, solid lugs and hallmarked with Glasgow import for 1937. The original black painted dial with painted gilt Arabic and baton numerals was once luminous but now that feature (along with the original dial signature) has disappeared into the mists of time.
Model 3116
Case no. 42329
Hallmark Glasgow 1937

The back of a sterling silver screw back and front "Dennison" transitional wristwatch, showing the first appearance of the milling seen on every Oyster.
Circa 1917

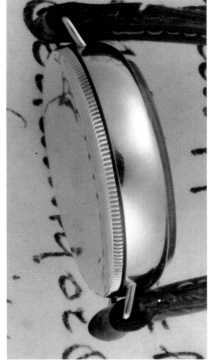

9kt pink gold screw back watch with luminous black and white porcelain dial. Note the skeletonized luminous hands. The two-piece case has a threaded back, and this is the earliest known Rolex with this feature. It is important to note that the milling is on the vertical face of the case back edge unlike the later Dennison screw backs and of course all subsequent oysters.
No model no.
Case no. 759130
Circa 1918

1934 steel and red gold "hooded" Oyster. Despite appearing to be a bubbleback, this is a three-piece case, manual wind watch. It also uses the "open face" movement seen earlier. The dial bears the name and location of the original retailer "Hopkins & Hopkins Dublin."
Model number indistinct
Case No. 31206

14kt yellow gold cushion Oyster with double sunken porcelain dial featuring gothic style numerals and unusual "double Roman" hands. Note wire lugs, milled bezel and oversized crown.
Circa 1937

Chapter 3
THE PERPETUAL
Perpetual Motion

We now turn our attention from the development of the Rolex case to their movements, particularly the movement that was just as important to Rolex as the Oyster case had been: the first Perpetual. From the very beginning, Rolex had used movements made by Aegler. The first model was the Rebberg movement which came in sizes from 7''' up to 13'''. The major advantage of this movement was the high level of interchangability of parts. When repairing a Rebberg movement, a watchmaker could pick a new balance staff from a box of spares and use it, confident that it would fit. Today we take this for granted, but it was not always so. The Rebberg came in a wide variety of sizes and also in versions with a subsidiary seconds hand, with a sweep seconds hand, or without any seconds hand. It was a simple, robust, and versatile movement that served Rolex well for almost 20 years. In 1922 Aegler introduced a new movement, first in only a 10-1/2''' size, and called it "The Hunter." This movement was, to a very large degree, an updated version of the Rebberg. The very first production run used a version of the complete winding gear from the similar earlier model, but the Hunter was built to much finer tolerances. The movement was given the name "Hunter" because its layout resembled that of a hunter pocket watch, having the winding crown at "3" and the seconds dial at "6". This is distinct from an open face pocket watch where the crown is at the "12".

When Rolex first produced a sweep seconds version of the 10-1/2''', the movement it used moved the winding crown 90° from the "3" position to the "12", making it much simpler to add a sweep seconds. To keep the nomenclature correct, this revised movement is signed, under the dial, with the name "10-1/2''' Open Face." Apart from the ultra rare free-sprung versions, this is by far the rarest version of the 10-1/2''' movement.

Over the next fifteen years the Hunter movement was made in sizes from the original 10-1/2''' size through subsequent miniaturization into 9-1/2''', 8-3/4''', and, finally, 7-3/4''' versions. When originally made it used a conventional balance wheel and came in three grades named "Prima", "Extra Prima" and "Ultra Prima." These three grades were usually just 15 jewel movements, but the two upper grades could be modified to make them adjustable to temperature and positions. In fact, the Hunter was the first Rolex movement that was regularly capable of being timed to chronometer precision. These chronometer movements were often either 17 or 18 jewel models with capped escape jewels, and were always inscribed "Chronometer" on the movement and often on the dial. The specific movements which obtained bulletins[12] from the Swiss official timing offices are recognizable by the fact that they are numbered on the movement plate.

In 1935 Rolex patented a new balance wheel design under patent 188077 and quickly brought it to production as the "Superbalance." In this design the balance adjustment screws were recessed from the external surfaces of the balance wheel, which had all its exposed surfaces rounded to present a much more streamlined form. It was thought that the streamlined form would be able to rotate without causing disturbance to the air around it. Conventional balances, the theory went, would dis-

British advertisement for the "bubbleback." Note that it is described as "The Masterpiece of Watch Craftmanship 1934."

turb the air as they passed through it on their revolutions, thereby creating little eddies of air which caused increased air resistance to the balance wheel on its return swing. How successful this was is open to conjecture, but it is worth remembering that the mid-1930s was known as the "Streamline" age. Streamlined steam locomotives were the icon of the time, and the trend toward smoothness extended to such diverse objects as refrigerators, diners, radios and toasters, simply because it was the thing to do. So, while the streamlining of the balance may have had some effect, it was probably more important as a promotional device. The real reason for the higher accuracy obtained from the superbalance was that it was a much heavier balance and with this weight came greater accuracy.[13]

The Hunter was to serve Rolex royally through almost forty years of revisions and updating. It was finally laid to rest as the 600 calibre in the 1960s, but the Hunter was present through many of Rolex's finest moments. It was in the first Oyster models, it was in the 140+ Kew A class chronometers that Rolex built in 1949, and it was the basis of the very first Perpetual.

The quest for a self-winding watch had been going on for almost as long as there had been watches. In the 18th century the world's greatest watchmaker, Abraham Louis Breguet had experimented and built a few prototypes, but the quest for a practical model continued until the early 1920s. As John Harwood tried to find a way to make a watch dirt- and dust-proof, he came upon the notion of doing away with the winding stem, where he thought most of the dust entered. Like most brilliant ideas it raised certain problems, beginning with the winding and setting of the watch. Harwood solved these problems by setting the hands via a rotating bezel and by winding it through the first viable self-winding wristwatch movement. Harwood's movement used a semi-circular winding weight pivoted at the center of the movement and swinging through an arc of almost 300 degrees. This weight was connected by a train of wheels and pinions to the main spring. To prevent overwinding, Harwood invented a simple blade spring which held a friction plate in position.

He had solved the main problems of a self-winding watch, but was unable to convince anyone to put the watch into production. He was saved by the intervention of Harry Cutts, a

First known advertisement for the Oyster Perpetual, 1934. Note that the watch is not signed "Oyster."

First post-War American advertisement for the bubbleback.

businessman (we would now call him a venture capitalist), who invested enough money in Harwood's company to save it. Cutts then set about exploiting the patent in a most unusual way. Instead of setting up a factory to build the watch, he first launched the company on the stock market and then set about selling exclusive national manufacturing franchises for the watch. This resulted in Fortis making the watch for sale in most of Europe, Blancpain building it for France and Spain, and the Perpetual Watch Company manufacturing it for the U.S.A and Canada. Harwood and Cutts made their money from the initial fees paid by the licensees and from royalties on each watch subsequently sold. The problem was that this arrangement was designed to make the maximum money for the investors, not to promote the watch.

The Harwood watch also had problems. It was made in only one size for both men and women and it was a round watch at a time the fashion demanded tonneau or rectangular shapes (although later Blancpain did make some shaped models).

Despite their less than massive success the company continued. Its possession of a patent which encompassed both a centrally mounted rotor and a functioning anti-overwinding mechanism made it impossible for anyone else to utilize either of these obvious solutions to the problem of the self-winding watch. Because of this, and in an attempt to follow fashion, all of the other companies who attempted to manufacture automatic wristwatches during the 1920s did so using a square or rectangular case and movements which relied on alternative winding systems. These ranged from such bizarre ideas as winding actuated by the expansion and contraction of the wrist (Wyler and Autorist), movements which rolled inside the case (Rolls and Wig Wag), pendulums (Perpetual Watch Co.) and winding weights which moved vertically on rails (Pierce).

While all of this was technically very interesting, the simple truth was that the very presence of the Harwood patent was restricting the development of the self-winding wristwatch via the most logical route. All of this changed overnight in 1929, when Harwood finally declared bankruptcy. The patent was still there of course but there was no-one to defend it.

Rolex did not follow the herd, and made no major attempt to produce an automatic watch which avoided Harwood's patent. There is a 1932 Swiss patent, no. 161351, showing a bizarre rectangular watch in which the dial remained fixed but the entire tonneau movement rotated through 30°, however, to the best of our knowledge Rolex never produced this watch. They had remained true to the concept of manual wound watches. In fact, this time when competitors all around them were introducing wild and wacky automatics, was the period when Aegler was developing the movement that would be known as the Prince.

Aegler S. A. Fabrique des montres: Brevet N° **161351**
Rolex & Gruen Guild A.   *1 feuille*

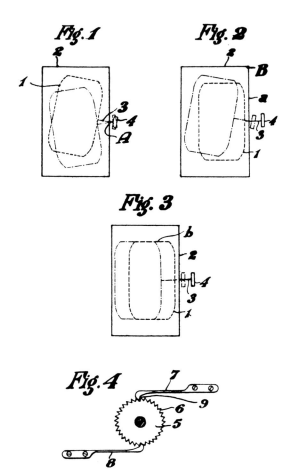

Patent No. 161351

For Rolex, the Oyster was proving very successful and met or exceeded all of Wilsdorf's hopes for it. There was one problem: the customers. People kept forgetting to rescrew the crown onto the tube after they had wound the watch, allowing water to seep in. Inevitably they would return to the jeweler who sold them the watch and say "This watch is supposed to be waterproof!" Wilsdorf realized that there were only two reasons to use the crown: to set the watch to time and to wind it. He knew that his watches were now accurate enough not to need hand setting more than once or twice a month, but he needed to get around the problem of daily hand winding. He needed an automatic system.

Once again Rolex (or Aegler, who actually patented it) took the road of patient modification of an existing product rather than the "blue sky" approach of so many of its rivals and of course it was Rolex who survived while the others are lost in history. The first perpetual models are simple subsidiary seconds 8-3/4‴ Hunter movements with the "Auto-Rotor" mechanism bolted straight onto the back. Because the rotor mechanism was on top of the regular movement, the winding stem could pass into the movement at a level where it did not obstruct the rotor. This allowed the rotor to revolve through a completely unobstructed 360°. This arrangement had three additional benefits. First, the watch could be wound manually if needed[14]. The facility to wind an automatic watch is now something to be taken for granted, but it is interesting to note that all of the early automatics listed above were incapable of manual winding. Apart from its use during periods of enforced idleness, the manual facility on the Rolex Perpetual gave confidence to those people who still did not trust the watch yet.

Second, because the rotor swung through a full rotation, there was nothing for it to hit (unlike all previous automatics), so the movement received no shocks through the sudden braking of a weight. This also meant that the watch was much less noisy than competing automatics, a benefit Rolex advertised heavily with the slogan "Rolex the silent self winder."

Finally, the arrangement allowed the watchmaker to remove the whole of the automatic mechanism with only two screws and then be faced with a simple manual movement that he would instantly recognize. This is an important factor in the success of Rolex's automatic movement. Many people believe that it was objections by the watchmaking trade that doomed the Harwood and its contemporaries. Rolex took this lesson well to heart and on the first Auto Rotor models the back of the mechanism is covered with instructions to the watchmaker on all three levels and in both French and English[15].

The principle disadvantage of the design was the thickness of the watch. In the end this disadvantage was far outweighed by the massive benefits the new design conferred: it was the first usable everyday automatic, waterproof wristwatch. In the end it was this additional thickness that gave the watch its nicknames; the "bubbleback" in English and the "Ovetto" (or little egg) in Italian.

The Auto-Rotor movement used a semi-circular rotor, pivoting from the center of the watch and geared through intermediate ratchets and pinions on to a specially modified raised winding wheel. The main breakthrough was a tiny flat spring, shaped like a three-armed swastika. This spring sat on the rotor and transferred the motion to the winding gear, a tiny click and spring prevented the rotor from unwinding the watch on its return. As with most Rolex innovations, it was a brilliantly simple device and one that made history. The simplicity of the device was rewarded with Swiss patent no. 160492 on May 16, 1933 and was first fitted to a model 1852 wristwatch in late 1933 and launched as "The watch sensation of 1934."

The exact date of the launch of the Rolex Perpetual would seem to be in late 1933 or early 1934. No earlier Rolex advertisement makes any mention of the watch. While the very first advertisements talk of "the watch sensation of 1934," a watch

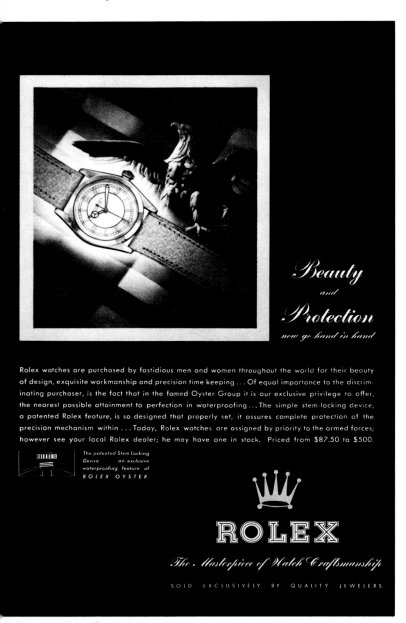

Canadian flyer for the bubbleback.

First known U.S. ad for the Oyster Perpetual, 1944

shown on page 82 bears an import hallmark dating from 1933, and its serial number of 29562 dates it as one of the very first Perpetuals[16]. More proof for this theory can be gained from an article written for "Practical Watch & Clock Maker" in late 1932. It describes a visit to the Rolex factory in Bienne and talks about "a factory which has produced *two* such interesting watches as the Rolex Oyster and the Rolex Prince..." but does not mention the Perpetual. The only problem with the date of 33/34 is that Rolex, in all their advertisements, from 1945 onward, talk about a date of 1931 as the introduction of the first Perpetual.

Interestingly the early automatics mentioned earlier seem to have soured the Swiss watch industry on the whole idea of automatic watches. In a review of the 1934 Basel Spring Fair "Watch & Clock Maker" reported that "Little can be said on the score of extreme novelty. Self-winding watches are shown in very few instances, the Wyler being the most prominent. Whether this is an indication that there is no particular demand for an automatic watch generally, or whether it is that the Swiss local market does not want them, is not apparent"[17]

The first Perpetuals were produced in a case that was another classic Rolex modification, simply the first model Oyster Royal with a new, deeper back. These cases were designed to take a manual 10-1/2''' hunter and the frame that held the rotor and winding gear on top of the 8-3/4''' movement was made to exactly the same diameter as the original 10-1/2''' hunter. This meant that the same case could be used, except for the new deeper back. They were all three-piece cases, like the very first Oysters. The movement and dial were held in an externally screwed brass movement ring, which, in turn, was located inside the main case by a pin at the "9 o'clock" position and by the winding stem. The separate bezel and back would then screw onto the movement ring.

These first models were all subsidiary seconds watches and almost always had very simple dials. They were the very first Rolex watches to be issued in a new stainless steel material called "Steelium." The Perpetual also was the first Rolex watch to introduce a style that has become synonymous with the company; a mixture of steel and gold[18]. The early watches were advertised as being available in "steel with gold bezel and winder." Perhaps the most amazing thing about the Perpetual was that, at its introduction in 1934, it cost 30% more in 18kt than an 18kt Prince; such was Rolex's belief in the watch's prestige.

Despite our modern perceptions, the watch was not an immediate success. Too many people, consumers and watchmakers alike, had had bad impressions of other early automatic wristwatches, and these prejudices were initially hard to overcome. However the usual Rolex processes of modify, improve and simplify were already at work.

The first thing to be improved was the movement. These second generation movements are readily recognizable by the chamfered edge to the rotor. Much more importantly they are no longer signed on the dial plate as 8-3/4''' Hunter, but by their own designation as calibre NA. The movements had several major improvements. The base calibre was now the 9-3/4''' Hunter, which was often fitted with a micrometer regulator. The most visible advance was that the movement was capable of driving a center seconds hand. These new models were still signed on the winding weight as "Auto Rotor," but almost all of the watchmaker's instructions, which had so distinguished the first model, were no longer there. The new models were available in a much simplified case. Though it looked almost identical the new case did away with the separate movement ring and removable bezel, consisting of only two parts: a case unit consisting of the center and bezel together and a removable back. Otherwise there was precious little difference in the look of the case. It still had a plain bezel and a flat back, but it did have a new model number 3131. This model was simpler (and cheaper) for Rolex to make and, with the new sweep seconds hand, it looked a lot more modern.

These two-piece case models retained the flat back of the original Perpetuals with both the case and model numbers stamped on the outside of the case back. For today's collector this has caused certain problems with the gold versions where over-polishing (or just plain wear over 50+ years) has caused the numbers to wear off. It is important to note that the gold versions of both the 3131 and 3132 were still made as three-piece case models and were made for more than another ten years.

On most models of this period Rolex charged a significant premium for the sweep seconds hand. In most of the promotional literature of the time it is described as "being suitable for the medical profession." These second model Perpetuals came in two different models, the 3131 sweep second model and the 3132 subsidiary seconds model, with the latter being approximately 5% less expensive than the former. 1940 saw the introduction of the model 3372, which was the first to feature the classic "machined" bezel with double batons at "12", single batons at the other quarters and circles for the remainder of the hour markers. The other interesting thing about the 3372 was that it reverted to three-piece case construction, even for the steel models.

What we must not forget is that in introducing the Perpetual in 1934, Rolex chose to do so in the depths of the worst depression the industrial world had seen, and as the world's economies began to improve the Perpetual was waiting.

---

[12] In English they would be known as "timing certificates."

[13] It is interesting to note the similarity between the "Superbalance" and the much later Patek, Philippe "Gyromax" balance.

[14] While the self-winding watch has proved to be one of the major success stories of this century, the necessity for it can still be argued. The energy needed to impart a swing to the balance is only four times that used in a flea's jump. It has been calculated that 1 horsepower would be sufficient to operate 270 million watches. So while the self-winding watch may have many benefits, it is difficult to argue that labor saving is one of them.

[15] This very first model Perpetual has the nickname in Italian of "Didatico" or "educator," due to these instructions on the movement.

[16] Taking the start of the Oyster serial numbers to be 20,001, would mean that Rolex had only produced around 9,500 Oysters before the introduction of the Perpetual.

[17] Rolex was not present at this Basel fair and did not show until much later.

[18] The mixture of steel and gold was called "Rollesor" by Rolex, who registered the name and the mixture on April 1, 1933.

The Harwood, the world's first automatic wristwatch with a centrally pivoted rotor. The rotor does not go all the way around and is, in fact, buffered by the small protrusions at the end of the weight.
Circa 1925

82 The Perpetual

9kt yellow gold Oyster Perpetual. Silvered three-tone dial with luminous Arabic and baton hour markers, subsidiary seconds and parallel luminous hands. The earliest known datable bubbleback, the hallmark inside the case dates the watch as 1933 and the case number is only 29562. Unusually not only is the watch very early, but it is in excellent condition and is shown in its original box.
Hallmark Glasgow 1933

Stainless steel and gold Rolex Oyster Precision with applied numeral Arabic dial subsidiary seconds. Notice three-piece case construction and movement fully signed with watchmaker instructions.
Case no. 48439
Circa 1937

9kt gold three-piece case Rolex Oyster Perpetual. Subsidiary seconds and note movement, signed "Autorotor" but without watchmaker instructions.
Model no. 3130
Case no. 48013
Circa 1937

84 The Perpetual

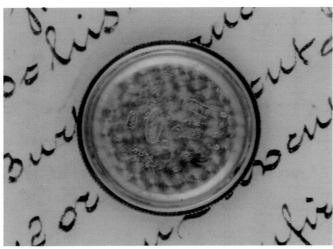

9kt gold three-piece case Rolex Oyster Perpetual. Early movement with full signature. Metal dial with applied Roman numerals. English made Oyster case signed ATO (Alan T. Oliver) but numbered in correct date sequence.
Case no. 36750
Circa 1936

Stainless steel two-piece bubbleback, original black dial with Arabic numerals and hands, signed "Chronometer" around subsidiary seconds. Movement signed "Rolex Auto Rotor" and numbered 91244
Model no. 2764
Case no. 96449
Circa 1939

Rolex Oyster Perpetual 18kt gold three-piece case, dial with subsidiary seconds and painted Arabic numerals. Notice this dial is not signed Oyster.
Circa 1935

Stainless steel and pink gold bubbleback. Subsidiary seconds, original dial, three-tone pink dial with luminous numerals. Note original two-tone expanding oyster bracelet. Two-piece case.
Model no. 3132
Case no. 45424
Circa 1939

86 The Perpetual

Stainless steel and yellow gold three-piece case bubbleback with painted subsidiary seconds dial. Signed "Rolex Perpetual." Like the watch opposite and the one in the advertisement shown on page 78, it is not signed "Oyster," though it obviously is one.
Circa 1936

14kt gold two-piece case bubbleback with Arabic and dot applied numeral dial.
Model no. 3131
Circa 1947

Stainless steel two-piece case bubbleback, subsidiary seconds dial, applied Arabic numerals and triangular batons.
Model no. 2764
Case no. 549803
Circa 1946

Stainless steel two-piece case bubbleback with retailer's name Mappin on the pink luminous number dial. Movement signed "Autorotor."
Model no. 2940
Case no. 145329
Movement 96694.
Circa 1941

88 The Perpetual

Stainless steel two-piece case, subsidiary seconds bubbleback, pink applied Roman numerals and batons.
Model no. 2764
Case no. 244975
Circa 1943

Gold and stainless bubbleback with smooth gold bezel with black half-Arabic half Roman dial. © Sotheby's Inc 1990
Circa 1947

Stainless steel round end bubbleback with subsidiary seconds and luminous even numbers and Mercedes hands.
Circa 1938

Stainless steel and gold two-piece bubbleback, bezel with machined hour markers. Luminous dial with inner minute track.
Circa 1938

Stainless steel bubbleback with pink gold bezel and winder, black and pink luminous dial and associated non-Rolex bracelet.
Circa 1945

18kt yellow gold three-piece case bubbleback with original black dial with luminous hour markers and luminous hands. Sweep seconds, dial signed "Rolex Oyster Perpetual." "Chronometer" above the six and with the retailer's name, "Beyer," above below the twelve. Movement signed "Autorotor"
Model no. 3131
Case no. 37793
Circa 1935

90 The Perpetual

Stainless steel and yellow gold two-piece case bubbleback, with yellow gold bezel and winder, applied numeral and baton dial.
Circa 1945

Stainless steel and pink gold two-piece case bubbleback. Bezel with hour markers, metal dial with applied gilt baton markers.
Circa 1948

Pink 14kt gold three-piece case bubbleback with later refinished luminous two-tone pink dial. Notice on this watch the double hour marker which should normally be at the twelve has been turned around and is at the three.
Circa 1947

18kt yellow gold three-piece case bubbleback, metal dial with applied Arabic and baton numerals, sweep seconds. Original 18kt expanding Oyster bracelet.
Model no. 3372
Case no. 348306
Circa 1946

The Perpetual 91

Two-piece stainless steel case with refinished black dial and luminous Mercedes hands. Interestingly this case does not seem to have a case number. Merely the number 10807 on the end opposite the model Number.
Model no. 2940
Circa most likely 1947 or 1948

18kt gold three-piece case. Movement signed "Rolex Perpetual Chronometer."
Circa 1947

92 The Perpetual

18kt gold three-piece case bubbleback, with metal dial and applied Arabic numerals, skeletonized Mercedes hands, sweep second hand with red arrow tip. Original 18kt gold Oyster expanding strap.
Model no. 3372
Case no. 387216
Circa 1946

14kt gold two-piece case, subsidiary seconds, bubbleback. Applied Roman numeral and baton dial, dial signed "Chronometer." Movement signed "Rolex Oyster Perpetual."
Circa 1947

Stainless steel two-piece case Rolex Oyster Perpetual, metal dial with applied gilt Arabic and baton numerals, sweep seconds. Movement signed "Rolex Oyster Perpetual."
Circa 1948

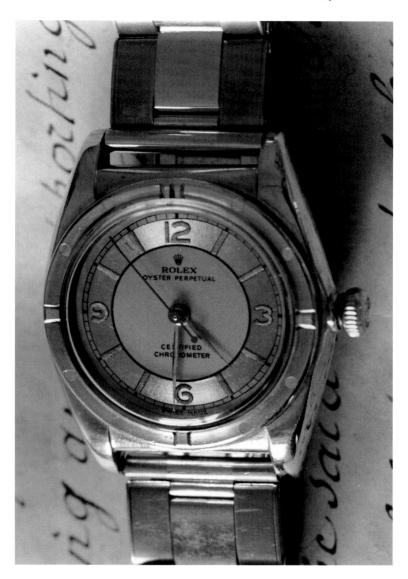

18kt pink gold three-piece Oyster Perpetual with refinished dial, with two-tone pink and luminous numerals. With original expanding Rolex bracelet.
Circa 1942

Stainless steel two-piece case bubbleback with original silvered Roman-Arabic dial and steel Mercedes hands. With original expanding bubbleback bracelet.
Model no. 2940
Case no. 529163
Movement signed "Rolex Perpetual Chronometer" and number N57350
Circa 1947

Stainless steel two-piece Rolex Oyster Perpetual signed chronometer on the dial with luminous hands and dial.
Model no. 2940
No Case no.
Circa 1941

94 The Perpetual

Stainless steel (steelium) bubbleback, pink gold machined bezel with hour markers and original black dial with mixed Roman and Arabic numerals, sweep seconds, Mercedes hands.
Model no. indistinct.
Case no. 50861
Movement signed "Rolex Autorotor Chronometer" and numbered 91644
Circa 1939

Stainless steel two-piece case bubbleback, sweep seconds dial with crown at twelve and applied luminous batons. With matching expanding Rolex Oyster bracelet.
Model no. 2940
Circa 1943

Stainless steel and pink gold hooded bubbleback with mixed Roman-Arabic luminous dial and Mercedes hands.
Circa 1945

96 The Perpetual

Stainless steel and pink gold hooded bubbleback with horizontal ribbed hoods. Metal dial with applied Arabic numerals, applied minute track and sunken subsidiary seconds.
Model no. 3595
Case no. 46079
Circa 1939

Stainless steel and yellow gold, hooded bubbleback with horizontally ribbed hoods, two-tone gilt and black dial with stick hands and sweep seconds. Notice original beaded steel and gold Rolex Oyster bracelet.
Model no. 3595
Case no. 231988
Circa 1943

Steel and pink gold hooded bubbleback, black dial with luminous mixed Roman Arabic and baton numerals. Inner 13-24 hour markers, red sweep seconds, and Mercedes hands.
Circa 1942

Stainless steel and pink gold bubbleback, two-piece case, applied Arabic numerals and sweep seconds and unusual bezel with double marker at twelve and single bars for the remainder.
Model no. 5015
Case no. 620760
Circa 1948

Stainless steel and yellow gold bubbleback. Unusual bezel configuration with bar at twelve and dots for the remainder.
Model No. 5015
Case no. 633960
Circa 1948

The Perpetual 99

Stainless steel and pink gold hooded bubbleback with luminous dial and hands, sweep seconds.
Model no. 3065
Circa 1948

Steel hooded bubbleback, pink dial with subsidiary seconds, luminous numbers and signed chronometer around the subsidiary seconds, Mercedes hands. Movement NA style bubbleback. Original Oyster beaded bracelet.
Model no. 3599
Case no. 275252
Circa 1945

Stainless steel and pink gold hooded bubbleback. Original black dial with luminous Arabic markers, signed "Self Winding" in a arc between seven and five on the dial. Fitted with original expanding bubbleback Oyster bracelet.
Model no. 2940
Case no. 36102
Circa 1936

14kt pink gold bubbleback hooded model with original black dial (slightly spotted) with luminous full Arabic numbers and Mercedes hands with red sweep seconds.
Model no. 3065
Case no. illegible
Movement number N86580
Circa 1941

Steel and pink gold hooded bubbleback with silver dial with applied Arabic numerals and batons, sweep seconds. This watch is in "as new" condition.
Model no. 3065
Case no. 345024
Circa 1946

Steel and pink gold hooded bubbleback. Original Roman/Arabic dial in black with pink gilt Mercedes hands. Notice original expanding pink gold and steel Oyster bracelet.
Model no. 3065
Case no. 302459
Circa 1945

102 The Perpetual

Boy's size stainless steel bubbleback with metal dial, with Roman quarter numerals, engine-turned bezel and 8-3/4''' Perpetual movement. Notice original model Jubilee style bracelet.
Circa 1945

Stainless steel, two-piece bubbleback with pink three-tone dial with separate hours minutes and seconds graduations. Signed "Rolex Oyster Perpetual Scientific."
Circa 1945

The Perpetual 103

18kt pink gold boy's size bubbleback with luminous dial, sunken subsidiary seconds and Mercedes hands. Original pink beaded bracelet.
Model no. 3767
Case no. indistinct
Circa 1939

Stainless steel boy's size Lifesaver bubbleback with extra wide bezel with engraved hour markers. Subsidiary seconds dial with sunken baton and square hour markers.
Model no. 3348
Case no. 153914
Circa 1941

Boy's size Oyster Perpetual Lifesaver model with sunken subsidiary seconds dial. Luminous and square hour markers also signed "Serpico y Laino," the Caracas, Venezuela retailer for Rolex. This watch is fitted with the original bamboo style Rolex Oyster bracelet.
Model no. 3348
Case no. 153990
Circa 1941

18kt gold round end bubbleback, two-piece case, original black dial luminous numbers, Mercedes hands and red sweep center seconds. Movement signed "Rolex Perpetual Chronometer." Watch has original 18kt gold expanding Rolex Oyster bracelet.
Model no. 5048
Case no. Indistinct
Circa 1948

Steel and pink gold scalloped hooded bubbleback with original pink luminous dial with subsidiary seconds.
Case no. 46920
Circa 1939

Stainless steel giant bubbleback, metal dial with painted Arabic numerals and sunken subsidiary seconds. Movement 10-1/2''' bubbleback.
Model no. 5026
Circa 1944

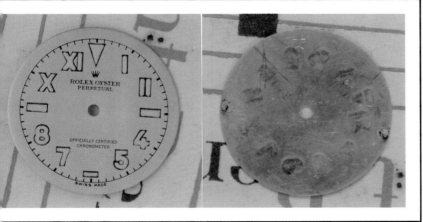

Roman and Arabic bubbleback dial, pink with luminous writing. Notice how on these dials the indentations for the luminous are punched through so they are visible from the back.
Circa 1947

Stainless steel Rolex Oyster Perpetual with sweep seconds. Screw-down crown, silvered matte dial, black Arabic numerals, tonneau water-resistant-type bubbleback, case. Case, dial and movement signed. Diameters 31mm.
© Sotheby's, Inc., 1990.
Circa 1945

106 The Perpetual

A gold front, stainless back "tropical" bubbleback with classic redone "California" dial. © Sotheby's, Inc. 1990.
Circa 1945

A steel and pink gold hooded sweep seconds bubbleback, with original luminous Arabic dial.
© 1990 Sotheby's Inc.
Circa 1940

Stainless bubbleback, with unusual dial.
© Sotheby's Inc 1990.
Circa 1945

A 14kt pink gold bubbleback, engine-turned bezel with hour markers. © Sotheby's, Inc., 1990.
Circa 1945

A small stainless self-winding wristwatch, steel bezel with hour markers and pink dial. © Sotheby's, Inc., 1990.
Circa 1947

Gold and stainless hooded self-winding Oyster Perpetual with unusual fluted gold hood.
© Sotheby's Inc., 1990.
Circa 1935

18kt gold Oyster Perpetual, black dial with applied baton numerals, three-piece case. Ref 3372. © Sotheby's, 1994.
Circa 1945

Rare boy's size bubbleback in 18kt with pink dial signed "Rolex Oyster Perpetual Chronometer." © Sotheby's, Inc., 1989.
Circa 1945

14kt gold Rolex Oyster Perpetual, late "round end" bubbleback, black matte dial with gilt applied Arabic even and triangular odd numerals. Ref. 4392. © Sotheby's, Inc., 1992.
Circa 1955

Rare Oyster perpetual self-winding cushion shaped watch with bracelet, Ref. 4961. 30mm © Sotheby's, Inc., 1992.
Circa 1940

Stainless steel and yellow gold hooded bubbleback, silvered matte dial with applied triangular and Arabic numerals. Ref 3065. © Sotheby's, 1992.
Circa 1947

108 The Perpetual

A gold and stainless steel sweep seconds, hooded Oyster Perpetual. Lever movement, screw-down crown, silvered matte dial, Arabic and baton numerals, tonneau water-resistant-type case. Bezel and hood are pink gold, the remainder stainless steel. Case, dial and movement signed. Diameter 32mm © Sotheby's, Inc., 1990.
Circa 1945

A stainless steel self-winding hooded Rolex Oyster Perpetual. Nickel lever movement, patented Superbalance, screw-down crown, pink matte dial inset with black matte chapter rings, Arabic and lozenge numbers, subsidiary seconds, water-resistant-type case, tonneau with articulated hood. Case, dial and movement signed. Diameter 32mm © Sotheby's, Inc., 1990.
Circa 1935

18kt gold square Rolex Perpetual, Super Precision. White matte dial, applied baton and square numerals, subsidiary seconds, square case with scalloped sides. Ref 8126. © Sotheby's, 1994.
Circa 1945

Stainless steel sweep seconds Oyster Perpetual. Ref. 2940. © Sotheby's, 1994.
Circa 1944

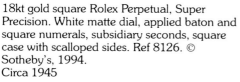

A gold and stainless steel, hooded, sweep seconds Rolex Oyster Perpetual. Screw-down crown, lever movement, silvered matte dial, luminescent Arabic and Roman numerals. Cylindrical water-resistant-type case, hood in pink gold, the remainder stainless steel. Case, dial and movement signed. Diameter 32mm © Sotheby's, Inc., 1990.
Circa 1935

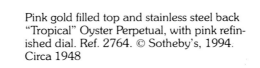

Pink gold filled top and stainless steel back "Tropical" Oyster Perpetual, with pink refinished dial. Ref. 2764. © Sotheby's, 1994.
Circa 1948

A 14kt gold Oyster Perpetual, Ref. 3130. Lever movement, screw-down crown, silvered matte dial, applied Roman and baton numerals, subsidiary seconds, tonneau water-resistant-type bubbleback case. Case, dial and movement signed. Diameter 32mm © Sotheby's, Inc., 1990.
Circa 1945

14kt yellow gold Oyster Perpetual in new old stock (NOS) condition. Note the shine to the bezel and the matte finish to the rest of the case. Silvered dial with applied Arabic and baton numerals, baton hands.
Model 3131
Circa 1942

Pink 9kt gold Oyster Perpetual bubbleback, pink dial with luminous Arabic numerals, sweep seconds and luminous leaf hands. Three-piece case with machined bezel with baton and dot hour markers.
Model 3372
Circa 1946

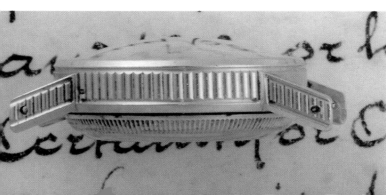

Very rare 18kt yellow gold Rolex Oyster Perpetual "Empire" model. Silvered dial with painted Arabic and baton numerals, subsidiary seconds. Two-piece case features "coin edge" milling all around the periphery of the case and lugs.
Model no. 3877
Case no. 57819
Movement no. N2026
Circa 1942

Chapter 4
SOME DAY MY PRINCE WILL COME

It is difficult to talk about the birth of the Prince in any great detail without looking at the relationship between Rolex and Aegler in more detail. As noted earlier, Rolex used movements from Aegler exclusively, but Aegler sold to many other watch companies. This meant, of course, that in the relationship Aegler had the strong hand.

Rolex wanted to change matters. To accomplish this they made Hermann Aegler a director of the company on the August 25, 1919. On April 12, 1920, Wilsdorf and Davis gave 6,960 unissued shares (worth £1 each) in the Rolex Watch Company to Hermann Aegler. This now meant that there were now three shareholders in the company: Wilsdorf, Davis and Aegler. Aegler was appointed chairman of the Rolex Watch Company for a period of twelve years and this appointment was backdated to July 1, 1919. In return for these shares and the appointment Aegler agreed to "use his best endeavors to promote the interests of the company (Rolex) and in particular shall use his best endeavors to cause the societé (Aegler) to sell its watches to the company exclusively and to no other person, firm or company in the United Kingdom and to sell the same to the company (Rolex) at the lowest current wholesale prices for watches of the like nature and type." At the same time Rolex began to advertise that their "technical department and movement factory" were in Bienne, though these facilities were in fact Aegler's factory. It is safe to assume that there was an exchange of shares between the two companies, with Rolex taking some part of Aegler and Aegler taking the 6,090 shares of the London Rolex parent company. Also at this time the name of Aegler's company was changed and became "Aegler, Société Anonyme, Fabrique des Montres Rolex & Gruen Guild A.," or Aegler Incorporated, Manufacturer of Rolex & Gruen Guild A Watches. It is obvious that by the mid-1920s the relationship between Rolex and Aegler had moved from being customer and supplier to being of associated, but still separate companies.

Slowly, but surely, the nature of Rolex was changing. The three directors now all had Swiss addresses: Aegler lived in Bienne, Davis had moved to Montreux in 1919, and now Wilsdorf lived in Geneva. More important to the evolution of the company was the gradual reduction of the role and share holding of Alfred Davis. Before Aegler came onto the board each of the two partners held 23,100 shares each comprising the initial 14,500 shares issued when Wilsdorf & Davis Ltd. was sold to Rolex and a further 8,600 shares issued in 1919. Only three years after receiving the additional 8,600 shares Davis sold 5,000 of them to Wilsdorf. The following year he sold another 5,000 shares to Aegler and the remainder of his holdings to Wilsdorf and his wife May, who appears on the company documents for the first time in 1924. What happened to Davis after his departure from the company he founded is unclear. It is not even known if he stayed in Switzerland, but his last known address was Grand Hotel des Alpes, Montreux. In the early years of this century tuberculosis was a major killer and prior to the introduction of antibiotics the only respite from it was prolonged convalescence at one of the mountain top Hotels in the Alps. While it is only conjecture, there is a strong possibility that Davis chose Montreux for this reason.

The company was in the hands of Aegler and Wilsdorf when, on August 26, 1926, a patent was filed with the Swiss Patent Authorities. Just over a year later on the October 1, 1927, it was granted the patent number 120849 for the movement that was to be named the Prince. The patent described the watch as being a "shaped watch movement with a seconds dial" and went on to state that the main advantage of the watch over normal watches was that it enabled the use of a much larger, and therefore much more visible seconds hand. The main advantage of the movement design was that, by placing the winding barrel and the balance at opposite ends of the watch, they could each be much larger than if they were arrayed closely to each other, as would be the case in a normal round watch. The

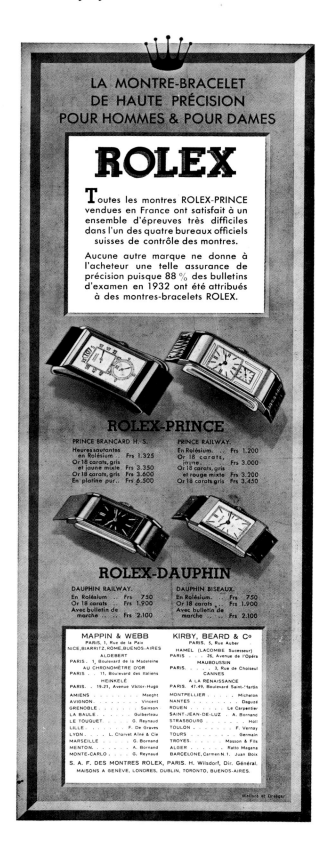

as the Duo-Plan. In this movement the winding and balance wheel were placed on two different levels.

The larger balance did its job; the Prince proved to be one of the most accurate wristwatches made to date. The accuracy of the watches comes from the very high quality balance wheel, which most unusually for a wristwatch, used solid gold screws to add extra weight and, thus, momentum. The watch also had a very high quality finish to the whole escapement, even the wheels. All of these efforts went to ensuring the accuracy of the Prince.

In 1928, when the watch was first introduced, it came in two styles: the Classic model 1343 and the Brancard[19] model 971. The Classic was a plain rectangular model and the Brancard was what is now normally known as the "flared" model. The Brancard was always nearly 10% more expensive than the Classic. The initial production of the watches was available in either 18 or 9kt gold and in sterling silver. The Brancard was available from the start in two color gold, though only in 18kt only. The silver was unusual in that it was chromium plated and was always advertised as "untarnishable silver." This fact has led some unsuspecting vendors of old Prince watches to sell their watch as chromium plated nickel when in fact they were sterling silver.

Two years later Rolex introduced the 18kt model with stripes in the case top, sometimes known as the "tiger stripe" model. That year they also introduced the most expensive watch they had ever made, the Brancard model in solid platinum. At its introduction in 1930 this watch was priced at £65.00 (worth $260.00 at the time), over three times the cost of an 18kt striped model. This was at a time when a small family car would have cost only £85.00.

The introduction of the "tiger stripe" model brought the number of Brancard models to three. Confusingly all had model number 971. The drawings in the catalog page of the time (*see illustration*) show the differences. The 971 had slightly curved end-pieces while on the 971U they were faceted. The end-piece of the 971A was also faceted, but this designation was reserved for the "tiger stripe" model.

In 1934, simultaneously with the launch of the Perpetual, Rolex brought out the Prince in steel. Interestingly, because of the difficulty of working in steel, at its introduction the steel model was more than 10% more expensive than the sterling model. The introduction of steel also enabled Rolex to bring out what was described as "Steel with gold mounts." These steel models with added gold have proved to be one of the easiest models for the unscrupulous to modify or enhance.

The Prince is often known as the "Doctor's" watch. Most people take this to mean that it was often used by doctor's because of its large seconds hand. While this may well have been true for some doctors, the real reason is that the watch was often given to doctors as a gift by grateful patients. Rolex always publicized their sweep seconds watches as "being particularly suitable for the medical profession," remembering that Rolex was making sweep seconds watches as early as 1914 (*see the 1930 catalog on page 44*).

As always Rolex was not content to leave a product alone; its policy of constant modification and updating continued. In the year after the launch of the steel watches they introduced both a new style and a new movement adaptation. The new case was a variation of the Classic model but with stepped sides. It was named the "Railway" Prince and given the model no. 1527. It did not prove to be one of the more popular models, lacking both the simplicity of the Classic and the unique styling of the Brancard. Its subsequent rarity makes it now one of the more collected models.

Simultaneously with the launch of the Railway Prince, Rolex (or Aegler) introduced the H.S. movement (H.S. meaning "Heures Sautantes" or Jumping Hours). In this model the hour hand was deleted and replaced with a slot in the dial at the 60

size of both the balance wheel and the winding barrel meant that proportionately the watch would be much more accurate and that it would run for much longer on a single winding. Both of these turned out to be true. Almost all of the Princes sold in Britain were sold with chronometer certificates. Rolex publicity of the time made much of the fact that the watch "goes for 58 hours on one winding." The idea of separating the balance and winding gear giving a large balance wheel in a quite small watch was not solely the preserve of Aegler. It was being developed simultaneously at Jaeger-le-Coultre where it surfaced minute position. Under the dial was a wheel on which the numbers 1 through 12 were printed, as the minute hand reached the 60 minute position the wheel would jump to a new position and the correct hour would be displayed through the slot in the dial. This additional complication had two disadvantages: thickness and difficulty in reading. The watch was now some 0.8mm thicker in order to accommodate the wheel and its attendant gears. While designed for ease of reading, it was, in fact, more difficult to read than a conventional dial. When its premium price is added to the mix, these factors mitigated against its success. Once again we have a watch that was a commercial failure when launched and is now a very desirable collector's piece[20].

The Prince name became so well known that, by 1933, Rolex began to trade on it. When the new HW tonneau-shaped movement (can it be any coincidence that those are Hans Wilsdorf's initials?) was launched, the range of watches using it was called the "Prince Dauphin" or Junior Prince. If any justification is needed for using the Prince name, a glance at the winding trains of the two will show that they are almost identical. Their line of higher cost lady's watches was called the "Princess."

Rolex manufactured some strange versions of the Prince, but none were stranger than the model called the "Sporting Prince." This model used either the conventional or the jumping hour movement in a hunting-style rectangular pocket case. When the front of the case was pushed, it would spring open and an inner case bearing the movement and dial would then pop up enabling it to be easily read. The rationale behind the watch was that, prior to the introduction of shock-resistant balances, it was inadvisable to wear a wristwatch while engaged in any form of sporting activity and most definitely not while playing golf[21]. The impact of the club hitting the ball would be transmitted up the shaft to the golfer's wrist and could very easily dislodge or break the pivots of a conventional balance staff. So began the rise of the sporting or purse watch. Every manufacturer from Movado through Longines to Vacheron et Constantin got into this market.

Rolex also used the technology invented (and patented under British patent nos. 333,853; 334,491 and 334,492) for the Sporting Prince to make a very few hunting cased Prince wristwatches model no. 1599. These are by far the rarest Prince models of all; less than a dozen are known. They are often recreated today by fakers using the Sporting Prince case with lugs soldered on. However it is quite easy to spot this modification as the back of the correct wristwatch is concave and the back of the Sporting Prince is convex; that is to say, the correct wristwatch case is curved to the wrist while the pocket watch back is bowed outward just like the front of the watch.

In 1940 two more case styles were added. The first was the asymmetric or "wedge shaped" case model no. 3362, which was the standard dial model, and its variation, model no. 3361, which was an asymmetric case with a sweep seconds movement. Strangely, this meant that the area where the seconds dial had originally been was now blanked over by the case. In some of these models a shield shaped piece of gold was added so that a family crest or a monogram could be engraved. The reason for the wedge shape was to allow for the additional depth of the center seconds pinion and wheel. The other case style introduced at this time was the "curled lug" model no. 3937. This was a style much more in the current vogue for the "streamlined" look. It became the basis of one of the most desirable of all Princes, the "Quarter Century Club." These watches were awarded to employees of the Eaton Company, a major Canadian department store chain, who had worked for the company for twenty-five years. They were 14kt gold and beautifully engraved on the back, but were most unusual in that the dial was custom-made for the Eaton company. Instead of the normal dial numerals, it had the letters "1/4 C E N T U R Y C L U B" inscribed and enamel-filled and in place of the name Rolex was "Eaton". Most of these watches, when sold by their original owners in the 1980s, had the dials refinished because they were perceived as being worth less than a regular Prince. Because so many were refinished, now, of course, a watch with a correct 1/4 Century Club dial is worth more than a standard one. Ironically people are now making reproduction 1/4 Century Club dials. So the world turns.

---

[19] "Brancard" means "stretcher" in French.

[20] In the Rolex literature of the time, the HS model was known as the "Automatic Hours" model.

[21] Then, as now, a high status pastime and therefore one likely to appeal to Rolex Prince owners.

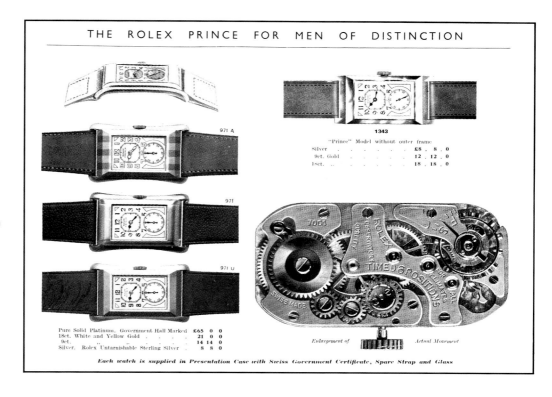

Three different Prince models from a 1930 catalog.

Before leaving Princes, we need to take another look at the name of the Aegler company. They manufactured for both Rolex and Gruen, including a Gruen watch that is nearly identical to the Prince. The question everyone asks is "Why was the identical watch sold as both a Rolex and a Gruen?" The answer is quite simple, when the watch first hit the market in 1928, Rolex was still a London based company selling almost exclusively in the British Empire and Europe, they had no distribution facilities in the United States, then as now, one of the largest markets in the world. Aegler had invented the watch and put time and money into it. They wanted to recoup their investment from as large a market as possible, and so Rolex was given the "franchise" for the British Empire and Europe, and Gruen was given the "franchise" for the United States. These were the only markets anywhere in the world worth selling an expensive watch to.

— 22 —

N° 58875. 7 juin 1937, 18½ h. — Ouvert. — 2 modèles. — Montres-bracelets avec dispositif protège-glace. — Hans **Wilsdorf**, Genève (Suisse). Mandataire: A. Bugnion, Genève.

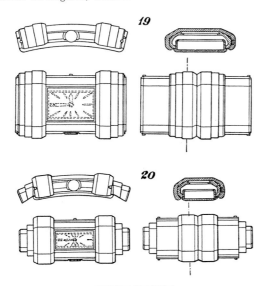

N° 58876. 7 juin 1937, 18½ h. — Ouvert. — 1 modèle. — Boîte de montre-bracelet. — Hans **Wilsdorf**, Genève (Suisse). Mandataire: A. Bugnion, Genève.

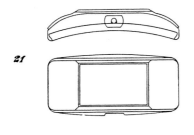

In 1937 Rolex designed a version of the Prince with hoods which became dial shutters. This unproduced model seems to be their answer to the recently introduced Reverso.

Rolex advertisement from Asprey's, London, featuring Princes.

Liste der Muster und Modelle
Liste des dessins et modèles — Lista dei disegni e modelli

N° 10

Zweite Hälfte Mai 1931
Deuxième quinzaine de mai 1931 — Secondo quindicina di maggio 1931

II. Abteilung — IIe Partie — II^a Parte

Abbildungen von Modellen für Taschenuhren
(die ausschliesslich dekorativen Modelle ausgenommen)
Reproductions de modèles pour montres
(les modèles exclusivement décoratifs exceptés)
Riproduzioni di modelli per orologi
(eccettuati i modelli esclusivamente decorativi)

Nr. 47741. 15 mai 1931, 18 h. — Ouvert. — 3 modèles. — Cadrans de montres. — Hans **Wilsdorf**, Genève (Suisse). Mandataire: A. Bugnion, Genève.

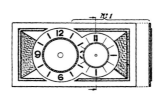  

Design drawings for the Sporting Prince.

Nr. 47776. 23 mai 1931, 15 h. — Ouvert. — 1 modèle. — Cadran de montre à heures sautantes. — Hans **Wilsdorf**, Genève (Suisse). Mandataire: A. Bugnion, Genève.

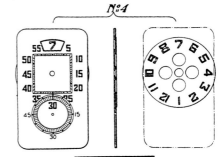

Design drawing for the jump hour Prince.

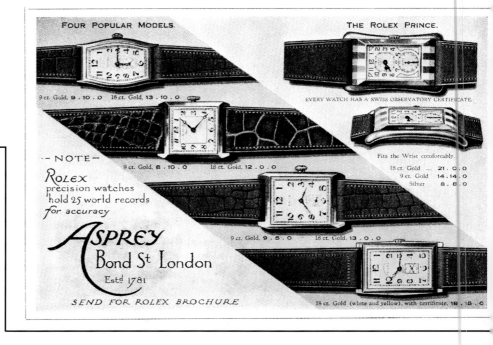

Someday My Prince Will Come 115

English Rolex advertisement for 1933 listing six dealers in India.

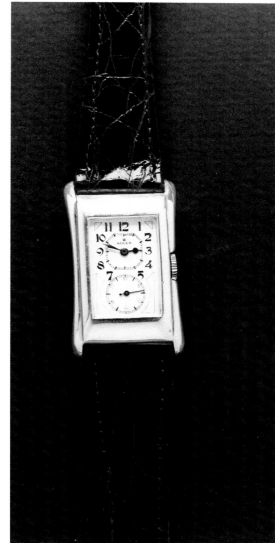

Sterling silver Rolex Prince. Flared case.
Model no. 971
Circa 1937

Stainless steel and gold straight cased Prince. Three-tone dial. This case is all stainless steel apart from a thin bar of gold down each side of the dial and the lugs themselves which are solid.
Model no. 1862
Case no. 012911
Circa 1929

Someday My Prince Will Come

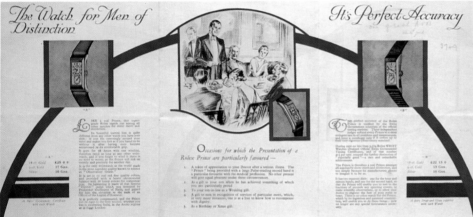

Promotional literature for the Rolex Prince from the early 1930s.

9kt gold Rolex Prince, straight sides, refinished dial.
Circa 1933

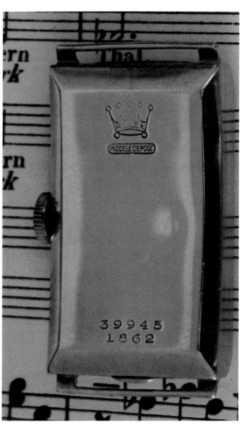

Straight-sided 9kt gold Rolex Prince. Original two-tone dial signed "Chronometre."
Model no. 1862
Case no. 39945
Circa 1936

Someday My Prince Will Come

Double stepped case stainless steel Rolex Prince. Note rare original black dial.
Model no. 1768
Case no. 010040
Circa 1928

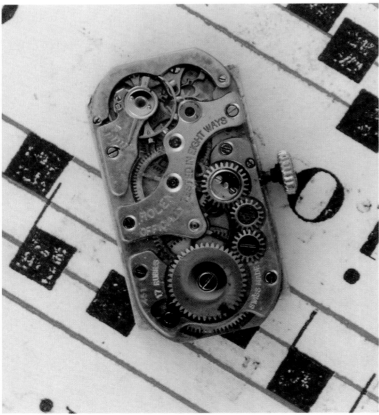

9kt gold flared, two-tone Rolex Prince. Yellow center, white sides. Movement 17 jewels with protective balance bridge.
Model no. 1491
Case no. 33761
Circa 1937

Someday My Prince Will Come 119

Two color pink and white 18kt gold railway, jump hour Prince. Notice the aperture at twelve for the hour indication, and the alternating colors of the step-sides on this watch.
Circa 1938

18kt gold flared single color Prince. Notice the sharp edge to the top case.
Circa 1936

18kt gold one color yellow step-sided Prince, original two-tone dial with retailer's name which has been inexpertly removed.
Circa 1937

18kt gold railway Prince, two colors, two-tone dial.
Circa 1938

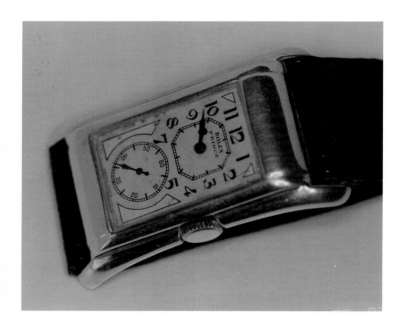

Sterling silver flared case Prince.
Model no. 1491
Circa 1935

9kt gold two color flared Prince. Solid lugs.
Presentation inscription dated 1937.
Model no. 1490
Case no. 8219
Circa 1937

14kt gold flared case, Rolex Prince, single color, yellow. Sold to T. Eaton & Co. and given by them to members of their staff who had served 25 years or more, thereby becoming eligible to join the Quarter Century Club. This watch was presented to George Sergeant who served from 1921 to 1946. The dials on these watches are always sterling silver. They are engraved with the numerals which are then filled with black enamel before being fired.
Model no. 1490
Case no. 1309294
Circa 1940

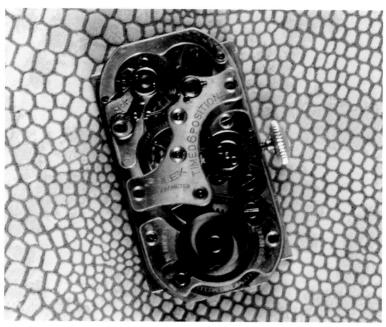

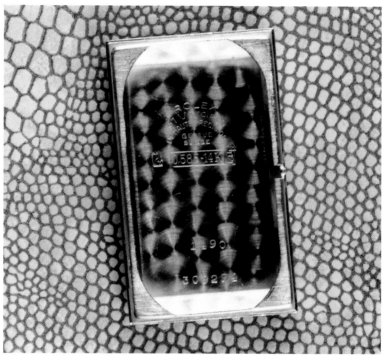

Two color striped 18kt gold flared Prince. This watch is one of the rare later models with the patent protective cap, which was a complete movement cover. The watches fitted with them have movements in excellent condition.
Model no. 971A
Case no. 1036
Circa 1938

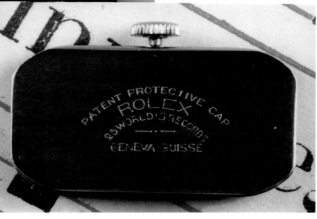

9kt gold single color, flared Prince with engraved sides. Presentation engraving dated 1931.
Model no. 971
Case no. 65269
Circa 1931

Rare flared two colored Prince with engraving along the sides of the dial.
Model no. 971
Case no. 71932
Circa 1931

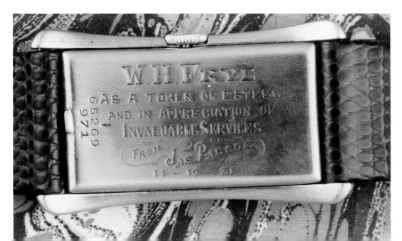
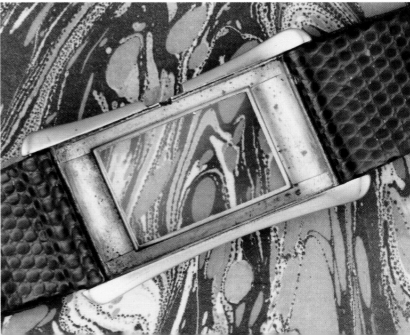

124 Someday My Prince Will Come

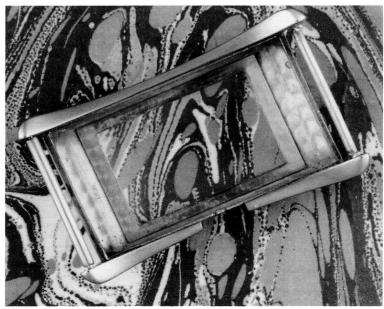
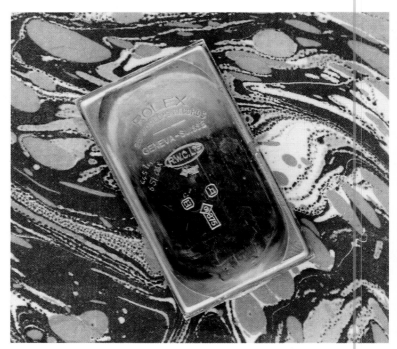

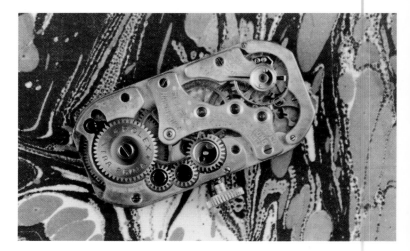

9kt gold, two-tone, flared Prince, fitted with a patent protective cap.
Model no. 971 U
Case no. 73309
Circa 1937

Someday My Prince Will Come

Stainless steel flared Prince with solid bars.
Movement fitted with Patent Protective Cap.
Note solid strap bars even on a steel watch.
Model no. 1490
Case no. F4150
Movement 73271
Circa 1938

Rolex Prince, 18kt two color gold with striped top. Movement fitted with patent protective cap. This watch still has the original winder signed "Rolex Patent," a very rare event.
Model no. 971A
Case no. 1036
Movement number 73800
Circa 1932

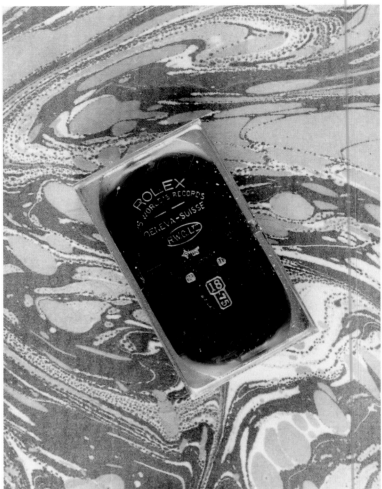

Someday My Prince Will Come 127

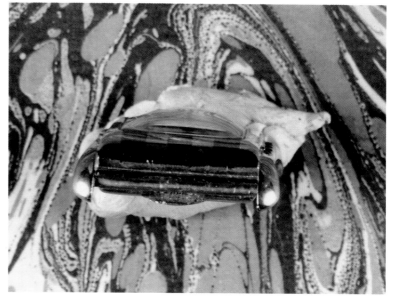

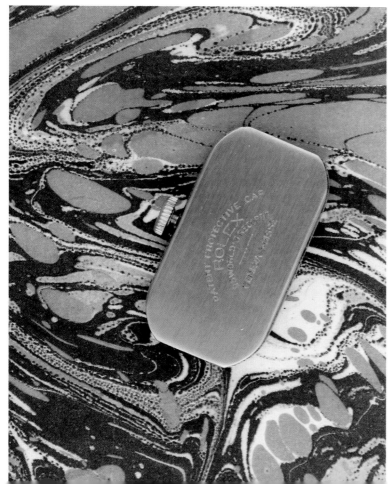

14kt gold Rolex Prince, sold to the T. Eaton Company, Quarter Century Club. Presentation engraving "1909 to 1934."
Case Style 1490
Case no. 309608
Circa 1932

Someday My Prince Will Come 129

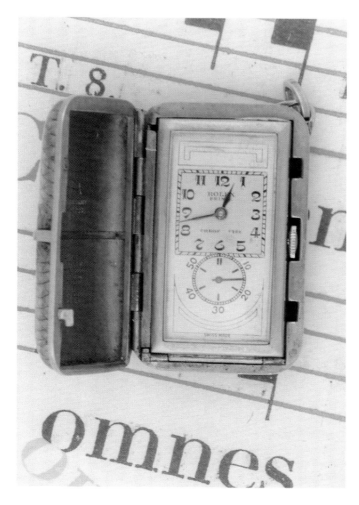

Steel hunting case Prince. Carried as a sport watch, these cases were normally covered in either in animal skin or lacquer. They are in fact much rarer than the standard Prince, but unfortunately no where near as desirable.
Circa 1937

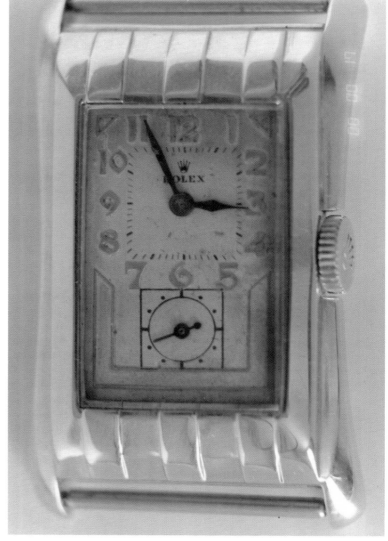

Single color flared Prince with rare stepped striped top. Applied Arabic numerals.
Circa 1938

Two-tone gold Rolex Prince. Almost the last model Prince made, with applied Roman, Arabic, and dot numerals. This was an attempt to bring the Prince into the modern age, but, unfortunately, it failed.
Circa 1952

9kt yellow & white gold square wristwatch with flared sides, once again showing the influence of the Prince. Two-tone quartered or "bow-tie" dial, subsidiary seconds and cathedral hands. Note once again, the solid case bars, a feature of most Rolex watches sold in Britain up until the outbreak of World War II. Stamped "7 World's Records."
Model no. 912
Case no. 55937
Hallmark Glasgow 1928

18kt gold, flared rectangular cased watch. While this is not a Prince it obviously displays the influence and styling of a Prince. Circa 1937

Square rectangular 9kt wristwatch with original black dial. HW movement, solid wire lugs. Once again another watch showing the influence of the Prince on Rolex design of the period.
Circa 1937

Someday My Prince Will Come 131

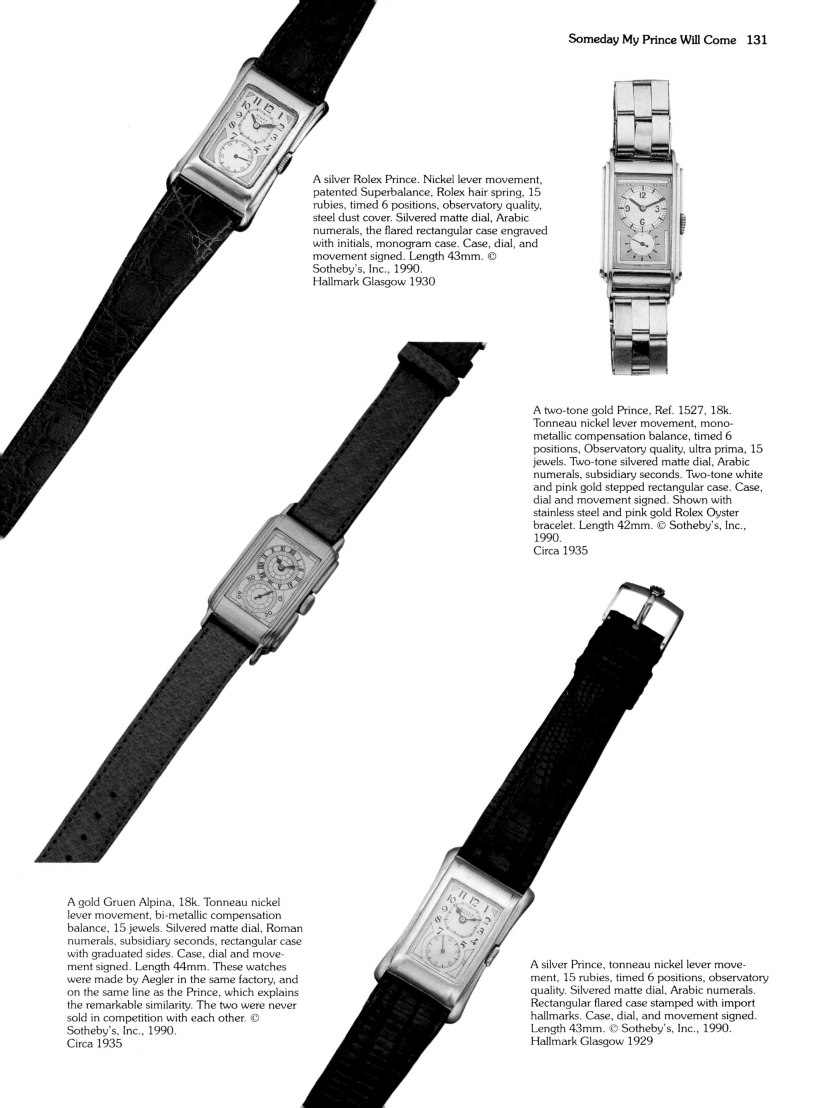

A silver Rolex Prince. Nickel lever movement, patented Superbalance, Rolex hair spring, 15 rubies, timed 6 positions, observatory quality, steel dust cover. Silvered matte dial, Arabic numerals, the flared rectangular case engraved with initials, monogram case. Case, dial, and movement signed. Length 43mm. © Sotheby's, Inc., 1990.
Hallmark Glasgow 1930

A two-tone gold Prince, Ref. 1527, 18k. Tonneau nickel lever movement, mono-metallic compensation balance, timed 6 positions, Observatory quality, ultra prima, 15 jewels. Two-tone silvered matte dial, Arabic numerals, subsidiary seconds. Two-tone white and pink gold stepped rectangular case. Case, dial and movement signed. Shown with stainless steel and pink gold Rolex Oyster bracelet. Length 42mm. © Sotheby's, Inc., 1990.
Circa 1935

A gold Gruen Alpina, 18k. Tonneau nickel lever movement, bi-metallic compensation balance, 15 jewels. Silvered matte dial, Roman numerals, subsidiary seconds, rectangular case with graduated sides. Case, dial and movement signed. Length 44mm. These watches were made by Aegler in the same factory, and on the same line as the Prince, which explains the remarkable similarity. The two were never sold in competition with each other. © Sotheby's, Inc., 1990.
Circa 1935

A silver Prince, tonneau nickel lever movement, 15 rubies, timed 6 positions, observatory quality. Silvered matte dial, Arabic numerals. Rectangular flared case stamped with import hallmarks. Case, dial, and movement signed. Length 43mm. © Sotheby's, Inc., 1990.
Hallmark Glasgow 1929

132 Someday My Prince Will Come

An 18kt pink gold Prince "Chronometre," Ultra Prima with two-tone pink dial, 42mm. © Sotheby's Inc 1990
Circa 1936

A two-tone 14kt gold Prince Chronometre. Nickel lever movement, with chased geometrical engravings. © Sotheby's Inc 1990
Circa 1940

A two-tone gold Prince Railway model with Deco dial. 42mm. © Sotheby's Inc 1990
Circa 1936

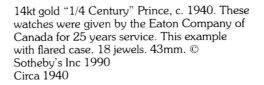

14kt gold "1/4 Century" Prince, c. 1940. These watches were given by the Eaton Company of Canada for 25 years service. This example with flared case. 18 jewels. 43mm. © Sotheby's Inc 1990
Circa 1940

Someday My Prince Will Come 133

A 18kt gold and steel Prince with rare wristwatch hunter case design. Case signed "Rolex" with Glasgow hallmarks. © Sotheby's, Inc. 1989.
Circa 1931

14kt white gold Prince. Silvered two-tone refinished dial with Arabic numerals. White gold Rolex watches of this period are very rare, although they seem to be much more popular in the Gruen equivalent model. Ref 1527. © Sotheby's, 1994.
Circa 1935

18kt gold rectangular wristwatch, 10½'" Hunter, 17 jewel movement, silvered matte refinished dial with subsidiary seconds, elongated curved rectangular case with hooded stepped lugs. This watch is 4mm longer than a Prince. Ref 829. © Sotheby's, 1994.
Hallmark Glasgow 1928

18kt gold asymmetric sweep seconds Prince. Calibre TS, 15 jewel, movement timed to 6 positions with micrometer regulator. Black matte dial with applied gilt triangular numerals. Sloping rectangular case, the bottom has an applied shield for a monogram. Ref 3361. © Sotheby's, 1994.
Circa 1945

Chapter 5
THEY ALSO SERVE...
Other Models

By the mid-1930s Rolex had almost achieved the position they are in today. The Perpetual was in production, the Oyster was *the* waterproof watch, and the company was positioning itself in the market place as the watch of the sportsman. It is worth remembering that in this era sport was not a universal pastime. It was still primarily an endeavor undertaken by the well-to-do classes. Most of the rest of society was too concerned with working or looking for work. Sport was still a pursuit of the leisured classes and Rolex was the watch of this sporting, mon-eyed minority.

This was good but the company wanted more and so Rolex began to use the name Wilsdorf had registered at the patent office in Bienne many years before on the January 24, 1911: Marconi. The name Marconi was, in fact, the fourth brand name that Wilsdorf had registered. Rolex was registered in 1908, Omigra later the same year, Elvira the following year, and then Marconi Lever, registered as a name for watches and watch parts. There is no record of the other names being used.

In fact it was nine years later, in 1920, before the name Marconi was used on a watch. Marconi was, of course, named after Gugliemo Marconi who successfully sent radio signals over the Atlantic Ocean in 1901. He was awarded the Nobel Prize for physics in 1909 in recognition of this achievement. At the time that Wilsdorf registered the name Marconi was regarded as a "cutting edge" technological figure, much as Bill Gates would be today. He would be an appropriate figure after whom to name an advanced modern watch. There is another point about Marconi which would have made Wilsdorf interested in him. Like Wilsdorf, Marconi was born outside Britain but patented his inventions there and established his company in London. It was from London that Marconi oversaw the expansion of his company to the farthest reaches of the earth.[22]

When Wilsdorf finally got round to launching a Marconi watch, it was because expansion was not proceeding as fast as he would have liked. This was due to the inability of his (and Aegler's) Geneva factory to produce sufficient watches at a lower price.[23] To solve the problem, he decided to buy movements from other ebauche houses. These movements were purchased in a completed stage, so there was no necessity to finish them in any way. They were simply cased in inexpensive cases, usually chromed nickel, and sold through parallel channels, which meant dealers who were not Rolex agents.

It would be wrong to think that the watches were a major success. The problem was that there was a huge number of brands of watches out there, all using ebauche movements. There was, frankly, very little to tell one from the other.

By the time the Marconi brand hit the market, it had one small problem: Marconi was a figure whose time had passed. His name carried little of its former power, so it was probably the wrong name to use.

All was not lost, however. Wilsdorf had gone through a phase of registering company and brand names in 1919 and 1920. On March 17, 1919, to celebrate the end of the "war to end all wars" and the foundation of the League of Nations (the predecessor to the U.N.) in Geneva, he registered the name LON (League of Nations) as a brand name. At the same time, just so that the journey to the office would not be wasted, he also registered the names Brex and Unicorn Lever. It was the Unicorn brand that would take over from Marconi and a few years later it was joined by Rolco, a simple contraction of "ROLex COmpany. The names are almost completely interchangeable. Shown here are a couple of watches which bear all three of the above names on one watch. One of them is even signed "Rolex Watch Co." as well, making a total of four brand names on one watch, probably a world record.

It should be noted that the correct brand names of both Marconi and Unicorn incorporate the word "Lever" as part of their title, and it is interesting to note that almost all of the dozens of watch names that Wilsdorf registered in this period also had "lever" in their name. This was to differentiate these watches from the much cheaper and inferior cylinder or pin pallet movements which were often fitted to the less expensive watches of the day.

If you examine the number of brand names that Wilsdorf registered in the period before World War II, it is obvious that he was unhappy with his secondary lines and was floundering to find a name and a model that worked. Ironically the answer was right under his nose, it was the Oyster. As early as 1932, the cushion Oyster had been made as a Rolco or Unicorn model and fitted with an ebauche movement from FHF (Fontmelon) and a case made of "Snowite," a chrome-plated, nickel-based material. It was no major problem to put an ebauche movement from FHF into the 2240 (boy's size) model. Most of these watches did without any of the subsidiary company names on their dials, they were simply signed "Oyster Watch Company" or "Oyster Junior Sport." The problem Rolex had had with all of the subsidiary lines was how to promote them without damaging the prestige of the main brand.

These watches were produced in large quantities until the end of World War II, when Rolex dropped all of the subsidiary brands. In 1945, the year Rolex celebrated their 40th anniversary or jubilee, it introduced its most expensive watch, the "Datejust." At the same time Rolex launched its final attempt at penetrating the lower cost watch market: the Tudor. The Tudor was one more name in Rolex's long list of "Regal" names which started with the Prince and continued through the Imperial and the Royal, and was epitomized by the Rolex crown (more correctly a coronet). The first symbol used for the Tudor was a stylized rose, representing the Tudor rose. By the 1960s the rose was replaced with a stylized shield.

The management at Rolex has seemingly always made sure that the Tudor line, which many watch aficionados erroneously consider to be a "cheap-Rolex," is up to par both in style and in quality with the watches that also bear the Rolex name.

In fact, an advertisement in Britain's Horological Journal, dated August, 1947, states that Tudors are "manufactured under strict Rolex supervision and bear the Rolex label of guarantee."

---

[22] It is worth remembering that the radio operator on the Titanic did not work for the shipping company; he worked for Marconi's Wireless Telegraphy Company Ltd.

[23] The Marconi watches were not the first watches from Rolex that can be described as "second class." In the very early years of the company, until 1920 or so, the Rebberg movement used in all Rolex watches was available as a 15 jewel or as a 7 jewel version.

136 They Also Serve...Other Models

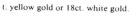
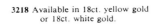
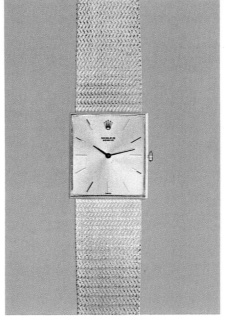

t. yellow gold or 18ct. white gold.
ld.

3218 Available in 18ct. yellow gold or 18ct. white gold.

3221 Available in 18ct. white gold or 18ct. yellow gold.

3224 18ct. gold.

Dress watches from 1966 including one of the early Cellini models (far right).

Advertisement for the Rolex "Girouette." 1937.

Oyster Tudor shock resisting, manual wind, stainless steel watch. This is one of the earliest of the Tudor series and uses the basic "boy's size" case.
Circa 1946

They Also Serve...Other Models 137

Brochure for the Canadian market.

1930 Catalog page featuring Portfolio or Traveling Clocks, including an eight day alarm model.

Sterling silver octagon Unicorn, 13''' Movement.
Case no. 104105
Model no. 73
Hallmark, Glasgow 1925

Still, since its introduction in 1945, Rolex has advertised the Tudor line as a "workingman's" watch. A Tudor advertisement in the Horological Journal's May, 1953 issue stated that not only could you "punish" your Tudor watch "without mercy," but that you should immediately go out and purchase one if your "aspirations are higher than your bank balance". Indeed, the people most often observed wearing a Tudor are the blue-collar worker, police officer, or anyone who leads an active lifestyle requiring a sturdy watch with good timekeeping abilities. In the two-Rolex family (one where both husband and wife wear them) the Tudor is often the first watch for their offspring.

In the U.S.A the Tudor line, which first appeared around 1946, is quite often given as a graduation gift as a "starter Rolex" of sorts. The lady's Submariner has proven to be enormously popular with the more active 1990s woman, making some wonder when Rolex will come out with a version signed Rolex. While the Tudor movements may have a different ebauche or fewer jewels than the traditional Rolex, they are, for the most part, just as good at time keeping as their more expensive "sister" watches and also just as durable. The early models of the Tudor, once disdained by collectors now are quite sought-after, particularly among those new to the field.

Some of the current models, like the Tudor Cosmograph, have become extraordinarily popular in the American markets and there is an almost rabid demand for them in Japan. This self-winding model, in fact, predated the "Rolex" Cosmograph Automatic version by 5 years! The Tudor Cosmograph is also the first Rolex/Tudor chronograph model to display a date function, with the exception of the scarce "full calendar" models such as the late 1940s era Ref. 6036 and 4768. Some of the earlier versions of the Tudor Cosmograph are collectible because of their varied and colorful dials. This of course contradicts the image that many less informed consumers have about the viability and originality of the Tudor line. Rumors of the demise of the Tudor line have been rampant for years, but Rolex S.A. appears to be committed to what they have referred to as the "undoubted future of the lower-priced field."

Other lesser regarded Rolex products are the portfolio and pocket watches. This is strange since the most accurate watches the company ever made were in fact pocket watches. In the 1930s and 1940s Rolex manufactured a wide line of folding watches. Often called purse watches today, they were originally known as portfolio watches and functioned as sport watches prior to the introduction of shock proof balances. They have some of the most interesting cases ever seen in any Rolex and have been unfairly neglected.

The pocket watches were the company's first products and remained major sellers until the start of World War II. As late as 1930 the catalog showed five standard pocket watches in open-face, semi- and full-hunter styles. These watches were available in a variety of materials ranging from rolled gold and silver to 18kt gold. There was also a line of eight slim dress pocket watches, many of them available with engraved and chased cases.

Two years later, in 1932, the company introduced a new pocket watch movement and launched it in a newly designed case as the "Rolex Prince Imperial." This new movement was to prove one of the most accurate watches ever made by Rolex, and its accuracy was well reflected in the price that the company chose set on it. At £36.15 an 18kt version was almost 30% more expensive than an 18kt Prince wristwatch.

Despite the introduction of these watches, Wilsdorf believed that there was room for an even more exclusive pocket watch. A year later he produced a small batch of very slim (and expensive) watches under his own name. These "Wilsdorf Watch Co." watches seem to have been made only in 18kt white gold and are signed on the case, movement, and dial. There seem to have been very few of these watches made. All that have surfaced, to date, bear serial numbers in a range from 102 to 122, so it can be assumed that less than 50 were made.

The reason for their lack of popularity and subsequent rarity is clear. Rolex was able to make a name for itself in wristwatches because it had no competition working at the same level. In the prestigious pocket watch market, however, the 1930s saw competition galore.

So it was that the Wilsdorf watch went to the wall, as did one of the most bizarre watches the company ever introduced: the cushion Oyster pocket watch. The only one we have ever seen bore a Glasgow 9kt hallmark for 1926. It was a perfect, four times life size version of the original cushion wristwatch Oyster, right down to the milled bezel, back and screw down winding crown (which was at 12 o'clock). The demand for a waterproof pocket watch must have been quite small. While one can imagine a lifestyle in which one's hands and wrists get immersed in water, it is difficult to imagine circumstances in which a vest pocket would be submerged!

It seems strange now to realize that pocket watches were actually included in the Rolex catalog until well into the 1960s. One of the very last to be so listed was the dress watch contained inside a U.S. $20 gold coin. This style of watch had first been introduced by a famous Parisian jeweler in the 1930s. By the time that Rolex introduced their version, a large number of companies ranging from Corum to Vacheron et Constantin were making wristwatch versions. It is ironic that the last pocket watch made by Rolex was simultaneously being made as a wristwatch by people who had in fact made their names in pocket watches.

A rarer example of a Canadian market Rolex. This one the Centregraph with inner 24 hour dial. Gold filled top and steel back.
Model number 3478
Case # 161118
Circa 1939

They Also Serve...Other Models 139

Rectangular white metal Rolex with dial signed "Rolco," movement signed "Marconi" and case signed "RWC LTD" Model number 1937
Case # 902871
Circa 1938

Early textured dialed Rolex Tudor with horned lugs. This one marked "Rotor" and "Self-winding."
Circa 1956

140 They Also Serve...Other Models

Tonneau shaped Rolex Marconi cased in Snowite metal with 10-1/2''' round movement. Notice that the case, dial and movement are all signed Rolex.
Model number 1370
Case #823778
Circa 1934

Canadian market, subsidiary seconds Oyster Lipton with unusual dial and wide bezel. This is a fine example of the many varieties of watch made exclusively for the Canadian market.
Model number 3136
Case #102249
Circa 1939

They Also Serve...Other Models 141

Canadian Solar Aqua anti-magnetic with subsidiary seconds hand. It has a threaded back, but is not a true Oyster watch, because it does not have screw down button, and is not, truly, waterproof. Circa 1948

Gold filled top, steel back, Canadian rectangular Rolex standard. To circumvent heavy tariffs, most of the non-Oyster Rolexes had their cases manufactured in Canada.
Circa 1941

142 They Also Serve...Other Models

Sterling silver small size "nurse's" pendant watch with sweep seconds and porcelain dial. 10-1/2''' Rebberg with the earliest known sweep seconds.
Circa 1915

Nurse's fob watch with subsidiary seconds, porcelain dial and sterling case and fleur d'lys hands. The design of this case was the subject of one of Wilsdorf's earliest patents.
Circa 1917

They Also Serve...Other Models 143

High grade pocket watch signed "H. Wilsdorf Geneve." 18kt white gold with sterling silver dial and enameled numbers. Case and movement numbered 102. Movement is extremely high grade, is ultra thin, has 18 jewels and is stamped twice with the seal of Geneva. Little is known of the Wilsdorf watch company but it is assumed that it represented an attempt to break into the premium watch market.
Circa 1935

Rolex pocket watch in 9kt gold case made by Dennison. Movement is high grade 21 Jewel, signed "Calibre 550," "Observatory Quality," "Class 'A' Certificate" and numbered 641. Movement and dial are signed "Camerer Cuss." Micrometer regulator, porcelain dial with sunken subsidiary seconds and Roman numerals. Watch is shown on its original timing certificate.
Circa 1931

Stainless steel Tudor Prince Oysterdate Submariner, waterproof to 200 meters or 660 feet. Blue bezel and dial. Note that the dial does not have white metal settings for the luminous indices, but otherwise the watch looks almost identical to the current Rolex Submariner.
Model 79090
Current Model

They Also Serve...Other Models 145

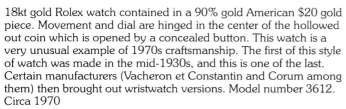

18kt gold Rolex watch contained in a 90% gold American $20 gold piece. Movement and dial are hinged in the center of the hollowed out coin which is opened by a concealed button. This watch is a very unusual example of 1970s craftsmanship. The first of this style of watch was made in the mid-1930s, and this is one of the last. Certain manufacturers (Vacheron et Constantin and Corum among them) then brought out wristwatch versions. Model number 3612. Circa 1970

146 They Also Serve...Other Models

Sterling silver and polychrome enamel "portfolio" watch. Enameled green and blue Art Deco patterns with ribbed and engine-turned back. The engine-turned silvered dial with Arabic numerals is revealed by pressing a button in the band. © Christie's, London
Case no. 63384
Hallmark Glasgow 1928

Another portfolio watch, this one in 9kt gold. It has extensive chasing on the case and to the Gothic style pendant. The watch has an unusual small oval dial in the center of the dial aperture. This watch appears to date from the late 1920s, but bears the W&D cartouche which was thought to have been phased out earlier in the decade.
Case #55729
Circa 1928

They Also Serve...Other Models 147

Polychrome enamel on sterling silver portfolio or travel watch. It has a silver dial with faceted batons. 10-1/2''' Hunter movement and gothic shaped pendant.
Case # 63466
Hallmark Glasgow 1927

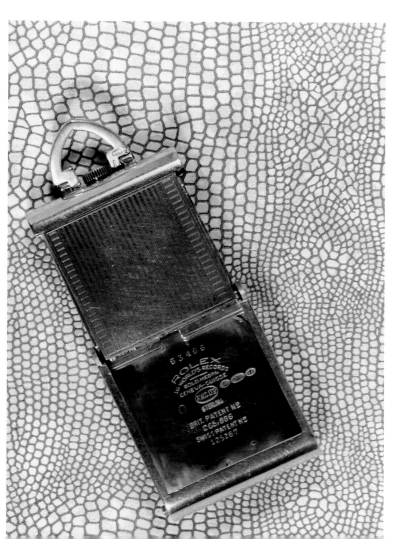

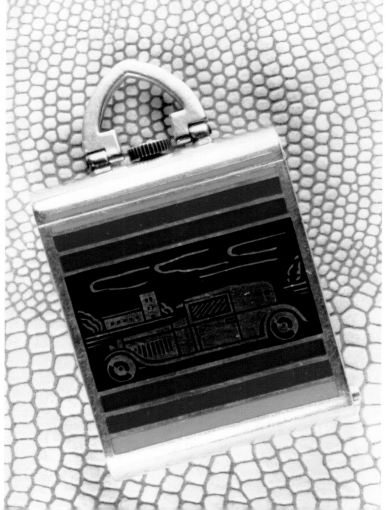

18kt white gold dress pocket watch, with original chain and cummerbund clip, which is also signed. Chronometer movement numbered L 52941.
Model no. 8437
Case no. 65766
Circa 1958

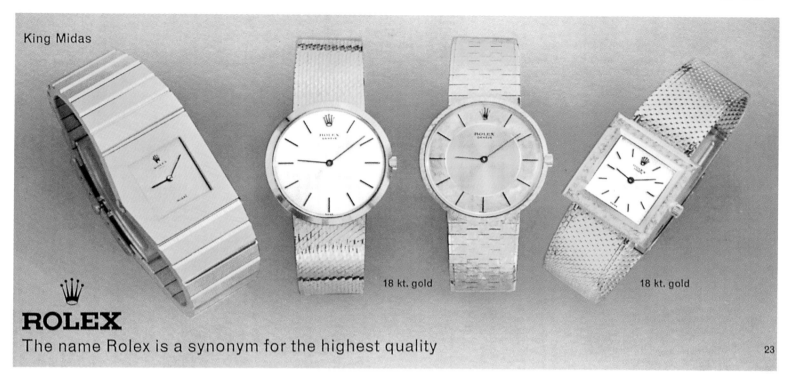

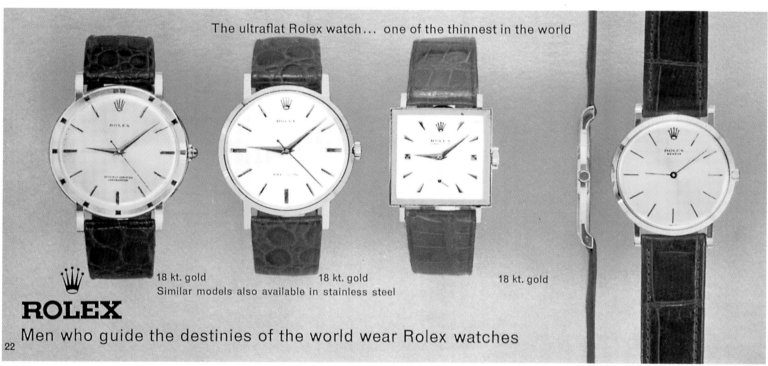

The "dress" (or non-Oyster) Rolex has continued to this day but still remains very much in the shadow of the Oyster Perpetual. Here are eight watches from a Bucherer catalogue of 1964. They include the King Midas model from the "Cellini" line although it is not identified here as such.

Advertisement promoting the Kew Observatory successes. 1938.

London Charivari   November 16 1938

No other wrist watch...

9 ct. gold with centre seconds hand for close timing
17 gns.

9 ct. gold with mobile ends **18 gns.**
18 ct. gold . . . **28 gns.**

Rolex Chronometers jewelled with 18 fine rubies. For the man who demands accuracy.

One pocket watch in one hundred, one wrist watch in ten thousand, may precisely be called a Chronometer. In accordance with the rules of the Chronometric Societies of Switzerland and France 'The title of Chronometer is permitted only if it is a watch of the finest quality and so precisely adjusted in all positions and extremes of temperature that it will pass a 14 days' continuous Observatory test.'

Every Rolex Chronometer can be supplied with SWISS GOVERNMENT CERTIFICATE

ROLEX

WRIST WATCH *chronometer*

Officially attested by Kew Observatory

AS THE

World's most precise wrist watch

THE ROLEX WATCH CO., LTD. (H. Wilsdorf, Managing Director)

GENEVA · LONDON · PARIS · BUENOS AIRES · DUBLIN · TORONTO
LONDON OFFICE: 40-44 HOLBORN VIADUCT · LONDON · E.C.1.

Chapter 6
CHRONOMETERS
Superiority in Accuracy

The logic Rolex used to justify making world record breaking watches was similar to that used in the 1960s by NASA in their exploration of space: "If we spend our time examining the frontiers of science, then we will, in so doing, extend those frontiers to the benefit of all mankind". NASA gave us the non-stick frying pan and Rolex gave us the wrist chronometer.

Any company or person searching for ways to improve on a world record must assume that the current record holder used every available bit of knowledge and skill to obtain the original record, "and to surpass that record new methods must be investigated and new problems solved. This is the important point. By this search for improved performance, for new models and technique, the difficulties are overcome one by one and a higher precision is the result. If, moreover, a firm is able to attain such wonderful records it is certain to have made extensive investigations in this field, and it is natural that it should endeavor to incorporate at least part of the results of the costly experience in the production of its ordinary lines. It is good to get a clear view of these records: and it is certainly a fact that these firms that take the greatest pains with single watches, have also high class quantity production. It is in this spirit that these records are made - and will be viewed by the clear sighted watchmaker." This was the view of Emil Borer, the technical director of Rolex, the inventor of the Rolex Perpetual and, later, a board director of the company. His views, from a 1935 essay, taken with Wilsdorf's marketing skills, express in the clearest possible way the route that Rolex had chosen to take.

One of the main problems involved in writing a detailed history of the early days of the Rolex Watch company, is that no-one knew how important the company was going to become. In the early days the company was just two businessmen trying to make a living in a tough environment. Nobody thought to build a company archive. They simply went about their business with no more thought for posterity than the rest of us do. When the company did become successful and Hans Wilsdorf had the luxury of hindsight he saw things through the rose colored spectacles that time always supplies.

This is not to lessen the contribution of Wilsdorf. He had one thing and had it in spades; he had vision. At a time when even very few women wore wristwatches, Wilsdorf could see this new form displacing the pocket watch entirely. When others ridiculed the idea of a waterproof wristwatch, Wilsdorf saw the potential and went for it. When everyone from Breguet to Blancpain had made self-winding wristwatches in the 1920s and failed to make a profit, Wilsdorf examined their failures, avoided them and made a fortune. All of the foregoing are the achievements we know Wilsdorf for and the ones that he is acclaimed for. But none of them were his own original ideas.

Wilsdorf's one great accomplishment and the one that was all his own idea is the one we all ignore. Wilsdorf was the man who made the wristwatch accurate, not just accurate enough to enable us to catch a train, but accurate enough to stand comparison with the finest contemporary mechanical chronometers in the world. This was Wilsdorf's obsession, his ambition and eventually his achievement.

There is a saying in Detroit stock car racing circles "Win on Sunday.....sell on Monday." People like to buy winners. Detroit spends millions producing very special vehicles looking similar to the standard production vehicles but with performance in an entirely different league. Rolex was doing this years before Detroit had even thought of it.

The Oxford English Dictionary defines a chronometer as "An instrument for measuring time: specially applied to time-keepers adjusted to keep accurate time in all variations of temperature. They differ from watches in having a more perfect escapement and a compensating balance, and are used for determining longitude at sea, and for other exact observations."

The dictionary also gives the Greek antecedents for the word: Chronos = time and Meter = measure. All of this seems to us to be blindingly obvious. Accurate time keepers on our wrists are a given in our lives. We take them as much for granted as automobiles and air travel. Wristwatches are something without which our lives could not be imagined. Yet all of these three items began in the early years of the century in which we live and did not assume the role we so naturally give them until the years after the Second World War.

In 1905, Hans Wilsdorf and Alfred James Davis formed Wilsdorf & Davis, they stated that the purpose of the company was: "To carry on all or any of the businesses of manufacturers, importers and factors of watches and other timepieces,........ "and to buy and sell and deal in (wholesale or retail),....and to let on hire, watches and clocks". Despite the use of the word "manufacture," the company manufactured nothing. Instead it imported finished watch movements from Switzerland, and assembled them in cases from either Switzerland or the U.K. In this aspect they were very little different from dozens of other small importers and Wilsdorf realized this. He wanted to make his company stand out from the crowd. He wanted to create a "U.S.P" (Unique Selling Proposition) for his company.

The first thing he did was to name the watches coming from his company. Wilsdorf and Davis was never going to be a name that rolled easily off the tongue; a Wilsdorf and Davis Explorer somehow just doesn't have the right ring to it. So on the July 2, 1908 he registered the name "Rolex" as a trade mark in Switzerland and began to use the name inside the watches on both the movements and the cases. Next he began to advertise the name so that people would ask for the watch by name.

This caused a problem common during the rise of brand names. His advertising (as most advertising does) listed the retail prices of the watches. He would sell a watch to a jeweler for a price x and the jeweler would then sell the watch for the advertised price y, leaving the retailer with a profit of the difference between these two figures. The problem arose because these two figures were both fixed. When a retailer sold a watch without a brand name, usually one he would brand himself, he could charge whatever price he chose and thereby set his own profit levels. The retailer, naturally, disliked having his margins fixed by someone else. This was the real reason that branded watches were slow to take off in Britain.

Wilsdorf saw that the way around this problem was to make the watches so desirable the customer would demand them. The jewelers would be forced to stock them, whatever the jeweler's policy. This is the first of many situations where we can see Wilsdorf's genius for facing a problem and solving it by looking at it from a different angle. He needed to convince the jewelers to stock his watches, when this proved impossible he decided to convince the customers. But, in order to make customers demand the watches, he had somehow to differentiate his watches from all others[24].

Up until the advent of quartz watches, the cost of a watch was in direct proportion to its accuracy. While the fixed costs of building any wristwatch mechanism are fairly constant, the costs of finishing of the movement and timing it vary in proportion to the degree of accuracy desired. Not only was there a great deal of time involved in the finishing and timing, but this work could be done only by the most highly skilled, and therefore highly paid, workers.

Wilsdorf looked at the problem from a new viewpoint and his answer was simple: don't make all of them accurate, just make one or two of them incredibly accurate and promote this fact. The company had always used the Rebberg movement from Aegler, which was one of the first industrially made movements from Switzerland. Its prime advantage was the easy interchangability of parts, not its outstanding accuracy. But if you throw enough money at it, anyone can make a silk purse from a sow's ear. Wilsdorf invested heavily in perfecting the Rebberg, and on July 15, 1914, Rolex was able to obtain a Kew A certificate for one of their 11''' movements.

It is worth looking at what that certificate meant. The Royal Observatory at Kew (a suburb of London) was an outstation of the Royal Observatory at Greenwich, where all the initial work

Philippe and Ulysse Nardin) could state that their watches were the most accurate in the world. This was another powerful sales promotion device and the major companies were prepared to spend time and money on building the special watches needed to win such certification.

All of the watches submitted to Kew were full size pocket watches, which often are known as "deck watches." So it was obvious that no-one would even attempt to put a wristwatch through such a test...but Wilsdorf did. We do not know who did the actual adjustment and timing of this first watch, or how long it took and at what cost. What we do know is, that whatever it cost, it was worth it. Everywhere, from the boxes in which the watches were shipped to the watch cases themselves, in advertising and all the other promotional materials, the message was pushed: Rolex was the only wrist(let) watch in the world to hold a Kew A certificate. Over the next few years more special watches were submitted to Kew and the successes were promoted widely.

As soon as a new Rolex movement was introduced, the factory would build a special version of it and submit the movement to Kew. This process began with a lady's oval movement "calibre O" which gained a Kew A certificate on June 14, 1925. Less than two years later, on March 16, 1927[26], another oval movement, "calibre T", also gained a Kew A certificate and on the following day a 10-1/2''' Hunter was awarded the same honor. These two watches were the first to be timed by Jean Matile, who went on to time all of the Rolex world record watches for almost 30 years. It can be argued that Matile was as important to Rolex as Emil Borer. When he had achieved near perfection with one example, Matile would then instruct all of the watchmakers on the production bench in how to adjust that model calibre in order to achieve the best results.

Jean Matile (Rolex, Bienne)

1925 Teddington, first-class certificate for a 5 3/4''' movement.
1927 Teddington, first Class A certificate at this Observatory with the mention « specially good » (86.5 points) for a 6 3/4''' chronometer.
1928 Geneva. 1929, at Neuchatel. 1932, Besançon, first-class certificates for 6 3/4''' chronometers.
1934, 1935 First class certificates at the Neuchâtel and Teddington Observatories, with ultra-flat 17''' chronometers.
1936 Teddington, best result in wrist-chronometers (87.6 points).
1945 Best result at Neuchatel with 8.4 points for a wrist-chronometer.
1948 Teddington 93.8 points. Shaped wrist-watch.
1949 Geneva. Record of single movement (859 points), with a 28.5 mm. movement.
1950 140 first-class certificates at Kew, 24 bearing the mention « particularly good » (more than 80 points) for 10 1/2''' wrist-chronometers.

Official Testing Stations : more than 75,000 Rolex wrist chronometers of all sizes have obtained the mention « particularly good ».

on accurate time keeping had been initiated. The search for a method of accurately finding longitude, the development of the first marine chronometer, the institution of an universal time standard, and the establishment of a central facility to test timepieces and certify them were all products of Greenwich. It became an institution so central to time measurement that today all the world runs to Greenwich Mean Time (although nowadays it is called Universal Time Co-ordinated).

The chronometer test at Kew was the toughest that a timepiece could be put through. Unlike the Swiss 15 day test[25], the Kew test lasted 44 days in six positions and three temperatures. Its principal purpose was to test (or rate) marine chronometers, whose accuracy was vital to marine navigation until the 1930s. The Kew tests were designed to time watches for the Royal Navy. If a watch scored in the top few percent of those tested, the Navy would offer to purchase it. This would then allow the maker to call themselves "Supplier to the Admiralty", another useful sales promotion tool. Beyond that, based on the chronometer tests at Kew major manufacturers (such as Patek

These successes were widely promoted. Starting around 1927 with the Prince, Rolex began to sign watches "Chronomètre" on both the movement and dial. Although these watches were not tested at any official timing station or observatory, Rolex was entitled to use the title Chronometer by virtue of a Swiss law that stated "A precision watch must be so regular at various conditions and temperatures that it can receive an official certificate of operation." What this meant was that manufacturers could call any watch a chronometer if it met or exceeded official timing standards while at the factory. Rolex took advantage of this facility to a greater extent than any other Swiss manufacturer. It must be emphasized that watches signed in this way were in no way inferior to those which held official timing certificates. In fact official timing certificates were available at extra cost for all chronometers sold by Rolex during the 1930s.

To understand the way Rolex dominated the chronometer field as early as the late 1920s, it is worth examining some of the annual returns from the Swiss testing stations. In 1927 the testing station at Bienne tested a total of 666 watches 204 of them being pocket watches and the remainder, 462, wristwatches. 423 of the wristwatches were from Rolex, or 92%.[27] By 1928 the figures were 1,190 watches in total, 1,049 of which were wristwatches. 891 of those gaining certificates (or 85%) were from Rolex[28]. The following year, 1929, 1,486 watches were granted certificates, 231 of them being pocket watches and the remainder, 1,255, wristwatches. 1,063 of the wristwatches were from Rolex, or 85%. The testing station also noted in its report for the year that "The immense majority of the certificates issued this year were in English". The following year provided similar results. Although the total number of pocket watches had risen by 50%, the total number of wristwatches had risen by over 300% to 4,113. 3,338 of them were from Rolex, around 81%. By 1931, the effects of the Great Depression were starting to be felt and the numbers were well down. Of a total 2,897 watches the number of pocket watches were up by another 100 or so to 454 and for the first time 25 of them were from Rolex. Wristwatches experienced a precipitous fall to 2,443 of which 1,831, or 75%, came from Rolex. Of the 2,897 certificates issued 2,107 of them were issued in English. By 1932[29] the path established the previous year was becoming well trodden, and now Rolex obtained 488 of the 910 pocket watch certificates, or over 50%, as well as 2,112 of the 3,007 wristwatch certificates, or 70%. By now the number of certificates issued in French now surpassed the number of those issued in English, 1,603 as against 1,181.

In addition to Bienne, there were many other observatories used by Rolex. Bienne's testing station was the closest to Rolex, but at the St. Imer observatory over the same years we see similar results. 1929: 80 Rolex watches from a total of 198. 1930: 236 Rolex watches from a total of 403. By 1940 Rolex were once again in the ascendancy and obtained 1,693 certificates from St. Imer out of the total of 1,851 issued, or 91%. In the same year the company obtained another world record when one particular watch being tested at Neufchatel observatory obtained a score of 13.08, the best score ever attained by a wristwatch.

The certificate Rolex included with new watches, noting the movement and case number.

The move toward increased precision in the 1930s was helped by three technical developments from Rolex. First, the Prince was a much more accurate wristwatch than had previously been seen. This was due to its large and heavy balance and, just as importantly, its extra long mainspring. While this was marketed as obviating the need for daily winding, its real advantage was that the watch mainspring was almost always under constant tension. This level of tension was in the middle area of the mainspring tension, the area where the most regular output occurs and, therefore, the one in which the most accurate timekeeping occurs. This of course had major advantageous effects on the timekeeping.

The second development from Rolex was the introduction of micrometer (or swan neck) regulators on many of their watches, beginning with the Prince, and appearing later on the 10-1/2''' Hunter movements. At the time these were the smallest production line movements ever fitted with this facility for regulation. The micrometer regulator allowed fine tuning of the hairspring, thereby allowing the watch to be set to a more accurate standard than had previously been available.

The final great leap forward was the invention of the "Auto-Rotor" movement. This perpetual movement meant that, as with the Prince, the mainspring almost always was kept at a constant tension.

These developments were not individually advertised or promoted to any great extent, but their results were. Rolex was now able to manufacture, on a large scale, high precision wristwatches. There were other companies that made more watches, and there were companies that made more accurate watches, but by the end of the 1930s Rolex had found their place in the market.

---

[24] Wilsdorf went to the extent of writing to the English journal "Practical Watch & Clock Maker" in June, 1928, to say "In no country of the world will the public buy a good watch unless they know the name of the factory," and "A shopkeeper may take the very greatest trouble with his silver, his plate, his diamonds and pearls but he will destroy his reputation by his *indifferent* watches." (Wilsdorf's Italics)

[25] The world's first timing contests for watches were run in 1869 at the Geneva Observatory. They were modified to be more impartial in 1873, when a system of competitions was instituted which ran for almost 100 years. Throughout the years, in 1879, 1882, 1890, 1908, and 1926, they were modified again, making the tests more and more severe. The format of these competitions was used to organize chronometrical services at places as far apart as Yale in the U.S. and Kew in the U.K.

[26] It should be noted that the 6-3/4''' watch which achieved a score of 86.5, ranking 39th among all watches tested at Kew in 1927, was obviously the smallest watch tested that year. The next smallest, other than a 10-1/2''' Rolex, was a 19'''.

[27] It is also worth noting that the balance of the wristwatches gaining certificates that year at Bienne came from Gruen.

[28] Once again Gruen produced the balance of the wristwatches.

[29] The technical director of Rolex in 1932, was C. Billeter, who was also the Senior Director of the School of Horology at Fleurier.

154 Chronometers

Timing Certificates from 1945, 1960, 1965.
Note that the last two have stickers stating that they had been retested at Rolex.

Chronometers

BUREAUX SUISSES
DE CONTRÔLE OFFICIEL DE LA MARCHE DES MONTRES
BIENNE, LA CHAUX-DE-FONDS, GENÈVE, LE LOCLE, ST-IMIER, LE SENTIER, SOLEURE

Épreuves pour montres-bracelet – Prüfungen für Armbanduhren
Prove per orologi da polso – Trials for Wristlet-watches

ref. 6605
S/458888

Bulletin de marche / Gangschein / Certificato di movimento / Watch Rate Certificate

Calibre:	Chronomètre	Diamètre du mouvement: 28,5 mm	Hauteur / Dicke des Werkes: 5,85 mm
Échappement:	ancre	Spiral: Breguet auto-compensateur	Balancier: non-magnétique

Manufacture des
MONTRES ROLEX S.A.
Bienne – Genève

1959	Jours - Tage Giorni - Days	Marches diurnes Tägliche Gänge Marce diurne Daily Rates	Variations des marches diurnes Differenz der täglichen Gänge Variazioni delle marce diurne Variations of the Daily Rate	Positions - Lagen - Posizioni - Positions		Températures
Sept.	23	+9		Verticale, 3 heures à gauche	Vertikal, 3 Uhr links	+20°C
	24	+12	3	Verticale, 3 heures en haut	Vertikal, 3 Uhr oben	"
	25	0				"
	26	+1	1	Verticale, 3 heures en bas	Vertikal, 3 Uhr unten	"
	27	+5		Horizontale, cadran en bas	Horizontal, Zifferblatt unten	"
	28	+8				"
	29			Horizontale, cadran en haut	Horizontal, Zifferblatt oben	"
	30	+11	1			"
Oct.	1	+10				"
	2	+13	3			"
	3	+15				+4°C
	4	+9				+20°C
	5	+16				+36°C
	6	+11		Verticale, 3 heures à gauche	Vertikal, 3 Uhr links	+20°C
	7	+12				+20°C

Résultats – Ergebnisse – Risultati – Summary

- Marche diurne moyenne dans les différentes positions / Mean daily rate in the different positions: +8,1
- Plus grande différence entre la marche diurne moyenne et l'une des marches dans les 5 positions: 8,1
- Variation moyenne / Mean variation: 2,2
- Variation par degré centigrade / Variation of rate per 1° centigrade: +0,03
- Plus grande variation / Maximum variation: 3
- Reprise de marche / Rate resuming: +1,5

We certify that this chronometer has, today, been once more thoroughly checked in our workshops and that its performance was found to be in accordance with this certificate.
Geneva 30 MAI 1960
THE ROLEX WATCH COMPANY LIMITED

BUREAUX SUISSES DE CONTRÔLE OFFICIEL DE LA MARCHE DES MONTRES
BIENNE, LA CHAUX-DE-FONDS, GENÈVE, LE LOCLE, ST-IMIER, LE SENTIER, SOLEURE

Extrait du règlement

Art. 4 Les Bureaux Suisses de contrôle officiel de la marche des montres désignés par l'abréviation B.O., reçoivent en dépôt les montres et les appareils horaires qui leur sont adressés, pour les soumettre à diverses épreuves et en contrôler la marche.

Art. 5 La marche de chaque montre est comparée toutes les 24 heures aux indications d'une horloge de précision, vérifiée chaque jour d'après le signal de l'un des Observatoires astronomiques et chronométriques de Neuchâtel ou de Genève. Les montres sont observées aux températures de +4°, +20° et +36°C. Ces températures sont maintenues dans une tolérance de ±1°C.

Art. 7 A la fin des épreuves réglementaires, et lorsque tous les résultats sont restés dans les limites prescrites, les B.O. délivrent sous le sceau officiel, un bulletin de marche. Ce document est d'un type unique pour tous les B.O.

Art. 8 Les montres ayant obtenu des résultats de marche satisfaisant aux limites avec mention prévues par le règlement, reçoivent un bulletin portant l'inscription «Résultats particulièrement bons».

Art. 20 Le signe + signifie de l'avance et le signe – du retard.

Une montre ayant obtenu un bulletin de marche a le droit de porter le titre de chronomètre.

Définition des genres de montres

Genre a: Montres-bracelet
Dans ce genre sont comprises les montres-bracelet simples, dont le diamètre est égal ou inférieur à 30 mm, ou dont la surface ne dépasse pas 707 mm² (mouvements de forme) et les montres-bracelet avec complications dont le diamètre est égal ou inférieur à 36 mm, ou dont la surface ne dépasse pas 1018 mm².

Genre b: Montres-bracelet de grandes dimensions
Dans ce genre sont comprises les montres-bracelet simples, dont le diamètre est supérieur à 30 mm, ou dont la surface dépasse 707 mm² (mouvements de forme), et les montres-bracelet avec complications dont le diamètre est supérieur à 36 mm, ou dont la surface dépasse 1018 mm².
Le mécanisme de remontage automatique n'est pas considéré comme complication.

Genre c: Montres de poche
Genre d: Montres de poche extra-plates ou avec complications
Dans ce genre sont comprises les montres de poche dont la hauteur entre le dessous de la platine et la pièce la plus haute ne dépasse pas 4,3 mm et les montres de poche compliquées: soit chronographes, montres à répétition, à quantièmes.

Limites pour l'obtention d'un bulletin

	Montres-bracelet (simples ou avec complications)		Montres-bracelet de grandes dimensions (simples ou avec complications)		Montres de poche		Montres de poche extra-plates (ou avec complications)	
	sans mention	avec mention	sans mention	avec mention	sans mention	avec mention	sans mention	avec mention
Marche diurne moyenne dans les 5 positions	−3 +12	−3 +12	−4 +10	−2 +5	−4 +10	−2 +5	−4 +12	−2 +6
Variation moyenne de la marche diurne dans les 5 positions	6	4	3	1,5	3	1,5	4	2
Plus grande variation entre deux marches diurnes consécutives dans la même position	10	7	5	2,5	5	2,5	6	3
Différence du plat au pendu			±10	±5	±10	±5	±12	±6
Plus grande différence entre la marche diurne moyenne et l'une des marches dans les 5 pos.	22	16	12	7	12	7	15	9
Variation par degré centigrade	±1	±0,7	±0,5	±0,25	±0,5	±0,25	±0,7	±0,35
Erreur secondaire			±9	±4,5	±9	±4,5	±11	±5,5
Reprise de marche	±10	±7	±5	±2,5	±5	±2,5	±8	±4

Toute imitation du présent bulletin sera poursuivie

Chronometers

BUREAUX SUISSES
DE CONTRÔLE OFFICIEL DE LA MARCHE DES CHRONOMÈTRES
Schweiz. Institute für amtliche Chronometerprüfungen
Swiss Institutes for official Chronometer tests · Oficinas suizas de control oficial de Cronómetros

Épreuves pour montres-bracelet – Prüfungen für Armbanduhren
Pruebas para relojes de pulsera – Trials for Wristlet-watches

Swiss made

Especially good results / Besonders gute Ergebnisse / Résultats particulièrement bons / Resultados particularmente buenos

Bulletin de marche / Gangschein / Boletín de marcha / Watch Rate Certificate No 1714573
Mouvement / Werk / Máquina / Movement No 4/39999

Genre / Uhrenart / Género de reloj / Type of watch	B
Particularités / Besonderheiten / Particularidades / Watch specialities	Automat. avec. sa rotore
Échappement / Hemmung / Escape / Escapement	ancre
Spiral / Spiralfeder / Espiral / Hairspring	Breguet auto-comp.
Diamètre du mouvement / Werkdurchmesser / Diámetro de la máquina / Diam. of Movement mm	28,5
Hauteur / Höhe / Espesor / Thickness mm	5,8
Balancier / Unruhe / Volante / Balance	non magnétique

Manufacture des
Montres Rolex S. A.
Bienne-Genève

Jours - Tage / Días - Days	Marches journalières / Tägliche Gänge / Marchas diarias / Daily Rates	Variations des marches journalières / Differenz der täglichen Gänge / Diferencias de las marchas diarias / Variations of the Daily Rates	Positions - Lagen - Posiciones - Positions		Températures / Temperaturen / Temperaturas / Temperatures
1.	M_1 + 4		Verticale, 3 heures à gauche / Vertikal, las 3 a la izquierda	Vertical, 3 o'clock left	+ 20° C
2.	M_2 + 3	V_1 1			"
3.	M_3 + 2		Verticale, 3 heures en haut / Vertical, las 3 arriba	Vertical, 3 o'clock up	"
4.	M_4 − 2	V_2 4			"
5.	M_5 + 2		Verticale, 3 heures en bas / Vertical, las 3 abajo	Vertical, 3 o'clock down	"
6.	M_6 + 2	V_3 0			"
7.	M_7 + 3		Horizontale, cadran en bas / Horizontal, esfera abajo	Horizontal, Dial down	"
8.	M_8 + 6	V_4 3			"
9.	M_9 + 6		Horizontale, cadran en haut / Horizontal, esfera arriba	Horizontal, Dial up	"
10.	M_{10} + 4	V_5 2			"
11.	M_{11} + 1				+ 4° C
12.	M_{12} + 4				+ 20° C
13.	M_{13} + 4				+ 36° C
14.	M_{14} + 4		Verticale, 3 heures à gauche / Vertical, las 3 a la izquierda	Vertical, 3 o'clock left	+ 20° C
15.	M_{15} + 3				+ 20° C

Date de la fin des épreuves: **5 MAI 1965**

Résultats - Ergebnisse - Resultados - Summary

Marche journalière moyenne dans les différentes positions / Mittlerer täglicher Gang in den verschiedenen Lagen / Marcha diaria media en las distintas posiciones / Mean daily rate in the different positions	+3,0	Plus grande différence entre la marche journalière moyenne et l'une des marches des 5 positions / Grösste Differenz zwischen dem mittleren täglichen Gang und einem der Gänge / Máxima diferencia entre la marcha diaria media y una de las marchas en las cinco posiciones / Greatest difference between the mean daily rate and any individual rate	5,0
Variation moyenne / Mittlere Gangabweichung / Diferencia media / Mean variation	2,0	Variation par degré centigrade / Gangabweichung pro Grad Celsius / Diferencia por grado centígrado / Variation of rate per 1° centigrade	+0,19
Plus grande variation / Grösste Abweichung / Máxima diferencia / Maximum variation	4,0	Erreur secondaire / Sekundäre Kompensationsfehler / Error secundario / Secondary error	
Différence du plat au pendu / Differenz zwischen liegend und hängend / Diferencia entre las posiciones horizontal y vertical / Difference between flat and hanging positions			

Toutes les valeurs indiquées sont en secondes / Alle Werte sind in Sekunden ausgedrückt / The values are indicated in seconds / Todos los valores están indicados en segundos

WE CERTIFY that this chronometer has been submitted to a final test in our workshops and that its performance was found to be in accordance with this Official Certificate.
THE ROLEX WATCH COMPANY, GENEVA

NOUS CERTIFIONS que ce chronomètre a subi un ultime contrôle dans nos ateliers et que sa précision est conforme à ce Bulletin Officiel.
MONTRES ROLEX S.A., GENÈVE

WIR BESTÄTIGEN, dass dieser Chronometer in unseren Werkstätten nochmals einer genauen Prüfung unterzogen wurde, und dass seine Präzision diesem offiziellen Gangschein entspricht.
ROLEX-UHREN AG., GENF

Extrait du règlement

1.21 Les B.O. reçoivent en dépôt les montres et les appareils horaires de fabrication suisse pour les soumettre à diverses épreuves et en contrôler la marche, sans distinction des mécanismes ou dispositifs utilisés.

1.23 Le titre de chronomètre (art. 1.22a) peut être délivré pour les genres de pièces suivants:
  genre A Montre-bracelet de petites dimensions,
  genre B Montre-bracelet de dimensions courantes,
  genre C Montre-bracelet de dimensions courantes, avec complications,
  genre D Montre-bracelet de grandes dimensions,
  genre E Montre-bracelet de grandes dimensions, avec complications,
  genre F Montre de poche,
  genre G Montre de poche, extra-plates,
  genre H Montre de poche, avec complications,
  genre J Appareil horaire mécanique 8 jours, utilisé dans une seule position,
  genre K Appareil horaire, utilisé dans une seule position, non compris dans l'un des genres précédents.
  Les dimensions sont données par le tableau 3.3.

1.24 Pour ces genres de pièces, différents critères de réglage sont établis. Pour chacun d'eux sont fixés deux champs de tolérances:
  a) soit avec mention «résultats particulièrement bons»,
  b) soit sans mention.

1.27 Les modèles des certificats de contrôle et des bulletins de marche individuels sont déposés au Bureau fédéral de la Propriété intellectuelle à Berne. Toute imitation ou falsification de ces documents sera poursuivie et pourra, en outre, entraîner une interdiction temporaire de déposer.

2.35 Une copie du certificat de contrôle ou du bulletin de marche individuel n'est accordée qu'après restitution de l'original et elle porte la même date que ce dernier.

2.36 Il ne peut être délivré qu'un duplicata. Il porte la mention «DUPLICATA».

3.11 Signification des symboles
  + = avance
  − = retard
  E = état de la montre
  M = marche diurne
  P = période (moyenne de période)
  T = température exprimée en degrés centigrades
  V = variation de la marche diurne
  m = marche instantanée

3.3 **Tableau des genres de pièces**

Genre	Désignation	Diamètre en mm		Surface en mm²		Hauteur en mm	
		mini	maxi	mini	maxi	mini	maxi
A	Montre-bracelet de petites dimensions	–	20	–	314	–	6,5
B	Montre-bracelet de dimensions courantes	20,1	30	317	707	–	8,5
C	Montre-bracelet de dimensions courantes, avec complications	20,1	36	317	1018	–	10
D	Montre-bracelet de grandes dimensions	30,1	–	712	–	–	10
E	Montre-bracelet de grandes dimensions, avec complications	36,1	–	1023	–	–	10
F	Montre de poche	–	50	–	1964	4,4	10
G	Montre de poche, extra-plate	–	50	–	1964	–	4,3
H	Montre de poche, avec complications	–	50	–	1964	–	15
J	Appareil horaire mécanique 8 jours, utilisé dans une seule position	–	–	–	–	–	–
K	Appareil horaire, utilisé dans une seule position, non compris dans l'un des genres précédents	–	–	–	–	–	–

3.6 **Tableau des limites** (Valeurs en secondes)

Nos	Critères	Genre A		Genres B et C		Genres D, E et F		Genres G et H	
		avec mention	sans mention	avec mention	sans mention	avec mention	sans mention	avec mention	sans mention
1	Marche diurne moyenne	−3 +12	−3 +12	−1 +10	−3 +12	−2 +5	−3 +8	−2 +7	−3 +10
2	Variation moyenne des marches	4	6	2,2	3,2	1,5	3	2	4
3	Plus grande variation	7	10	6	9	2,5	5	3	6
4	Différence du plat au pendu	± 10	± 14	± 8	± 12	± 4	± 7	± 5	± 9
5	Plus grande différence	16	22	12	18	7	12	9	15
6	Différence dans une période	–	–	–	–	–	–	–	–
7	Variation par degré centigrade	± 0,7	± 1	± 0,6	± 1,0	± 0,25	± 0,5	± 0,35	± 0,70
8	Erreur secondaire	–	–	–	–	± 4,5	9	5,5	11
9	Reprise de marche	± 7	± 10	± 5	± 9	± 2,5	± 5	± 4	± 8
10	Isochronisme	–	–	–	–	–	–	–	–

What is the purpose and value of an OFFICIAL TIMING CERTIFICATE

•

The Society of Swiss Chronometry says:

A watch is entitled to be called a Chronometer only if it is so carefully adjusted in 5 positions and 3 temperatures, that it will pass the severe tests of Observatories or Government Controlling Stations, and obtain an **official Timing certificate.**

Therefore, the name Chronometer upon your watch and the official certificate jointly testify to fine quality. The description given officially says:

Under the auspices and the control of the State, official institutes were created for the purpose of ascertaining the degree of precision attained.

All watches to which an official certificate has been awarded enjoy the benefit of an official
    guarantee of quality,
    guarantee of sound construction,
    guarantee of precise timekeeping.

Important: Fine grade watches show the same variations of seconds every day, if carried always in the same manner, and can easily be adjusted to show perfect time, by your watchmaker. Once adjusted to your habits, which should be done only a few weeks after purchase, your watch will give permanent satisfaction.

ROLEX
31 FIRST HONOURS GAINED FOR HIGH PRECISION BY BRACELET WATCHES

Special advantages of ROLEX movements:

1. They are entirely made in our own factories, under closest technical supervision, from the rough metal up to the finished movement.
2. ROLEX depend upon no outside sources and are thus in a position to really guarantee the quality of their productions.
3. ROLEX have specialised in making small watch-movements since 1878, they are one of the principal pioneers of the bracelet-watch. Their great accumulation of experience gathered since, is being constantly added to, so that Rolex remain one of the chief leaders of the industry.
4. ROLEX movements are made in two grades, i.e. 17 Rubies and 18 Rubies. The finer of these two grades is also available with Swiss Government certificates of quality and construction.
5. The hairsprings of ROLEX are anti-magnetic, which is an important feature nowadays.
6. The aero-dynamic balance-wheels to be found in ROLEX movements are their own invention. They are of very solid construction and of larger diameter than others, ensuring full swinging-oscillations and better timekeeping.
7. Every ROLEX watch has been regulated in 5 different positions by expert craftsmen.
8. All winding buttons used by ROLEX are of their own patented devices; they keep out dust and dirt.

Montres ROLEX S. A.
GENÈVE (Suisse)

From 1912 to this day the ROLEX factories have won a great number of honours at the famous Observatories of Geneva, Neuchâtel (Switzerland), Besançon (France) and Kew (England) as well as at the Swiss Testing Stations, which are under Government Control.

They were obtained for highest precision by wrist watches of all sizes and shapes, from the smallest Baguette, used for diamond set Ladies watches, up to 28½ millimeters for Men's wear.

ROLEX were the first in history to manufacture a small wrist watch Chronometer that was awarded the much coveted Kew A certificate already in 1914, after a 45 days test in 5 positions and 3 temperatures. Since then many other distinctions have been added by Rolex, so much so that chronometers worn on the wrist are now being produced in their factories regularly and serially for sale throughout the universe. For instance, 500 Gold & Platinum Rolex Prince watches made to order to commemorate the silver jubily of King George V reign, in 1936 obtained, each one of them, a Swiss official certificate with distinction "Especially Good".

Rolex factories at Geneva and Bienne

Distributing offices at: London - Paris - New-York
Buenos-Aires - Dublin - Toronto

ADVICE TO OWNERS
of good watches

The movement of a watch is an infinitely small and delicate piece of machinery and its many parts play jointly an important role in the effort of obtaining precision timekeeping.

The fine pivots revolve within highly polished ruby-jewels of first quality.

The balance wheel and escapement regulate and control the absolutely even unrolling of mainspring power.

Hardly any other machine is called upon to perform unceasingly day and night, every second of the year, as watches are expected to do.

One must therefore not be surprised that they require periodically to be most carefully oiled to prevent parts from wearing.

Science has not solved yet the problem of preserving oil in Men's watches for more than 18 months, and in Ladies watches for 12 months.

You will therefore remember that unless you hand your fine watch to an experienced watchmaker regularly within these periods for careful cleaning and oiling, it is bound to suffer damage and will disappoint you.

How to take care of a good watch:

Wind it every morning regularly so that the full mainspring-power is available for day-time service. It is immaterial whether you wear your watch day and night, or whether you remove it while you sleep.

But you must always behave the same way. If you take your watch off during the night you must place it upon the table, **always in the same position.**

The public seem unaware that watches behave differently and vary according to the positions in which they are made to function.

For that reason Official Testing Stations and Observatories control all Chronometers in 5 different positions and 3 temperatures.

Please remember:

1. Your watch gains a few seconds during the night if placed flat upon the table dial upwards.

2. It loses a few seconds in the vertical position, winding button downwards.

3. It loses still a few seconds in the vertical position, winding button upwards.

When you wish to correct a few seconds variation per day just avail yourself of this information, instead of asking your watchmaker to correct the watch. The result is very satisfactory.

General remarks

Very good watches require to be handled with care, but ordinary watches stand rougher treatment.

You cannot, however expect the same accuracy and dependability from an ordinary watch as from your own fine watch.

It will be your life-long friend if you follow the ways suggested by ROLEX. Punctuality helps you to success.

It is preferable to appeal to a firm who are appointed ROLEX agents for technical help rather than others; they have been carefully selected from among the best in each country.

ROLEX watches are sold and can easily be repaired in all countries of the world.

GUARANTEE
GIVEN BY ROLEX FACTORY

We certify that

ROLEX WATCH No.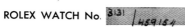

was produced entirely in our factories from first class materials, carefully selected and scientifically examined by our experts. It underwent very close tests throughout its many processes of manufacture, and was adjusted in 5 positions for accuracy.

Rolex guarantee this watch to be technically perfect in all its details.

MONTRES ROLEX S. A.
GENÈVE - SUISSE

Sold on:

By: Montres ROLEX S. A.
18, rue du Marché
GENÈVE

A brochure showing the purpose and value of an Official Timing Certificate.

CHAPTER 7
THE 1940s
Out of the Empire and Into the Fire

As the Perpetual and all gold Rolex watches were unavailable in Britain before 1949, Jewelers in Ireland would advertise such watches in British magazines.

The thirties saw several major changes for Rolex on fronts apart from the horological one. The most important happened on September 20, 1931 when the British Government of the day decided to abandon the Gold Standard. What this meant was that the British pound was no longer backed by gold and it immediately became worth 25% less than it had the previous day. Because Rolex made all of their watches in Switzerland and then imported them into London, they were very badly hurt. It was this one event that finally convinced Hans Wilsdorf to move the administrative offices of the company from London to Geneva. While the company had had a presence in Geneva almost from the start, and Hans had moved there in 1919, Rolex had always been run from London.

Later in the decade Rolex made another major move with the beginning of exports to the United States through its newly established American subsidiary. This was a very astute move, for just as Europe was sinking toward war, America was rising from the dust of the Depression under Roosevelt's New Deal and the economy was turning round. Even with the benefit of hindsight it is impossible to say whether it was good luck or good judgment, but Rolex once again was in the right place at the right time.

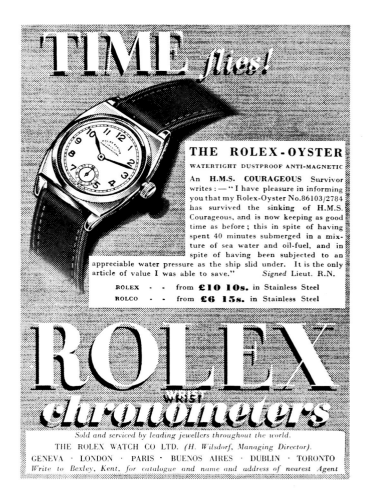

War time advertisement for Rolex.

The late 1930s marked the period when Rolex began to move toward mass production of chronometer certificated timepieces. Under the watchful eye of Emil Borer, their talented technical director,[30] Rolex technicians began to research into the basic principles of timekeeping. Lengthy investigations into such subjects as friction, types of oil and wearing surfaces were undertaken and the results published regularly in "Journal Suisse d' Horlogerie et de Bijouterie". It was only through this elemental research that the specialists at Rolex were able to discover ways in which to improve the timekeeping of their watches by small, incremental amounts.

The end of the thirties saw Rolex firmly ensconced in their traditional markets with a very successful range of products. The Oyster had transformed itself from the original two models to over a dozen in both men's and women's sizes. The Prince was still the ultimate wrist status symbol and the Perpetual had firmly taken hold. It was from this position of strength that Rolex embarked upon the most fruitful period of their entire corporate existence, before or since.

Because the future of the company was then safe, one could speculate that the designers felt free to let their creative juices flow. It was the designer's time to express themselves, and during this period Rolex made some of their most flamboyant cases and dials. Almost all of this creative effort went into the Oysters, both manual and Perpetual. It was there that the company's future was headed. As long as the patents held, these models were the company's unique selling points.

One of the first of the new Oyster models was the "Plage"[31], introduced in 1933. A tonneau-shaped model, no. 3359, it is often referred to by the name it was later given by Rolex: the "Viceroy." The watch came in men's and women's sizes and was the first non-perpetual Oyster to be offered in steel and gold, as well as the usual steel or gold. The very early versions of the watch were three-piece cases, but most of those made after 1935 were two-piece models.

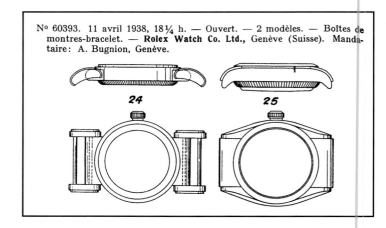

Original drawing for a hooded bubbleback.

The Viceroy was followed shortly by the model no. 2518, called the "Egyptian." This model was the first really styled Oyster. It has a simple round case with a double stepped bezel and with stepped hooded lugs. Introduced in 1934 and produced only until 1938, it was the first Rolex Oyster to wear hooded lugs, which were a stylistic peculiarity soon to figure heavily in the Oyster's future. The hooded (or concealed) lug style soon spread from the Egyptian, appearing next on the original cushion Oyster, which was issued in a hooded version in 1935 under the new model number 3096. This watch was not very successful, as by now the cushion Oyster style had been in production for ten years and was now regarded as old hat. This lack of success resulted in the hooded cushion becoming probably the rarest cushion model other than the ultra rare white gold model.

The models mentioned above are commonly considered to be the first hooded models from Rolex. In fact, the company had designed a whole range of exotically styled non-Oyster watches at the end of the 1920s. The range included twelve different models, all hooded. The original factory drawings (which were used to register the designs in June, 1928) show some very adventurous designs. To our knowledge, only a few of the proposed designs ever made it to the production line.

1934 advertisement featuring the Imperial and the Plage.

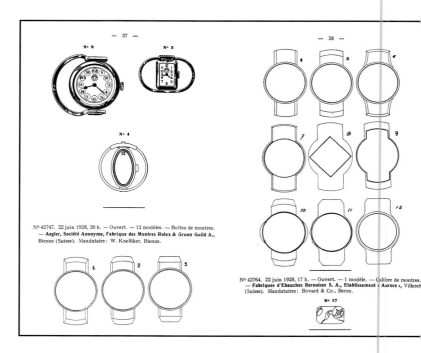

A range of case designs, 1928.

On the 11th of April, 1938, Rolex submitted design drawings of two new models to the patent authorities in Geneva. The designs had neither the names nor the model numbers the watches would bear when they were finally launched later that year. Instead they bore the prosaic design office numbers 24 and 25. The watch with the number 24 was a manual Oyster with a wide stepped bezel and large mobile lugs. Its companion watch, however, was destined to become the more famous when introduced, the following year, as the model 3065 hooded bubbleback.

The wartime period was a strange one for Rolex. Being Swiss the company was obviously not directly involved, but with its main markets of Britain and the British Empire cut off it was essential that Rolex find alternate markets. They supplied the Italian retailer, Panerai, with some custom built watches for the Italian Navy divers, but even when the Italians increased their order to supply some other segments of the Navy, it did not cover the loss of their principal markets.

Making a shift in their strategy Rolex began to think laterally and, in common with other Swiss watch companies, began to ship to Britain via neutral Spain and Portugal using other neutral flagged ships and planes. The new methods of shipment proved so effective that Switzerland was able to export more during the second half of 1940 than in the similar period the previous year. The system began operation in early 1940, just in time to meet a pent up demand for Rolex Oysters from the pilots of the rapidly expanding Royal Air Force. These pilots were supplied by the RAF with government issue watches, usually Longines, Omega, IWC or Movado, but the young Spitfire and Hurricane pilots wanted nothing to do with government issue watches.

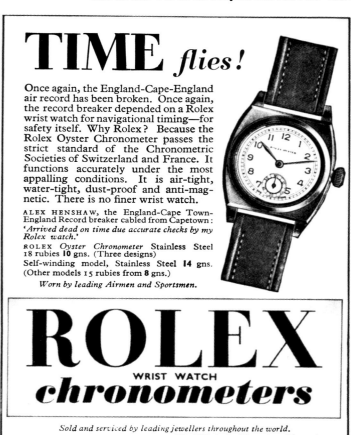

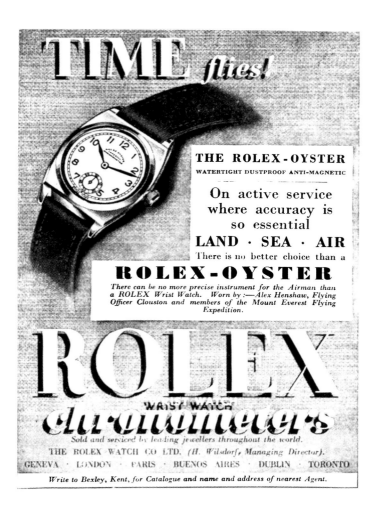

Advertisement with an endorsement from Alex Henshaw.

Another Henshaw endorsement.

In the period immediately prior to World War II Rolex had been very actively involved in the supply of watches to pioneering British long distance fliers. Alex Henshaw, the first to fly from London to Capetown, South Africa, and back wore one. So did the members of the RAF team members who were the first to fly over Everest. Flying Officer Coulston, who made the pioneering round trip flight between London and New Zealand, cabled the Rolex London office immediately after landing to praise the performance of his watch. Incidentally, Coulston had purchased his watch privately some six years earlier and was forced to use it for navigation purposes after his aviation timepiece expired. All of these exploits had been promoted by Rolex in the late 1930s with a series of advertisements under the title "Time flies." What this all meant was that any self respecting pilot wanted a Rolex Oyster, and suddenly there were a very large number of pilots.

The Battle of Britain was won by pilots wearing "Boy's size", manual wind Oysters with sweep seconds (where possible) and luminous dials. These watches were the unsung heroes of the 1940s for Rolex. While all the publicity and the focus was on the Perpetual and, later, on the Datejust, Rolex produced the 2240, the 4220, the 3121 and the 5056 in hitherto unheard of volumes. None of the pilots mentioned above or the soldiers or sailors who fought in World War II would have ever thought of wearing a self-winding watch. That was new and, as far as they were concerned, untried technology and in wartime you want something that you know will work every time. They chose the manual Oyster. With their own money, members of the British armed forces purchased every Oyster in the country. In truth there could be no higher compliment to a product. The compliment was returned by Rolex later in the war when British officers, who had been taken prisoner by the Germans, had only to write to Rolex in Geneva and ask for a replacement watch (their own had been appropriated by their captors) and it would be sent to the prison camp, with the understanding that it would be paid for when the war ended. While the vast majority of the "Boy's size" oysters had the simple dials described above, many were sold into civilian markets in the parts of the world still untouched by the war and these watches had some spectacular dials.

With dials, as with everything else, Rolex made few major changes, preferring gradual ones. When the "bubbleback" was first introduced in 1934, the dial was essentially a variation of the manual dial then current. It featured the words "Rolex Oyster" on one line with the word "Perpetual" centered beneath. There was no crown or "chronometre" anywhere on the dial. As we have seen earlier, all these first model watches were subsidiary second models and some of them bore the inscription "Self Winding" around the perimeter of the seconds dial. All the dials bore the legend "Swiss Made" and some bore the additional French translation "Fabrique en Suisse" below.

The first change was three years later when the new NA movement was introduced. As well as being slightly deeper, the new movement brought the benefits of a sweep seconds hand to the bubbleback. Despite the loss of the subsidiary seconds dial, Rolex (or to be more correct, their dial suppliers) still printed the curved "Self Winding" on some of the dials. However, the greatest change was the appearance of the word "Chronomètre" on the dial below the center post on these watches, marking the ability of the new movement to be timed to chronometer status.

The next main dial change did not occur until 1946 when the top lines of text on the dial were reformatted to read "Rolex" on the top line and "Oyster Perpetual" on the second line, the whole surmounted by the Rolex crown. The reformatting involved much more than just re-arranging the three words. The word "Rolex" remained in the same heavily serifed[32] typeface as before, while "Oyster Perpetual" was set in a new sans serif typeface. The size of the word "Rolex" was made 50% larger than the other words on the dial. The word "Chronometer" was now in English. This style remained the prevalent one for bubblebacks until 1949 when Rolex began to mark all their chronometer watches with the new legend, "Officially Certified Chronometer." All the watches which were not so certified were marked "Precision."

While Europe was sliding into World War II, Rolex was expanding its operations both in North and South America. It was for these markets that Rolex developed some of its most flamboyant models. They were required for these particular markets because the U.S. companies, such as Hamilton, Wittnauer/Longines and Gruen, had been producing their own flamboyant styles for quite some time and had succeeded in making the watch into a fashion item for men. The South American market also required much more radically styled watches than were favored in Europe. These markets were, at the beginning of the 1940s, the last flourishing consumer markets left in the world, and Rolex had to be there.

The next styling innovation was a simple variant of the 3065 hooded bubbleback. Introduced with the model number 3595 in 1941 it featured a wider, flatter bezel and five horizontal flutes across the hoods. Originally offered as a subsidiary seconds model, it (like all Rolex Perpetuals of the period) was available as a sweep seconds model for an additional SF 10. The model was the first watch Rolex introduced that was available in steel and gold only, with the customer's choice restricted solely to whether he wanted the hoods, bezel and winder in yellow or pink gold.

The 3595 was never popular and the following year Rolex introduced another variant on the hooded bubbleback. Starting with the style of the 3595, the new model simply turned the direction of the fluting from horizontal to vertical and increased the number of flutes from 5 to 13. Given the model number

1950 British advertisement showing one of the last "bubblebacks," which was also one of the first Rolexes to be imported after the end of World War II.

3598, it proved to be no more popular than the 3595 and is subsequently one of the most sought after Rolex models of the period.

It is not, however, as eagerly hunted as the next hooded bubbleback, the model 3599 all steel version. This was in fact a much simplified version of the 3598, featuring the wide bezel and only two raised vertical bars on the hoods. It was an attempt by Rolex to introduce a lower priced version of the hooded model. What Rolex finally faced was that the Oyster, whether perpetual or manual, was viewed by the public as a sports watch and that they weren't particularly interested in a stylish sports watch. This comparative lack of popularity has made the hooded bubblebacks much sought after today. It also accounts for the "normal" bubbleback being so widely produced.

Rolex got the message and, in 1944, registered the design for the machined bezel for a two-piece case, which had initially been introduced on the 3372 three-piece case model. This bezel design with very fine machined concentric circles, double raised bars at "12", single raised bars on the other quarter marks and raised circular indices at the other hour marks, became the Rolex signature bezel until the arrival of the fluted bezel ten years later. The first variation on this machined bezel was introduced only one year later as model 5015 in steel and gold. It featured a flat bezel with double bars at "12" and single bars for the remainder. This format was used later the same year on the model 5011, but the bezel was sloped toward the edges of the case instead of being flat.

Perhaps the rarest of the bubbleback bezels was the one used later on the model 5015, when the watch was relaunched for the 1949 season. The flat bezel featured a single bar at "12,"

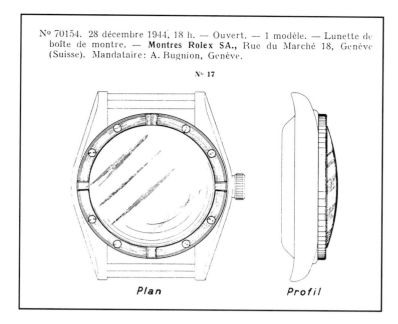

The Rolex Oyster again comes to England

THE OYSTER, pride of the Rolex Company of Geneva and first waterproof watch in the world, will soon be arriving from Switzerland in small quantities. The thousands who bought Oysters before and during the war know that this is a watch that, for all its elegance of design, is as strong and unfailingly accurate as a ship's chronometer—a watch as right for the drawing-room as for the golf course. The Rolex Oyster in stainless steel is a prized possession.

ANOTHER MEMBER of the Rolex family, the Tudor, also in stainless steel is being imported in small but increasing quantities. The Tudor is the perfect watch for those who want a genuine Swiss movement at a lower price. It is manufactured under strict Rolex supervision and every watch bears the Rolex label of guarantee, proof positive of its reliability.

QUANTITIES are still small, but the next few months will add to the number of watches and the variety of models. Meanwhile, leading jewellers *do* have Rolex watches, and may be able to satisfy your long-felt desire to own one of the finest watches ever made in Switzerland.

ROLEX, creators of the *first* wrist-CHRONO-METER and the *first* WATERPROOF watch, also perfected the *first* waterproof AND SELF-WINDING watch and the *first* waterproof, self-winding and CALENDAR watch, the last two not yet available here.

The waterproof crown was patented in 1926. To wind the watch, unscrew the crown—which automatically springs up and out. The watch can now be wound and the hands set.

When the crown is closed, two metallic surfaces are pressed tightly against each other, thus ensuring the permanent, hermetic sealing of the case, without the aid of perishable materials.

ROLEX
WRIST CHRONOMETERS

The Rolex Oyster in Stainless Steel with leather strap (incl. pur. tax) £23.15s.; The Tudor Oyster with leather strap (incl. pur. tax) £15.15s.; Steel Bracelet, when available (incl. pur. tax) £1.5s.
Prices are subject to fluctuation.

The Rolex Watch Co. Ltd., 1 Green Street, Mayfair, London, W.1
(H. Wilsdorf, Governing Director)

Owing to present-day conditions the repair service has been suspended. Its resumption will be announced.

The return of Rolex to Britain is heralded in this advertisement.

circles for the remainder of the hour markers, and, unlike any other contemporary bezel, it had radial minute markers.

The next development from Rolex was probably the most significant of their history. It started with the simple decision to introduce a larger version of the bubbleback. The new watch was powered by a perpetual version of the classic 10-1/2''' Hunter, a new calibre called the A295 (or calibre 720). It was initially used in a simple large bubbleback with the reference 5026. Launched in the midst of the war, in 1942, it was never a big seller. However three years after the introduction of the new movement Rolex made two additions to it, and renamed it the calibre 740. With these two modifications, a sweep seconds hand and a date disc visible through an aperture in the dial at "3," the definitive Rolex was created, the Datejust.

While Rolex was scaling up the self-winding mechanism before adding it to the 10-1/2''' Hunter, it was decided that it would be just as easy to scale the movement down a little and add it to the 8-3/4''' Hunter. This was done and the resulting movement named Calibre AR (later changed to Calibre 520). This movement was fitted to some of the most flamboyantly styled watches Rolex ever made. Known collectively as "Boy's size" bubblebacks, they were made for almost as long a period as the standard size bubblebacks, but were made in much smaller quantities. At the time they entered production, men's watches were, in fact, becoming larger. So the decision was made to emphasize the smallness of the watches rather than hide it. This decision brought forth a series of watches sometimes collectively known as "Lifesavers," small perpetuals with wide bezels and tiny dials.

The next watch to use the new AR movement was another exotically styled piece, the scalloped bubbleback (model 3353). It has hoods made from three wide scallops, and was available with small or sweep seconds and in steel with gold hoods or all steel. The scalloped bubbleback was made for only three years and is now eagerly sought after.

The forties were also a time of great tragedy for Rolex and for Wilsdorf himself. On April 26, 1944, his wife, May passed away and less than two months later his long time business partner Hermann Aegler also died. Wilsdorf must have been devastated, the foundations that he had built upon for forty years were gone in matter of days. The effect on Wilsdorf was to make him decide to build for the future. Seeing so much taken from him he cast his mind to the building of a living memorial, giving birth to the idea for a charitable foundation. May and Hans Wilsdorf were childless. If they had had children there is no doubt that they would now be running Rolex.

In 1941 World War II was in full swing and Rolex was still obtaining the vast majority of all the Chronometer bulletins issued by the official testing stations. In their home town of Bienne they obtained 2,632 certificates, over 80% of the 3,253 issued that year. Almost as impressive was the fact that they obtained almost 63% of all the certificates with the mention "Especially Good Results."

The 1945 results showed the steady improvement once again. Rolex obtained 1,231 certificates at Locle which represented 69% of the total 1,776. Over 669 of them were "Especially Good Results" or over 75% of the 886 watches so honored. At St. Imer the Rolex earned 275 "Good Result" mentions (almost 59% of those given) and 174 regular certificates (42% of those awarded). Out of the total of 898 certificates Rolex was awarded 449 or exactly 50%. In its home town of Bienne the company was awarded 13,214 certificates out of the total 14,449 delivered or an astounding 91.45% and the 9,084 "Especially Good Results" watches it won represented an even more amazing 92.46% of the watches qualifying. It was the perfect end to a perfect Jubilee year for Rolex.

The end of the war saw Rolex in what seemed to be the strongest position they had ever been. In truth the company's future looked bleak. Its major market, Britain and its Empire was gone, as, of course, was the rest of the European market. As the world clawed itself out of the ruins of World War II, so did Rolex.

164 The 1940s: Out of the Empire into the Fire

Though Britain had been on the winning side of World War II, it emerged economically devastated and the importation of Swiss watches was prohibited until 1947. When it slowly began in that year, only steel watches were permitted and only the less expensive ones at that. In fact the perpetuals were not imported into Britain until early 1949 and then only in steel. The advertisement shown here from 1949 talks about both the Perpetual and the Datejust, but states "the last two not yet available here." Rolex was in a desperate position. Their largest pre-war market could not import their latest, most expensive (and therefore most profitable) products.

It was for this reason that Rolex turned more to the ever expanding U.S. market. To enable them to crack the U.S. and international markets Rolex recruited a new partner, J. Walter Thompson, at the time the world's largest advertising agency. In many ways they were a perfect fit. Both companies, while being essentially creative, also were essentially conservative in outlook. Though they both were obvious issue of their respective birthplaces (Switzerland and the U.S.) both were international in outlook. The relationship between the two still lasts. In the fifty years since the first advertisement there have only been four main campaigns. The latest, showing famous Rolex wearers and sometimes bearing the line "You can tell by the men who wear them," was first used over twenty-five years ago in 1970.

[30] Borer was related to Hermann Aegler and, today, the Aegler/Borer Family Trust is the only shareholder in Rolex apart from the Wilsdorf Foundation.
[31] The word "Plage" means "beach" in French, it is *not* a misspelling of "plague."
[32] Serifs are the little twiddly bits at the end of letter strokes. ROLEX is in a serif typeface, while ROLEX is in a sans (or without) serif type.

Rolex steel watch original two-tone black dial with outer seconds track and inner 13-24 hour track, luminous dial and hands, sweep seconds.
Circa 1935

The 1940s: Out of the Empire into the Fire 165

9kt gold Rolex Oyster Imperial two-piece case, original two color criss-cross dial signed "Rolex Oyster Imperial Extra Precision." Movement 10-1/2''' open faced converted to sweep seconds with 17 jewels, adjusted to six positions signed "Ultra Prima."
Glasgow Hallmark 1935

166 The 1940s: Out of the Empire into the Fire

18kt pink gold hooded manual Rolex with scallop hoods, original black dial with diamond shaped markers, sunken subsidiary seconds. Movement 10-1/2''' Hunter Ultra Prima. Notice the different positions of the balance on this and the previous watch.
Model no. 3270
Case no. 43626
Circa 1938

Stainless steel Rolex Oyster Speedking Precision. Luminous hands and dial with unusual tear drop shaped luminous hour hand.
Model no. 5056
Case no. 630264
Circa 1944

The 1940s: Out of the Empire into the Fire

Stainless steel Rolex Oyster Royal two-piece case with luminous dial and "Mercedes" hands, sweep seconds. Non-original Bonklip-style contemporary bracelet.
Model no. 2595
Case no. 141310
Circa 1941

Rolex Oyster Speedking Precision. Luminous dial and classic parallel hands.
Model no. 4220
Case no. 385273
Circa 1942

Stainless steel Rolex Oyster Royal two-piece case. Applied gilt Arabic numerals and sunken subsidiary seconds with the legend "Shock Resisting" around subsidiary seconds dial. Note that the numbers of this and the previous watch are within 170 of each other.
Model no. 4220
Case no. 385106
Circa 1942

Rolex Oyster Speedking two-piece case, pink dial with Arabic and baton luminous numerals and luminous Mercedes style hands.
Model no. 4220
Case no. 228009
Circa 1941

Rolex Oyster Royal two-piece stainless steel case with pink dial with luminous Arabic even numbers and luminous baton odd numbers. Dial also signed "Precision" and with the name of the retailer, Burdocks of Capetown, South Africa.
Model no. 2280
Case no. 247632
Circa 1942

The 1940s: Out of the Empire into the Fire 169

Stainless steel two-piece case Rolex Oyster Speedking, silvered luminous dial, sweep seconds with outer seconds track.
Model no. 5056
Case no. 629567
Circa 1948

Rolex Oyster Speedking in two-piece stainless steel case with applied gilt Arabic numerals, sunken subsidiary seconds, signed "Precision" around the subsidiary seconds hand. Movement 10-1/2''' Hunter Super Balance.
Model no. 5056
Case no. 657120
Circa 1948

Rolex Oyster Speedking stainless steel two-piece case with original black dial with mixed Arabic, Roman and baton numerals. Original gilt colored stainless steel hands and the dial also has a gilt minute track.
Model no. 4220
Case no. 232298
Circa 1942

The 1940s: Out of the Empire into the Fire

Stainless steel and pink gold Rolex Oyster Viceroy with gold machined perfect bezel and gold winder. Original black dial with mixed Roman, Arabic and baton luminous numerals. Movement 10-1/2''' Hunter Super Balance with 17 jewels signed "Chronometer" with micrometer regulator and capped endstones. Movement number 15815.
Model no. 3359
Case no. 297107
Circa 1944

Rolex Oyster Viceroy Chronometer in 14kt pink gold with original pink dial with mixed Roman, Arabic and baton luminous numerals and Mercedes style hands. Movement signed "14741" 10-1/2''' Hunter Superbalance.
Model no. 3116
Case no. 56630
Circa 1939

The 1940s: Out of the Empire into the Fire 171

Rolex Oyster Perpetual in boy's size, stainless steel two-piece case with applied Arabic gilt numerals and luminous hands with sweep seconds. Signed above the six "Officially Certified Chronometer."
Circa 1947

Rolex Oyster Speedking, stainless steel two-piece case with silver metal dial. Luminous even numbers and luminous batons for the odd numbers. Parallel luminous hands, sunken subsidiary seconds dial with the word "Precision" around it.
Model no. worn off
Case no. 334654
Circa 1945

Stainless steel and pink gold Rolex Oyster Firefly with original pink metal dial with luminous Arabic numerals, sweep seconds and Mercedes style hands. The case seems to be stamped with two model numbers, 3121 and 4302, but no case number.
Circa 1944

Stainless steel square dress Rolex. Silver metal dial with luminous Arabic, baton, and dot numerals, sunken subsidiary seconds.
Model no. 4572
Case no. 495926
Circa 1944

172 The 1940s: Out of the Empire into the Fire

Rolex Oyster Royalite stainless steel two-piece case, silvered three-tone metal dial with luminous even numbers and luminous dots for the odd numbers. Outer 13-24 hour track and the minute track contained in between the normal hours and the 24 hour track. Dial is signed "Observatory" above the six. As with almost all watches which have the word Observatory on the dial, the Royalite was made solely for the Canadian Market and comes in a number of different dial designs. Circa 1941

Stainless steel two-piece rectangular Rolex wristwatch with unusual three dimensional chased lugs, which rise above the case to the level of the crystal. Silvered metal dial with applied Arabic and baton numerals and subsidiary seconds dial. Movement is the classic HW movement with the protective balance guard and capped endstones. 17 jewel movement.
Model no. 6565
Case no. 837610
Movement no. 565460
Circa 1946

The 1940s: Out of the Empire into the Fire

9kt gold "top hat" style dress rectangular wristwatch with applied metal numerals and yellow metal dial with sunken subsidiary seconds. Dial also signed "Bucherer" for the retailer. Movement HW with protective balance bridge and 15 jewels.
Model no. 2317
Case no. 29644
Circa 1943

9kt gold Rolex dress watch with stepped ends, solid strap bars, metal dial with criss-cross two-tone finish and subsidiary seconds. Movement HW with 15 jewels and protective balance bridge no capped end stones.
Circa 1931

174 The 1940s: Out of the Empire into the Fire

Rolex stainless steel Prince Dauphin with triple stepped case sides recessed for winding crown. Black dial with stylized numerals and sunken subsidiary square.
Circa 1934

Stainless steel two-piece Rolex tonneau-shaped dress watch. Stepped ends, solid strap bars and two-tone dial in which the luminous numerals have been washed off leaving just the outline. Movement HW, 15 jewel movement.
Circa 1935

The 1940s: Out of the Empire into the Fire

9kt gold Rolex rectangular wristwatch with movable lugs. Original metal dial with applied full Arabic numerals and sunken subsidiary seconds. Notice how the lugs are completely flexible.
Model no. 2943
Case no. 30864
Circa 1936

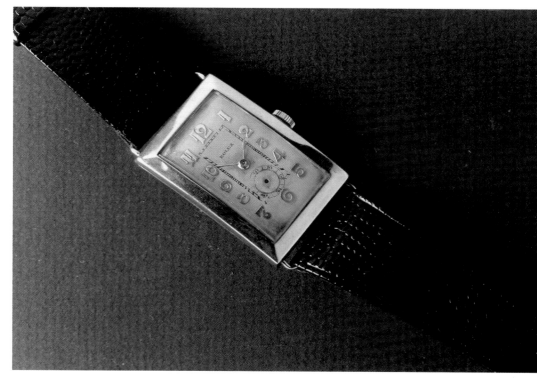

9kt pink gold curved rectangular wristwatch with applied exaggerated Arabic numerals and subsidiary seconds dial, although subsidiary seconds hand is missing. Solid strap bars.
Circa 1934

Rolex stainless steel two-piece cushion shaped dress watch. 10-1/2''' Hunter movement. Notice the solid strap bars, dial with applied Arabic numerals and sunken subsidiary seconds. The case is signed by the manufacturer, Dennison, made in England and marked "Denisteel," which was their first attempt at stainless steel.
Model no. 1045
Circa 1948

Rolex two-piece 9kt gold, cushion shaped non-Oyster dress watch. The construction of this case is unusual in that the back, the center of the case, and the lugs are all one piece and the rest is merely a removable bezel. The dial is metal with painted Arabic numerals, sunken subsidiary seconds. The movement is a 10-1/2''' Hunter which fits into a movement ring that drops straight into the back.
Circa 1937

The 1940s: Out of the Empire into the Fire

9kt gold Rolex two-piece Scientific non-Oyster model. Two-piece porcelain dial with Arabic and Baton numerals for hour markers, outer minute track and then further seconds track divided into 1/5th of a second. 10-1/2''' Hunter Ultra Prima Chronometer movement timed to six positions and with 18 jewels and capped endstones. This is one of the rarest Rolex models made during this period. It is believed that less than 500 were made and they were the highest quality movement from Rolex at the time. The two-piece dial construction itself is also very rare and is more often seen on pocket watches. The case is stamped with the serial number both inside and outside which is also very unusual. The large red sweep seconds hand is to ensure perfect visibility at any time.
Model no. 2942
Case no. 30198
Movement no. 77277
Circa 1936

178 The 1940s: Out of the Empire into the Fire

18kt yellow gold Rolex Oyster Impérial, retailed by Bucherer's. Unusual three color dial signed "Extra Precision." Blued steel stick hands with unusual red sweep seconds hand. Very rare 10-1/2''', 17 jewel, "open face" movement signed "Ultra Prima."
Model no. 2595
Case no. 36028
Circa 1937

Stainless steel and pink gold square dress watch with two-tone criss-cross dial, Arabic numerals, sunken seconds and unusual "fleur d'lys" blue steel hands. This watch is extremely unusual as it is the first simple steel rectangular watch with applied pink gold hoods to surface. It is evident that the hoods have not been applied recently as the fixed bars for the watch strap are immediately behind and attached to the hoods. 10-1/2''' Hunter movement.
Model no. 1872
Case no. 018530
Circa 1939

Rolex Oyster Perpetual stainless steel boy's size bubbleback. Original black dial with applied gilt Arabic and triangular markers, subsidiary seconds dial.
Style no. 6006
Case no. indistinct
Movement no. A7910
Circa 1939

The 1940s: Out of the Empire into the Fire 179

Stainless steel two-piece case Rolex Oyster Perpetual Precision. Metal silvered dial with painted Arabic numerals and outer minute track, sunken subsidiary seconds dial, with the word "Precision" around it. Blued steel leaf hands. This model used a 10-1/2''' Hunter version of the original bubbleback movement and is the missing link between the bubbleback and the first Datejust.
Model no. 5026
Case no. indistinct
Circa 1943

180 The 1940s: Out of the Empire into the Fire

Stainless steel rectangular watch with double stepped case ends, curved back, and fixed strap bars. Called the "Egyptian" in contemporary advertisements, it was one of the first watches to use the new HW tonneau shaped movement.
Model no. 1936
Case no. 019685
Circa 1937

Rolex 18kt gold three-piece case Oyster Perpetual Datejust with coin-edge style milled bezel. Original dial with applied triangle markers and aperture at three for the date. 10-1/2''' Hunter Perpetual, bubbleback-style movement numbered G23154.
Model no. 4467
Circa 1946

The 1940s: Out of the Empire into the Fire 181

9kt gold cushion dress watch with silvered metal dial, applied full Arabic numerals and subsidiary seconds. Signed with retailer's name "E.S. Campbell Ltd." The style of this watch is almost identical to the 2420, although it is, of course, not an Oyster. 10-1/2''' Hunter Superbalance movement, 15 jewels.
Model no. 7247.
Hallmark Glasgow 1938

Stainless steel Oyster "Army" two-piece cushion case, similar in design to the earlier 2420 gold model. Refinished black dial with luminous hands. 10-1/2''' Hunter Superbalance movement, 17 jewels.
Model no. 3139
Case no. 245057
Circa 1941

Rolex Oyster stainless steel watch with unusual round hidden lug case. 33mm © Sotheby's, Inc., 1990.
Circa 1940

A 9kt gold Viceroy Oyster Imperial with black dial indicating it was retailed by P. Orr and Sons of Madras and Rangoon, Rolex's principal agent in India. © Sotheby's, Inc., 1990.
Circa 1945

182 The 1940s: Out of the Empire into the Fire

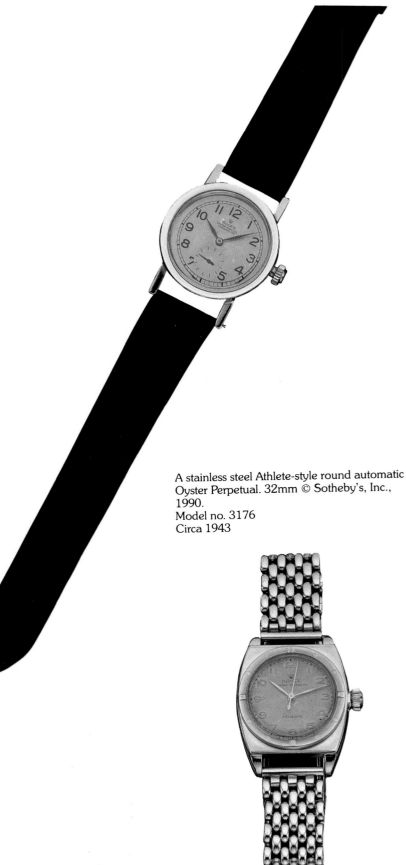

A stainless steel Athlete-style round automatic Oyster Perpetual. 32mm © Sotheby's, Inc., 1990.
Model no. 3176
Circa 1943

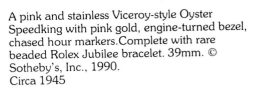

A pink and stainless Viceroy-style Oyster Speedking with pink gold, engine-turned bezel, chased hour markers. Complete with rare beaded Rolex Jubilee bracelet. 39mm. © Sotheby's, Inc., 1990.
Circa 1945

A pink gold cushion-shaped 14kt Rolex Oyster Majestic, 17 jewels, 28mm, sweep seconds. © Sotheby's, Inc., 1990.
Circa 1945

Rolex Oyster Imperial with two-tone dial, only 29mm.© Sotheby's, Inc., 1990.
Circa 1945

The 1940s: Out of the Empire into the Fire 183

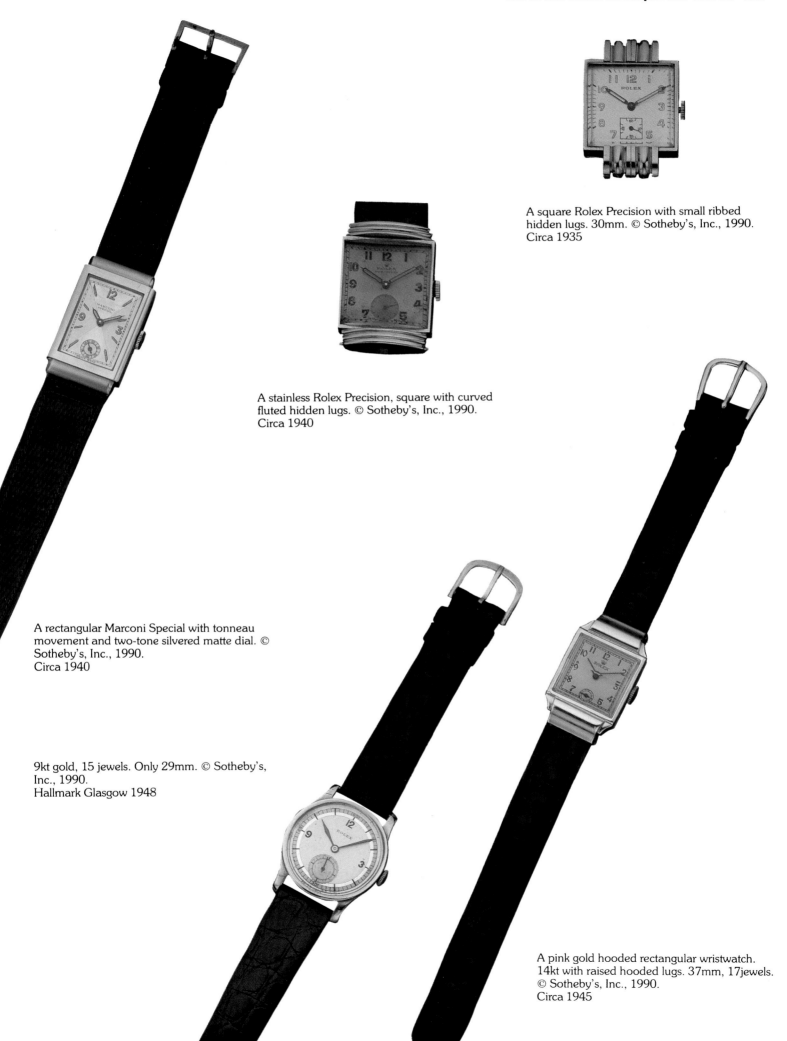

A square Rolex Precision with small ribbed hidden lugs. 30mm. © Sotheby's, Inc., 1990. Circa 1935

A stainless Rolex Precision, square with curved fluted hidden lugs. © Sotheby's, Inc., 1990. Circa 1940

A rectangular Marconi Special with tonneau movement and two-tone silvered matte dial. © Sotheby's, Inc., 1990. Circa 1940

9kt gold, 15 jewels. Only 29mm. © Sotheby's, Inc., 1990. Hallmark Glasgow 1948

A pink gold hooded rectangular wristwatch. 14kt with raised hooded lugs. 37mm, 17 jewels. © Sotheby's, Inc., 1990. Circa 1945

184 The 1940s: Out of the Empire into the Fire

Rolex Oyster Extra Precision retailed by Mappin of London. Two-tone steel and gold with hooded lugs, with Rolex steel and gold bracelet. © Sotheby's, Inc., 1991.
Circa 1945

Steel Rolex Pall Mall Observatory for Canadian market with black matte dial and sweep seconds hand. © Sotheby's, Inc., 1991.
Circa 1940

Small scallop-hooded manual wind Oyster, with black dial with pointed hour markers and subsidiary seconds. Ref. 3270. © Sotheby's, Inc., 1992.
Circa 1944

A stainless and pink gold boy's sized Precision with sweep seconds and hour markers on bezel. Ref. 3121.
© Sotheby's, Inc., 1991.
Circa 1940

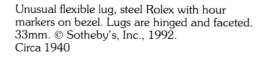

Unusual flexible lug, steel Rolex with hour markers on bezel. Lugs are hinged and faceted. 33mm. © Sotheby's, Inc., 1992.
Circa 1940

The 1940s: Out of the Empire into the Fire 185

Stainless steel & pink gold sweep seconds Rolex Oyster Viceroy. Black dial with luminous Roman/Arabic numerals. 17 jewel, 10-1/2''' Hunter movement, pink gold winder, and engine-turned bezel. Ref. 3359. © Sotheby's, Inc., 1992.
Circa 1946

A stainless steel Rolex Oyster Athlete. Sweep seconds, nickel lever movement, patented "Superbalance," 17 jewels. Screw-down crown, pink matte dial, Arabic and baton numerals, cylindrical water-resistant-type case. Case, dial and movement signed. Diameter 32mm. © Sotheby's, Inc., 1990.
Circa 1940

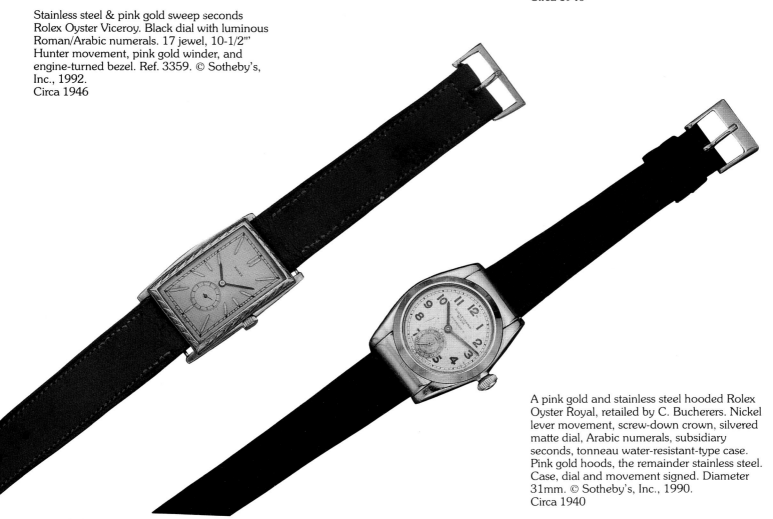

A pink gold and stainless steel hooded Rolex Oyster Royal, retailed by C. Bucherers. Nickel lever movement, screw-down crown, silvered matte dial, Arabic numerals, subsidiary seconds, tonneau water-resistant-type case. Pink gold hoods, the remainder stainless steel. Case, dial and movement signed. Diameter 31mm. © Sotheby's, Inc., 1990.
Circa 1940

14kt two-tone gold rectangular Rolex. 10-1/2''' Hunter "Superbalance," 15 jewel movement. Gilt matte dial with applied baton numerals and subsidiary seconds dial. Yellow gold bezel chased with scrolls, white gold band similarly chased, hinged yellow gold back. Ref. 758. © Sotheby's, Inc., 1994.
Circa 1945

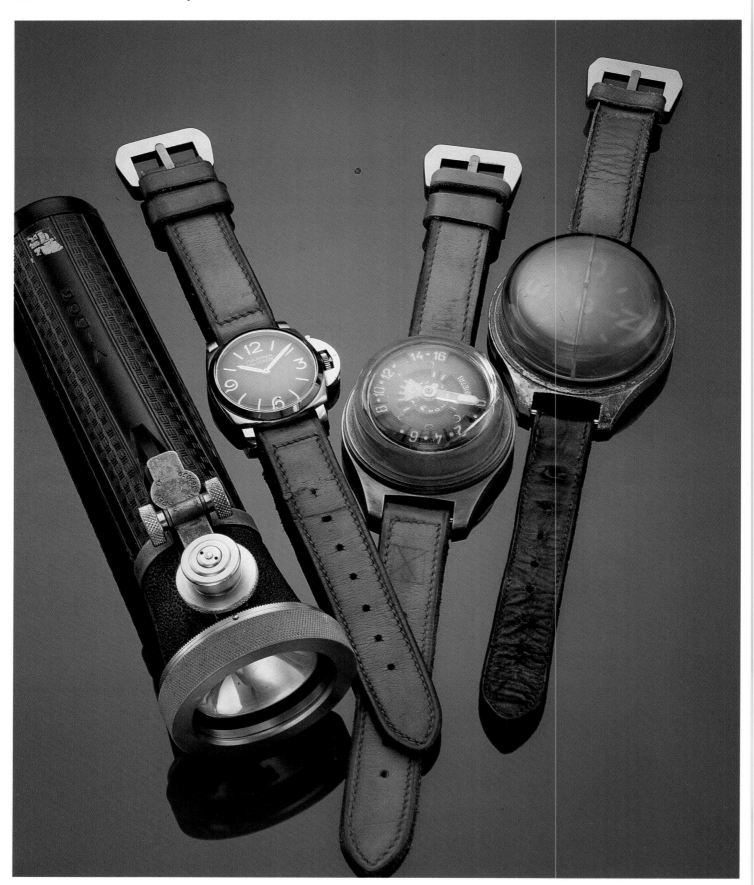

A stainless steel military Rolex, "Marina Militare," with compass, altimeter, flashlight and knife. Nickel lever movement, 17 jewels, mono-metallic compensation balance, dust cap. Matte black dial, luminous baton and Arabic numerals. Water-resistant-type cushion-form case with winding crown protector. Case and movement signed; dial signed Marina Militare. Liquid mounted compass in a stainless steel case, back signed "officina Panerai, Firenze." Similarly case altimeter is back signed "officina Panerai, Firenze Matr. #212 Brevatto," and the dial is signed "Luminor Panerai metr." Also, a military issue knife and a rubber mounted Vidor Flashlight. Ref. 6152. © Sotheby's Inc., 1990.

The Panerai Company of Italy commissioned Rolex to manufacture diver's watches during World War II for the use of the Italian Navy Frogmen. Several model names appeared on these watches: Marina Militare, Luminor Panerai and Radiomir Panerai. This group is interesting as it is rare to find what would appear to be a complete set.

Circa 1940

9kt yellow gold two-piece cushion case with 10-1/2''' Hunter movement, silvered dial with Arabic numerals and leaf shaped hands. The case on this watch is unusual in two aspects. First, it bears a Glasgow domestic hallmark for 1937. By this time almost all Rolex cases were Swiss made and the few that were made in England were hallmarked in London. This is only the second Rolex to be seen with a Glasgow domestic hallmark. Second, the case construction itself is also very unusual, in that the movement sits tightly in a deep ring inside the case back and the whole back then fits almost hermetically into the case top. The gap in the top of the case for the winder is protected by a small protrusion from the bottom half of the case.
Model 4890
Hallmark Glasgow 1937

"Athlete" bubbleback with the rare Roman/Arabic variation dial. Note the difference between this dial and the normal dial shown earlier. This dial has a luminous dot at "6" rather than the more normal baton. The perpetual version of the "Athlete" is much rarer than the manual version.

188 The 1940s: Out of the Empire into the Fire

Viceroy in stainless steel with pink gold machined bezel and original early Oyster bracelet. Unusual three color "scientific" style dial.
Model no. 3359
Case no. 284966
Circa 1945

Stainless steel "boy's size" Oyster Scientific. Very unusual black dial with luminous Arabic and baton hour markers and classic scientific outer minute, seconds, and 1/5 second divisions. This is the only black Scientific dial to surface to date.
Model no. 2280
Case no. 269561
Circa 1944

The 1940s: Out of the Empire into the Fire 189

The evolution of the "bubbleback" dial. **Left:** the first sweep seconds model bears the words "Rolex Oyster" on one line and the word "Perpetual" on another line. The word "Chronomètre" is spelled in the French mode. **Center:** the 1942 dial has the words "Oyster Perpetual" on the second line and the word "Chronometer" is now in the English spelling. **Right:** the 1947 dial bears the legend "Officially Certified Chronometer."

9kt yellow gold asymmetric wristwatch with hooded lugs. Nickel plated Extra Prima movement, black dial with Roman numerals and subsidiary seconds. © Christie's, London
Model 3260
Hallmark Glasgow 1938

The original Rolex box for the dials shown below. Note the stylized image of sails in the form of pairs of hands, while a Rolex pocket watch rises from behind the clouds.

Dial options from the 1940s. Rolex catalogs, until the 1970s always offered a variety of dial options for each watch. Here we see the range of dials for a model 1872, a square hooded 10-1/2''' watch. Four are luminous and two have applied gilt numerals. Note the little Rolex sacks each dial was contained in.

Chapter 8
THE 1950s
250,000 Chronometers
Can't Be Wrong

George Mallory was one of the first of the great Himalayan mountaineers, and when he was asked why he wanted to climb Everest he replied "Because it's there!" It was in this spirit that the watchmakers of Rolex (and most of the other great Swiss watch companies) approached the challenge of the 1950s. The War was over, evil had been defeated and the world was now a place fit for heroes. The technology that had helped to produce the machines that won the war was now available on the open market for all to buy, and the Swiss bought it by the train load and then they used it to its limits. Everything now seemed possible. The atom bomb (and its sole ownership by the United States) meant that now there could never be another war.

It was in this spirit of optimism bold new steps were taken. In the U.S., Detroit began to introduce the high-compression V8 engine and the first steps toward the longer, lower look in auto design were made. In Britain a new young queen was crowned, the birth of a new Elizabethan era. In Switzerland Rolex began the second half of the 20th century with a giant celebration, which they called the Rolex Festival. It was held to celebrate four significant anniversaries: the 20th year of the Auto Rotor automatic, the 25th year of the Oyster, Mr. Wilsdorf's 50th year in the watch business, and his 70th birthday. The festivities lasted four days and included a banquet, numerous cocktail parties, a luncheon excursion into France, a garden party and a festive voyage on Lake Geneva. The voyage was brightened up by one of Wilsdorf's classic publicity stunts. Prior to the start of the excursion six Oyster perpetuals were set to the correct time but not wound, they were then attached to a frame which was then thrown into the lake and towed behind the boat all day. At the end of the day the frame was withdrawn and all the watches were running perfectly and going to time. The movement of the waves and the momentum of the ship had provided the mechanism with the necessary winding energy and the Oyster cases had performed their part perfectly, too, by protecting the movements and dials. The following day the watches and their frame were proudly displayed in the window of Geneva's finest jeweler.

The 1950s were, for Rolex, a time for some of their greatest successes. It also was a time for some of their more impressive failures. As is often the way, it was the failures that came first.

Despite the success of the 25 year old patent for the screw-down Oyster crown, Rolex introduced the "Super Oyster" crown. This was a non screw-down crown whose impermeability was supposedly ensured by a system of gaskets and precision tolerances. Called in some of Rolex's advertisements by the bizarre name of the "phantom crown," it was introduced on the model 6062, (automatic/day/date/month/moon phase)[33] which was the first ever Rolex to show both the day and date together.

The watch and the crown both proved to be initially successful. It solved the problem encountered by many people still unaccustomed to unscrewing the crown when winding or hand-setting the watch. After completing the operation they would forget to rescrew the button into the secured position, with disastrous results when the watch was brought to water. That this really was a problem for Rolex is evidenced by the instruction

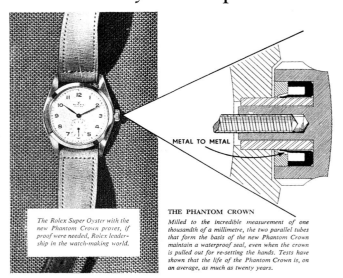

leaflets of the day which gave heavy emphasis to the rescrewing of the button. The Super Oyster aimed to solve this problem by using three gaskets which theoretically provided protection even when the winder was in the hand setting position. Please note the use of the word "theoretically." What in fact happened was that general wear and tear and the universal application of Murphy's Law made the Super Oyster less than Rolex Oyster waterproof within months, usually just after the expiration of the warranty. Rolex learned from the lesson and the Super Oyster lasted less than five years. It was perhaps their most notable failure, but it was not all lost; the interlocking gaskets, coupled to a conventional screw-down crown, lived on to become the "Triplock" crown system of the later Submariner model 6538 in 1957.

After its initial success and strong sales for its first year, demand for the "Cosmograph" 6026 began to tail off. Soon Rolex had so many unsold movements in stock that it decided to case

18kt gold Bombé Oyster Perpetual with unusual honeycomb finished dial.
Circa 1952

In 1950 the Datejust was made only in 18kt gold and was the only model with a date window. Rolex did not want to damage the prestige of the Datejust, but did want to exploit their patented date disc in a wider market, so, in late 1952, they introduced the Rolex Oysterdate. Placed in a stainless steel Oyster case, it had a manual wind 10-1/2''' Hunter (calibre 710) movement to which was added the complete calendar mechanism of the calibre 286 Datejust. The watch was available in two models and sizes, 6266 and 6466, and remained in the catalog until the late 1970s. It was the first new model to carry the company-designed shock proof system, which was then gradually introduced to the rest of the line.

Further technological steps taken in the 1950s resulted in the introduction of two of the most esoteric models the company ever produced: the "Tru-beat" (model 6556) and the "Milgauss" (model 6541). The "Tru-beat" was a standard chronometer grade non-date Oyster Perpetual with a major difference: it had a dead beat sweep seconds hand. Instead of making a continuous sweep of the dial, the second hand would stop at each seconds mark before jumping to the next one. In other words it performed exactly the way that analog quartz watches do today. Like early dead beat watches, the motion work that operated the seconds hand was on a special plate between the normal seconds drive and the automatic winding mechanism and needed a special pinion and numerous other special parts. Apart from this, though, it operated more or less identically to other movements.

Dead beat watches had been made since the days of Breguet and had always been seen as the mark of a very finely engineered watch. But in the 1950s this watch was surely the answer to the question no-one asked. The arrival of the "Tru-Beat"[34] on the international market was greeted with overwhelming apathy and it was withdrawn within five years. This has made an unaltered model one of the rarest and most desirable

Further technological steps taken in the 1950s resulted in the introduction of two of the most esoteric models the company ever produced: the "Tru-beat" (model 6556) and the "Milgauss" (model 6541). The "Tru-beat" was a standard chronometer grade non-date Oyster Perpetual with a major difference: it had a dead beat sweep seconds hand. Instead of making a continuous sweep of the dial, the second hand would stop at each seconds mark before jumping to the next one. In other words it performed exactly the way that analog quartz watches do today. Like early dead beat watches, the motion work that operated the seconds hand was on a special plate between the normal seconds drive and the automatic winding mechanism and needed a special pinion and numerous other special parts. Apart from this, though, it operated more or less identically to other movements.

These failures, frustrating as they may have been, pale in light of the successes that were to follow. At the 1954 Basel Fair, Rolex launched four major watch models, all in production to this day, the Explorer, Turn-O-Graph, Submariner and the Lady's Perpetual Chronometer. This knockout blow was followed two years later with the introduction of the watch destined to knock the "Datejust" off its pedestal as the company's flagship. This was, of course, the "Day-Date," a watch important enough to merit its own chapter a little further on.

Oyster Perpetual gold cap with steel back, honeycomb effect dial. Notice the officially certified chronometer above the six is written in German.
Model no. 6564
Case no. 141551
Circa 1956

The 1950s: 250,000 Chronometers Can't Be Wrong 197

Stainless steel manual Oyster with unusual black dial signed "Rolex Oyster Shock Resisting" and with the even numbered markers made up of differently sized Rolex crowns. Circa 1956

Rolex Oyster Royal manual, shock resisting, with gold filled top and stainless steel back. Very unusual dial with chevron finish and inner minute track.
Back dated 2/56

The 1950s: 250,000 Chronometers Can't Be Wrong

Oyster Royal, manual, stainless steel. Unusual honeycomb dial and triangle and arrowhead shaped markers.
Model no. 6244
Circa 1957

18kt pink gold dress watch with unusual two-tone matt and mirror finish 18kt pink gold dial. The Rolex signature below the 12 is engraved and then enamelled, with applied lozenge shaped markers, sweep seconds and leaf hands. 17 jewel movement numbered N92182. Original open link 18kt bracelet with deployant style clasp. © Christie's, London
Case no. R9004
Circa 1958

The 1950s: 250,000 Chronometers Can't Be Wrong

Rolex stainless steel Oyster Perpetual with honeycomb finished dial and luminous triangle markers. 10-1/2''' Hunter-styled Perpetual movement.
Circa 1957

200 The 1950s: 250,000 Chronometers Can't Be Wrong

Rolex Oyster Perpetual, mid-sized, in extremely rare platinum. Dial numbers one and three in green enamel and the remainder in diamonds.
Model no. 6548
Case no. 348891
Movement caliber 1130
Circa 1955

Rolex Oyster Perpetual Everest model. It dates from prior to the introduction of the Explorer.
Circa 1953

The 1950s: 250,000 Chronometers Can't Be Wrong

9kt gold Rolex dress watch. Shock resisting with horned lugs. Dial with Arabic twelve, three, six, and nine, and triangle markers for the remainder.
Circa 1958

Very rare 18kt gold Oyster Perpetual with engine-turned bezel and very unusual dial showing two soccer players contesting the ball. This dial is engraved and then painted.
Model no. 6085
Case no. 7303551
Movement no. F18896
Circa 1955

Stainless steel Oyster Perpetual Bombé with unusual subsidiary seconds dial. Signed "Officially Certified Chronometer" around the subsidiary seconds dial.
Model no. 08/16
Case no. 642017
Movement no. N71904
Circa 1953

The 1950s: 250,000 Chronometers Can't Be Wrong

Rolex Oyster Perpetual True Beat in stainless steel.
Circa 1957

18kt gold Rolex dress watch with blue and cream porcelain dial. Applied gilt bar numerals, sweep second. 10-1/2''' Hunter Precision movement.
Model no. 8382
Case no. 821801
Circa 1955

18kt red gold Oyster Perpetual Datejust with ultra rare purple enameled dial.
Model no. 1601
Case no. worn away
Circa 1957

Stainless steel dress watch with very unusual green and cream porcelain dial with applied gilt numerals.
Circa 1954

The 1950s: 250,000 Chronometers Can't Be Wrong

Stainless steel dress watch with blue enamel and cream enamel dial with applied gilt markers. 10-1/2''' Hunter Rolex Precision movement.
Circa 1954

18k Bubbleback with blue & cream enamel dial. Note similarity of this dial to the previous examples above. These watches were all produced at the same time. To this date only two bubblebacks with enamel dials have surfaced, this one and a 3372 with a similar dial in lilac.
Model No 3131
Case No 492502
Mvt No N89225
Circa 1955

18kt gold Oyster Perpetual Bombé, with original expanding Oyster bracelet. Unusual blue and cream enamel dial with applied gold hour markers.
Model no. 6090
Case no. 928030
Movement Number M74799
Circa 1955

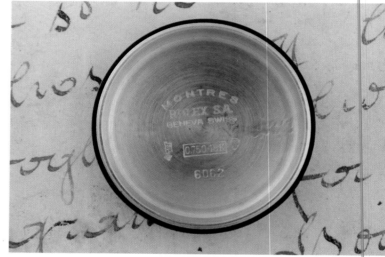

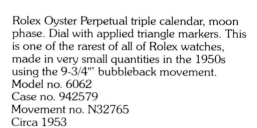

Rolex Oyster Perpetual triple calendar, moon phase. Dial with applied triangle markers. This is one of the rarest of all of Rolex watches, made in very small quantities in the 1950s using the 9-3/4''' bubbleback movement.
Model no. 6062
Case no. 942579
Movement no. N32765
Circa 1953

Rolex Oyster Perpetual, moon phase with star marker dial. The watches of this model with star dials are much rarer than the conventional dial. It is interesting to note that while the case for this watch was made prior to the previous one the movement was made after. Obviously they were all in stock and merely assembled at different times.
Model no. 6062
Case no. 911083
Movement no. N33638
Circa 1954

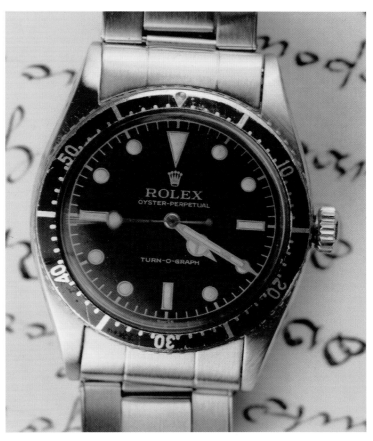

Rolex first model Oyster Perpetual Milgauss with unusual Honeycomb finished dial and lightning effect second hand. The bezel on this watch may look as if it rotates, but in fact it is fixed and is numbered 1 to 5 rather than 10 to 50 on the Submariner.
Model no. 6541
Case no. 412034
Circa 1955

Rolex Oyster Perpetual Turn-O-Graph. Stainless steel, honeycomb finished dial, rotating bezel numbered 10 to 50 with markers for every intervening minute. Textured dial with gold finished hands and markers.
Model no. 6202
Case no. 950508
Circa 1954

Rolex Oyster Perpetual Turn-O-Graph. Stainless steel with original Oyster bracelet.
Circa 1955

Rolex Oyster Perpetual Turn-O-Graph. Rare steel and gold model cataloged as the deluxe model with gold bezel, hour markers, and hands. The movement in this watch was also finished to chronometer standard rather than just a super precision as was the standard model.
Model no. 6202
Case no. 258899
Circa 1956

The 1950s: 250,000 Chronometers Can't Be Wrong

Rolex Oyster Perpetual Datejust in 18kt white gold with slightly dished dial, diamond cut bezel, and original 18kt white gold Jubilee bracelet. This was the third model of the Datejust but looks almost identical to the current model. It was the first model Rolex in which the dial was signed "Superlative Chronometer Officially Certified" rather than "Officially Certified Chronometer."
Circa 1954

18kt gold Rolex Oyster Perpetual, with calendar and moon phases. Screw-down crown, silvered matte dial, applied triangular baton numerals, subsidiary seconds combined with moon phases, windows for date and month, outer ring calibrated for date, tonneau water-resistant-type case. Case, dial and movement signed. Diameter 39mm. © Sotheby's Inc., 1990.
Circa 1954

A rare 18kt gold and enamel sweep seconds Rolex Oyster. Lever movement, Super Oyster crown, polychrome enamel dial with maps of North and South America set in cloisonne with similarly enamelled compass rose, birds, and fish, applied gold faceted baton numerals, all against a black enamel ground. Engine-turned gold bezel chased with hour markers, tonneau water-resistant-type case. Case, dial and movement signed. Diameter: 34mm. © Sotheby's, Inc., 1990
Circa 1955

The 1950s: 250,000 Chronometers Can't Be Wrong 207

18kt gold Rolex Oyster Perpetual with calendar and moon phases. Super Oyster crown, silvered matte dial, applied gold stars and triangular baton numerals, subsidiary seconds, combined with phases of the moon, windows for the day and month, outer ring calibrated for date. Tonneau water-resistant-type case. Case, dial and movement signed. Diameter: 38mm. © Sotheby's, Inc., 1990. Circa 1954

A fine 18kt gold Rolex Oyster Perpetual with calendar and moon phases. Ref. 6062. Lever movement, Super Oyster crown, black matte dial, applied gold star-form numerals, subsidiary seconds combined with moon phases, windows for day and month, outer ring calibrated for date. Tonneau bubbleback water-resistant-type case. Case, dial and movement signed. Rolex buckle. Diameter: 35mm. © Sotheby's, Inc., 1990.
Circa 1952

18kt gold square bracelet Rolex. Circular nickel lever movement, timed 6 positions, patented Superbalance, mono-metallic compensation balance, silvered textured dial, applied gold baton numerals, subsidiary seconds, with integrated Rolex 18kt bracelet composed of brickwork links. Case, dial and movement signed. Diameter: 27mm. © Sotheby's, Inc., 1990.
Circa 1950

Stainless steel Rolex Oyster Perpetual with calendar and moon phases. Ref. 6062. Lever movement, screw-down crown, silvered matte dial, applied triangular numerals, windows for day and month, subsidiary seconds combined with moon phases. Tonneau water-resistant-type case. Case dial and movement signed. Diameter: 36mm. © Sotheby's, Inc., 1990.
Circa 1955

Fine 18kt gold, black dial Rolex Oyster Perpetual, with calendar and moon phases. Ref. 6062. Lever movement, Super Oyster crown, black matte dial, applied gold star-form numerals, aperture for day and month, outer ring calibrated for date, subsidiary seconds combined with moon phases. Tonneau water-resistant-type case, engraved with monogram. Case, dial and movement signed. 14kt Rolex buckle. Diameter: 36mm. © Sotheby's, Inc., 1990.
Circa 1951

208 The 1950s: 250,000 Chronometers Can't Be Wrong

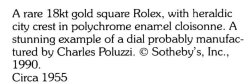

A rare 18kt gold square Rolex, with heraldic city crest in polychrome enamel cloisonne. A stunning example of a dial probably manufactured by Charles Poluzzi. © Sotheby's, Inc., 1990.
Circa 1955

14kt gold, square Rolex, ref. 4533. Subsidiary seconds, black dial. 26mm. © Sotheby's, Inc., 1991.
Circa 1950

18kt gold automatic Oyster moon phase, calendar watch with square number markers and 18kt Jubilee band. © Sotheby's, Inc., 1990.
Circa 1950

Stainless steel calendar/moonphase Oyster Perpetual. 35mm. © Sotheby's, Inc., 1990.
Circa 1950

A fantastic Rolex Oyster perpetual with unusual black star dial in 18kt. The black star dial is by far the rarest of all dials for these watches. © Sotheby's, Inc., 1990.
Circa 1950

The 1950s: 250,000 Chronometers Can't Be Wrong

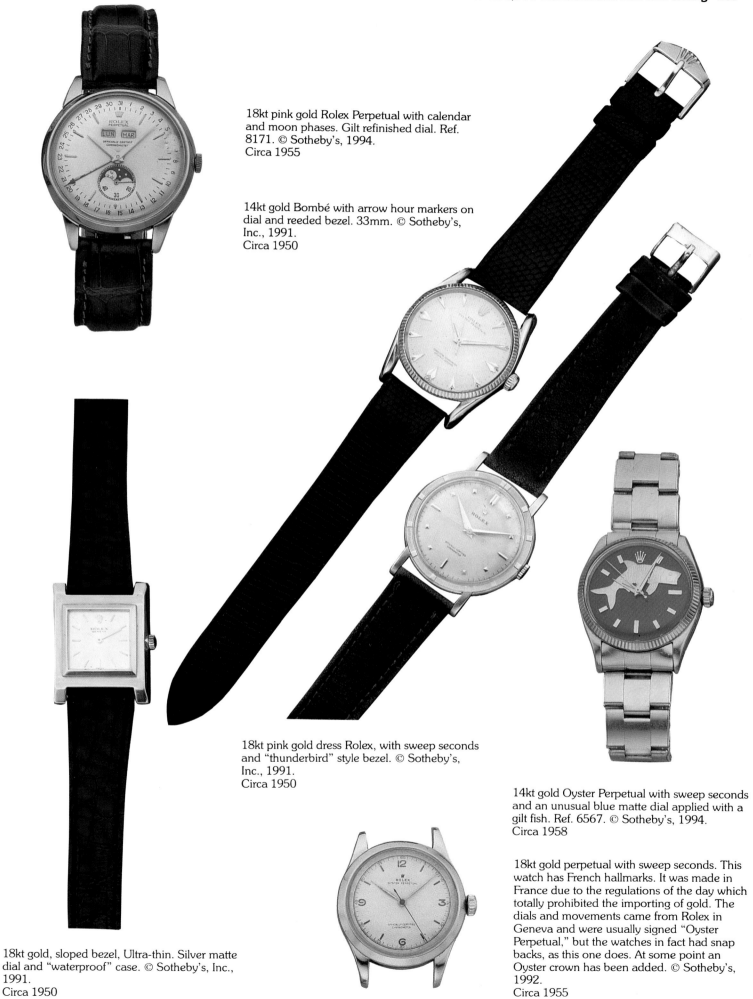

18kt pink gold Rolex Perpetual with calendar and moon phases. Gilt refinished dial. Ref. 8171. © Sotheby's, 1994.
Circa 1955

14kt gold Bombé with arrow hour markers on dial and reeded bezel. 33mm. © Sotheby's, Inc., 1991.
Circa 1950

18kt pink gold dress Rolex, with sweep seconds and "thunderbird" style bezel. © Sotheby's, Inc., 1991.
Circa 1950

14kt gold Oyster Perpetual with sweep seconds and an unusual blue matte dial applied with a gilt fish. Ref. 6567. © Sotheby's, 1994.
Circa 1958

18kt gold perpetual with sweep seconds. This watch has French hallmarks. It was made in France due to the regulations of the day which totally prohibited the importing of gold. The dials and movements came from Rolex in Geneva and were usually signed "Oyster Perpetual," but the watches in fact had snap backs, as this one does. At some point an Oyster crown has been added. © Sotheby's, 1992.
Circa 1955

18kt gold, sloped bezel, Ultra-thin. Silver matte dial and "waterproof" case. © Sotheby's, Inc., 1991.
Circa 1950

The 1950s: 250,000 Chronometers Can't Be Wrong

9kt gold manual wind dress watch with case made in England by Dennison for Rolex. 10-1/2''' Hunter 17 jewel superbalance movement.
Model no. 13874
Case no. 318425
Circa 1957

Stainless steel oversized dress watch, black dial, diamond cut batons and dauphin hands. 10-1/2''' Hunter movement with micrometer regulator and signed "Superbalance."
Model 8029
Circa 1952

Stainless steel manual Oysterdate Precision (non-chronometer). Silvered dial with steel baton markers and hands. This watch is unusual in that someone has gold plated the watch after manufacture.
Model no. 6694
Circa 1958

9kt pink gold manual wind dress watch with unusual curved solid case bars. Silvered dial with applied pink Arabic numerals and subsidiary seconds. Note original "dimpled" winding crown.
Model no. 4409
Case no. 318425
Circa 1957

The 1950s: 250,000 Chronometers Can't Be Wrong

Stainless steel mid-size Oysterdate, with honeycomb dial, gilt Arabic and baton numerals and red date disc. Original box.
Model no. 6266
Case no. 964789
Movement no. 08421
Circa 1955

1954 brochure for the model 6202 Turn-O-Graph. It is interesting in that Rolex go to great lengths to disparage chronographs in the explanation of the watch. This was done at a time when Rolex had a number of Chronographs in their line. The back of this brochure is the only place where any mention is made of the very rare two-tone version of the first Turn-O-Graph, one of which was shown earlier in this chapter.

Stainless steel Bombé, silvered dial with applied dagger markers, sweep seconds and triangle hands. Super Oyster non-screw down crown. Note that the dial is still signed "Officially Certified Chronometer" without the "Superlative" which came along after the demise of the Bombé.
Model no. 6090
Case no. 710776
Circa 1951

Stainless steel Oyster Speedking, in the larger (Air-King) size with machined bezel featuring radial hour markers. Silvered dial with applied quarter hour markers and painted batons for the remainder.
Circa 1957

18kt yellow gold non-Oyster manual chronometer retailed by W. Rösch of Berne. Engine-turned bezel with hour markers. Silvered dial with subsidiary seconds and painted baton markers. Very high quality Superbalance movement with 18 jewels, capped end stones and swan neck micrometer regulator. Movement contained in base metal movement ring.
Movement no. E11090
Circa 1952

The 1950s: 250,000 Chronometers Can't Be Wrong 213

9kt yellow gold dress watch with teardrop form lugs and double stepped bezel. Silvered dial with applied markers and sweep seconds.
Circa 1954

This first model Milgauss has a flat (rather than honeycomb) dial, but still shows the unique "lightning strike" seconds hand. Note the word "Milgauss" is in red.
Model 6541
Circa 1954

14kt yellow gold Oyster Perpetual (non-date) with silvered dial bearing enameled "25 years service Coca Cola." The rear of the case has a presentation inscription to "John B. Campbell for 25 Years Service to the Coca Cola Bottling Company of Boston."
Circa 1958

Chapter 9
THE CHRONOGRAPH
Stopping Time

The flag drops and the roar of the crowd is lost in the rattle of spoked wheels as the race cars speed down the cobbled surface. It is 1922 in Europe, and automobile racing is getting serious. Low income taxes and easy stock market investment opportunities have translated into more leisure time for more people than ever before. Many of the well-dressed people at the race wear watches on their wrists, but they don't use them to measure the lap times of the vehicles. These racing enthusiasts also carry a more practical timepiece, a cumbersome old-fashioned pocket sports timer.

It was to be another four or five years before anyone was able to combine the functions of the sports timer with the wristwatch to produce a completely new kind of timepiece: a wrist chronograph.

A chronograph was an ordinary wristwatch to which has adapted the principle of the sports timer. The recording hands function independently from the ordinary watch movement and also are capable of measuring the seconds, the minutes, and, occasionally, the hours of a desired event. While the chronograph and sports timer may seem similar, their operations are in fact very different. A sports timer is started and stopped by a brake acting upon the balance wheel. This would not work on a chronograph which is also required to keep accurate time. To do this the balance must be allowed to swing at all times. The first wrist chronographs used a switching wheel and levers to allow these two contradictory functions.

Rolex was not among the chronograph pioneers, choosing to focus their energies on the Oyster and then on the Perpetual. However by the late 1930s the Oyster had propelled Rolex into the position of being the "sportsman's" watch and sportsmen demanded chronographs. So Rolex began to fulfill that demand.

The 1934 Rolex catalog does not show any chronographs at all, but the 1937 one shows five. The company made the decision to follow most other companies producing chronographs and utilize a proprietary movement, rather than expending the time and money to develop one themselves. The movement they chose, and would use for almost 50 years, came from the renowned firm Valjoux[38]. Available in three sizes from 10-1/2''' to 14''' they were all "simple" chronographs, that is to say, one button models. These allowed users to time a single continuous event. The first pressure on the button started the sweep seconds hand and a second push would stop it, allowing the elapsed time to be read. A third push would reset the hand.

The popularity of the wrist chronograph caught the watch industry by surprise. The demand for the new devices came not only from animal and vehicle racing spectators, but from anyone whose hobby or profession required them to have an accurate measure of elapsed time. Production controllers used the wrist chronographs to measure rates of production and photographers used them to measure exposure and development times.

The one button chronograph was popular, but as more were used the limitations of this simple system became clear. Customers demanded the facility to intermittently stop the time. The real breakthrough came in the late 1930s with the ingenious two-button configuration. It signalled a new era in timekeeping and enabled wrist chronograph users to stop the timing for breaks in the action. Long-distance automobile racers, for example, could stop the chronograph when they were fueling or taking "necessary" breaks, thus easily determining the actual time of travel without subtracting break times from the overall elapsed time. Production foremen at large factories could more accurately determine the actual production output of their crew without taking into account lunch and coffee breaks. A soccer referee could stop the chronograph each time the ball went "offside" and restart it each time play resumed. This way he could accurately measure a 45 minute game period.[39]

One of the most interesting Rolex chronographs, a hybrid of sorts, is the "Zerograph," produced in very limited quantities at the end of the 1930s. This watch was an in-house conversion, if you will, that simply took a basic 10-1/2''' Hunter sweep seconds movement, added fewer than 10 additional parts and a protruding back, and turned an everyday watch into one of the most sought after vintage Rolexes on the market. The watch (model 3462) marked an important breakthrough in Rolex chronographs. It was the first model to feature the Oyster crown.

The modifications began to be added to other models. The first true Oyster chronograph was the model 4500, which never sold in large numbers due to fact that it was launched in the depths of World War II. The 6232 and 3668 models followed in short order, but proved to be just as unpopular. Both new models used the same 13''' Valjoux movement as the 4500 and, like it, had just one 30 minute register. While Rolex had introduced an hour register chronograph as early as 1942 it was a snap back model. Another seven years passed before the benefits of a third register and an Oyster case were to be brought together in the form of the model 5034. Over the next fifteen years the 5034 mutated into the 6034, which then became the 6234. This, in turn, became the 6238 before finally evolving into the 6239.

As always at Rolex the changes were almost imperceptible and seemingly trivial, but the final watch in the above progression revived a name Rolex had shelved in 1956 with the final moonphase watch. It was the "Cosmograph". The 6239 Cosmograph differed from the 6238 only in that it had a bezel with engraved tachymetric graduations, rather than having them on the dial. This alone gave the watch a much more modern look and caused it to outsell its companion model. But we should not think that even the Cosmograph was a popular watch in its day; it still lagged far behind its other Rolex stablemates in sales.

A story will give you some idea of the Cosmograph's "popularity." When the author was 25 or so, he decided to buy a GMT-Master. He walked into his local Rolex agent and examined one. He agreed to purchase the watch and produced his credit card. He paused for a moment and asked the salesman "If I pay with cash, do I get a discount?" The response spoke volumes: "Sorry sir, we never discount Rolex watches...except the chronographs. We will give you 20% off one of those."

The last major change to the manual chronograph came in the late 1970s when Rolex introduced the 6263 model Cosmograph. This was the first model to have truly waterproofed pushers. The earlier Oyster chronographs had simple round pushers with internal gaskets as the only sealing mechanism. They functioned perfectly at depths of 25 feet or less, but as Rolex began to emphasize the waterproof capabilities of all its other watches, the Cosmograph was in danger of being left behind.

The screw down pushers were added for another reason. People kept trying to operate the push buttons while the watch was underwater and in so doing let water into the movement. The locking buttons also functioned to prevent inadvertant operation of the pushers. The new 6263 was initially listed as being waterproof to 165 feet (or 50 meters), but 10 years later was listed as being capable of twice that depth, 330 feet (or 100 meters), although there seems little visible difference between the models. The 6263 was unusual in one very strange way: it was the only watch Rolex made where the quality of the movement differed depending on the case material. Putting it simply, watches with 18kt cases had movements which, though still manual wind Valjoux 72 ebauches, were timed to chronometer standard. The steel cased watches were not. This difference started with the earliest of the 6263s, although it was not publicized until the second series, when the dials of the 18kt watches bore the legend "Superlative Chronometer Officially Certified" under the "12". To the best of our knowledge these are the only manual wind Rolex watches ever to be so signed[40].

Despite all of this, sales of chronographs still lagged until 1986. That's when the surprising demand for Rolex Daytona Chronographs began. This was the start of the wristwatch craze in the United States and Great Britain. The vagaries of the international market were such that a Rolex authorized dealer in the U.S. would be happy to sell a watch for full retail price, but would often settle for selling one at a discount. The same watch would immediately be sold for double the list price in Italy. Many small-time collectors and "vest pocket dealers" began little cottage industries, burning up the phone lines rounding up Rolex Daytonas by the hundreds and immediately selling them to their individual Italian contacts.

Several U.S. dealers were sure that this incredible thirst for Daytonas would soon be quenched. Their mindset touched off a frenzy of activity between European collector/dealers and their American counterparts. As the market value of these watches increased rapidly, Italian dealers would occasionally balk at a price they claimed was too high. However, they never failed to buy the particular Daytona model which was made with an unusual red outer track on the dial, no matter how high the cost.

This model featured square markers on the subsidiary dials, and an outer track that was of the same color as the subsidiary dials. The Italian collectors/dealers began referring to this dial as the "Paul Newman" dial, though the origins of this nickname are obscure. The reason most often heard was that Paul Newman allegedly wore one in the movie "Le Mans." Subsequently, a picture of the watch, attached to Mr. Newman's wrist, was plastered all over Italy on movie posters. Another story cites Mr. Newman wearing a Cosmograph on the cover of Italy's most popular magazine as the reason for the name, and possibly for the popularity of Rolex Daytonas in general. (Mr. Newman still wears a Daytona Cosmograph, but his personal watch is the current stainless steel automatic version, and not a "Paul Newman" dial at all.)

This unusual dial, colorful and rather deco in appearance, was only produced for a short time after its introduction in 1970. At the height of the Daytona craze, the 18kt solid-gold Daytona, complete with Paul Newman dial, was reportedly selling in Italy for the U.S. equivalent of $35,000! Rolex, quick to capitalize on a trend, ceased production of the Daytona Cosmograph, promising to re-release it "soon" with a completely updated look. As collectors and speculators waited patiently for the new model to appear, older versions were snapped up at an incredibly rapid rate from authorized Rolex dealers worldwide. Entrepreneurs began placing advertisements in newspapers and magazines in a fervent attempt to flush out these elusive watches.

The basic stainless-steel plain-dialled Rolex Daytona Cosmograph rose in value from a meager $1,400 in the mid-1980s to some $7,000 in the mid-1990s. Rolex finally released the new version in 1991. Barely noticed was the fact that Rolex had abandoned the stalwart workhorse movements made by Valjoux and replaced them with Zenith movements. The newly redesigned dial featured a bold look with large subsidiary dials and an inner track in a contrasting color. Combinations included white on black, black on white, and even a gold on black configuration. Every dial sported the word "Daytona" in red, and all models had screw-down pushers.

The stainless steel version was an immediate success. Pandemonium would not be too strong a description for the attention it received both in collectors circles and with the general watch buying public. It immediately resold for much more than retail list price. People the world over put in orders for the watch and some waited as long as 6 years for delivery! Rolex was well aware that their biggest profit margin was in the gold and gold/stainless models, so in their haste to manufacture these much more profitable models, the stainless version was in very short

The only known advertisement showing both the "Paul Newman" and the standard dial.

supply. The 18kt gold model resold for close to its "retail" price, but the stainless and gold specimen proved to be unpopular, and remains the most difficult of the modern-day Daytonas to sell.

Rolex was not finished. The much-heralded 18kt Daytona, complete with leather strap and 18kt gold deployant buckle was released in 1992. It was an immediate success, generating interest comparable to the hysteria surrounding the stainless steel model. During this time, the lira was hovering in the 1,100/$1 range, the Swiss franc was close to $1.22 and the mark was in the low DM1.50s. Because the dollar was so weak, this watch was being purchased immediately at full retail in the United States, and shipped right back to Europe to be sold for 10 percent to 20 percent over Rolex retail list price. This hysteria was short-lived, however, as the watch market and the currencies reversed themselves with the devaluation of the lira in 1993.

The value of the lira fell from the 1,100/$1 level to a values in the 1,700/$1 range by mid-1995, in a long-overdue "correction." The ever-popular Daytona, at one time thought to be priceless, was at long last showing signs of weakness. Prices plummeted, led by the Italians' reluctance to pay what, for them, would have been close to double the previous worldwide market price.

Of course, the desire to own a Rolex Daytona was still strong, not only in Italy, but in Japan, Germany, and even the United States. With the combination of the shrewd marketing techniques by Rolex, the continued operation of the laws of supply and demand, the firming up of the lira and Swiss franc, and the continued hunger for Daytonas worldwide, the Rolex Daytona Cosmograph remains the most desirable production timepiece ever produced by Rolex.

---

[38] The name comes from the location of the factory in the Joux Valley, or, in French, "Val de Joux."

[39] This explains why many wrist chronographs have registers reading to 45 minutes. They are for football referees and spectators.

[40] The only other manual Oyster with a chronometer rating, the model 6512 "Veriflat," predated the Superlative designation and was signed simply "Officially Certified Chronometer."

Square Chronograph, stainless steel with two registers and "Anti-Magnetic" and pulsations dial. Square Chronographs were produced in very limited numbers in the late 1930s and early 1940s.
Circa 1942

Rare one-button Chronograph 18kt. yellow gold with unusual dial featuring two registers and blue outer track.
Model no. 2303
Case no. 14717
Circa 1935

218 The Chronograph: Stopping Time

A small stainless steel Chronograph with register, telemeter and tachymeter, Rolex Oyster, "Anti-Magnetic," Ref. 3184. Nickel lever movement, screw-down crown, mono-metallic compensation balance, 17 jewels, four adjustments. Two-tone silvered matte dial, applied Arabic numerals, two subsidiary dials indicating constant seconds and register for 30 minutes, outer ring calibrated for tachymeter and telemeter, engine-turned bezel chased with hour markers, tonneau water-resistant-type case. Case, dial and movement signed. Diameter 29mm. © Sotheby's, Inc., 1990.
Circa 1940

A stainless steel square Chronograph with register and tachymeter, Ref. 3529. Circular nickel lever movement, 17 jewels. Silvered matte dial, baton and Roman numerals, two subsidiary dials indicating constant seconds and register for 30 minutes, inner ring calibrated for tachymeter, square case. Case, dial, and movement signed. Width 26mm. © Sotheby's, Inc., 1990.
Circa 1945

A small 18kt gold Chronograph with register and tachymeter, "Anti-Magnetic," Ref. 3055. Nickel lever movement, mono-metallic compensation balance, 17 jewels. Silvered matte dial, applied gold baton numerals, two subsidiary dials indicating constant seconds and register for 30 minutes, outer ring calibrated for tachymeter, circular case. Case, dial and movement signed. Diameter 30mm. © Sotheby's, Inc., 1990.
Circa 1930

18kt "mono-bloc" domed back Oyster Chronograph with two-registers and two-color telemeter and tachymeter scales. Telemeter markings are in blue. On contemporary 9kt Rolex band.
Model no. 3525
Serial no. 386555
Circa 1950

A stainless steel Chronograph with register, telemeter and tachymeter, Ref. 2508. Nickel lever movement, bi-metallic compensation balance, 17 jewels. Pink matte dial, baton numerals, two subsidiary dials indicating constant seconds and register for 30 minutes, outer rings calibrated for telemeter and tachymeter, circular case. Case, dial, and movement signed. Diameter 35mm. © Sotheby's, Inc., 1990.
Circa 1945

The Chronograph: Stopping Time 219

Stainless steel two register Anti-Magnetic Oyster Chronograph with unusual pink gold milled bezel featuring baton and circular hour markers. Pink dial has tachymeter and telemeter tracks.
Model no. 3668
Circa 1942

Stainless and gold Rolex Oyster Chronograph. Two registers, anti-magnetic, telemeter and tachymeter.
Circa 1952

220 The Chronograph: Stopping Time

Stainless steel and pink gold Oyster Chronograph with two registers, applied numerals, and gold screw down crown.
Circa 1943

A 18kt gold Chronograph with register, telemeter and tachymeter, Anti-Magnetic, Ref. 4062. Nickel lever movement, mono-metallic compensation balance, 17 jewels. Black matte dial, applied gold baton numerals, two subsidiary dials indicating constant seconds and register for 30 minutes, outer rings calibrated for tachymeter and telemeter. Circular case with teardrop lugs, coin-edge band. Case, dial and movement signed. 18kt Rolex bracelet. Diameter 36mm. © Sotheby's, Inc., 1990.
Circa 1940

A stainless steel Chronograph with register, tachymeter and telemeter, Ref. 3484. Nickel lever movement, mono-metallic compensation balance, adjusted to three positions, 17 jewels. Black matte dial, baton numerals, two subsidiary dials indicating constant seconds and register for 30 minutes, outer rings calibrated for tachymeter and telemeter. Case, dial, and movement signed. Rolex bracelet. Diameter 33mm. © Sotheby's, Inc., 1990.
Circa 1940

The Chronograph: Stopping Time 221

Stainless steel "mono-bloc" cased Chronograph with three registers retailed by Beyer of Zurich. This dial has faded telemeter track.
Model no. 4048
Circa 1949

Reference 4500 stainless and pink gold Chronograph with telemeter and tachymeter.
Circa 1945

222 The Chronograph: Stopping Time

18kt gold Chronograph with registers, telemeter and tachymeter, Ref. 6238. Lever movement, screw-down crown, silvered matte dial, applied baton numerals, three subsidiary dials indicating constant seconds and registers for 30 minutes and 12 hours, outer rings calibrated for telemeter and tachymeter. Tonneau water-resistant-type case. Case, dial, and movement signed. 18kt Rolex bracelet. Diameter 36mm.
© Sotheby's, Inc., 1990.
Circa 1947

Stainless steel three-register Chronograph with blue telemeter track and triangular markers.
Circa 1957

18kt gold Oyster Chronograph, two registers, with telemeter and tachymeter.
Circa 1951

The Chronograph: Stopping Time 223

Stainless steel Chronograph with black matte dial, tachymeter and registers. Nickel lever movement, 17 jewels, mono-metallic compensation balance. Three subsidiary dials indicating constant seconds, and registers for 30 minutes and 12 hours. Tonneau water-resistant-type case. Case, dial and movement signed. Diameter 36mm. © Sotheby's, Inc. 1990.
Circa 1960

18kt Oyster Cosmograph Daytona with gilt "Paul Newman" dial featuring black registers and black enameled tachymeter bezel. Note: The gold "Paul Newman" watches have no red markings at all.
Model no. 6264
Case no. 2357483
Circa 1969

Stainless steel three-register Oyster Chronograph with triangle markers on the "Anti-Magnetic" dial, which seems to have discolored to brown.
Circa 1954

The Chronograph: Stopping Time

Stainless steel Oyster Cosmograph Daytona. "Paul Newman" dial with steel bezel featuring tachymeter. Applied square steel hour markers and red outer seconds track and stylized divisions on inner registers.
Case no. 2,005,380
Circa 1967

Stainless steel Oyster Cosmograph of the rare "Paul Newman" variety. It is rare to find the "Paul Newman" dialed Cosmographs with screw down pushers.
Model no. 6262
Serial no. 2788924
Circa 1971

The Chronograph: Stopping Time 225

Rare stainless steel Oyster Cosmograph Daytona with "Paul Newman" dial. This watch has a dial configuration that is the reverse of the previous watch. i.e. black register on white dial. Note the rare "Tiffany & Co." on the dial.
Circa 1967

Similar watch to previous, but this watch has the word "Daytona" in red.
Model no. 6263
Case no. 7627582
Circa 1980

Interesting Oyster Cosmograph without Daytona markings, this version in stainless steel.
Model no. 6265
Case no. 3485375
Circa 1977

Stainless Oyster Cosmograph Daytona with enameled tachymeter bezel and Daytona in red.
Model no. 6265
Case no. 6047495

226　The Chronograph: Stopping Time

Stainless steel Cosmograph Daytona with registers. Lever movement, silvered matte dial, inset with three black subsidiary dials indicating constant seconds, registers for 30 minutes and 12 hours, black enamel bezel calibrated for units per hour. Water-resistant-type tonneau case. Case, dial and movement signed. Diameter 41mm. © Sotheby's, Inc., 1990.
Circa 1970

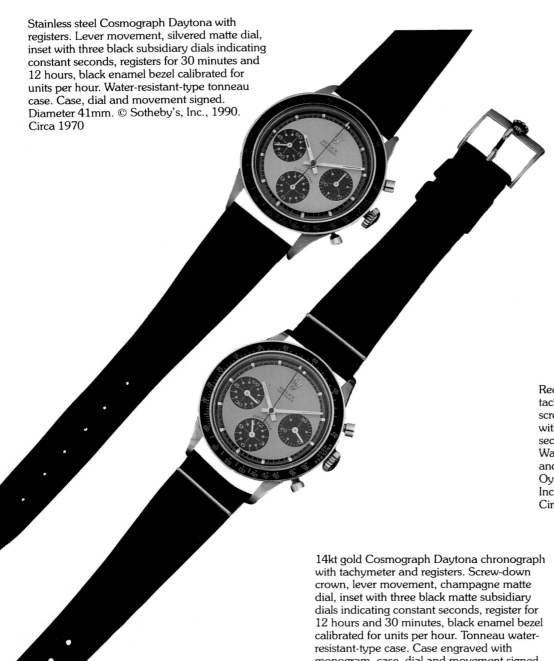

Recent stainless steel Chronograph with tachymeter and registers. Lever movement, screw-down crown, silvered matte dial, inset with three subsidiary dials indicating constant seconds, registers for 12 hours and 30 minutes. Water-resistant-type tonneau case. Case, dial and movement signed. Stainless steel Rolex Oyster bracelet. Diameter 39mm. © Sotheby's, Inc., 1990.
Circa 1974

14kt gold Cosmograph Daytona chronograph with tachymeter and registers. Screw-down crown, lever movement, champagne matte dial, inset with three black matte subsidiary dials indicating constant seconds, register for 12 hours and 30 minutes, black enamel bezel calibrated for units per hour. Tonneau water-resistant-type case. Case engraved with monogram, case, dial and movement signed. © Sotheby's, Inc., 1990.
Circa 1968

Stainless Chronograph with registers, tachymeter. Signed Cosmograph. 36mm. © Sotheby's, Inc., 1990.
Circa 1960

18kt non-Oyster pink gold Chronograph with refinished dial, 17 jewels. 35mm. © Sotheby's, Inc., 1990.
Circa 1945

The Chronograph: Stopping Time 227

18kt gold Chronograph Calendar with registers, 36mm. © Sotheby's Inc., 1990.
Circa 1945

Pink gold Chronograph with two registers, tachymeter, and black dial. Square buttons. 33mm. © Sotheby's, Inc., 1990.
Circa 1945

18kt pink gold Chronograph with unusual "Pulsemeter" dial, telemeter and register dial and oval buttons. "Anti-Magnetic." 33mm. © Sotheby's, Inc., 1990.
Circa 1935

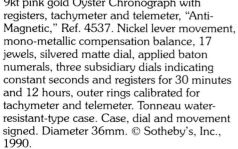

9kt pink gold Oyster Chronograph with registers, tachymeter and telemeter, "Anti-Magnetic," Ref. 4537. Nickel lever movement, mono-metallic compensation balance, 17 jewels, silvered matte dial, applied baton numerals, three subsidiary dials indicating constant seconds and registers for 30 minutes and 12 hours, outer rings calibrated for tachymeter and telemeter. Tonneau water-resistant-type case. Case, dial and movement signed. Diameter 36mm. © Sotheby's, Inc., 1990.
Circa 1960

Small stainless steel Chronograph with two registers and large smooth bezel, Ref 2508. 32mm. © Sotheby's, Inc., 1990.
Circa 1938

228   The Chronograph: Stopping Time

Rolex Oyster Chronograph with rare Pulsations dial retailed by Joyeria Riveria of South America. Round pushers, "Anti-Magnetic." 18kt gold. Ref. 4500, 36mm. © Sotheby's, Inc., 1990.
Circa 1946

A rare small, 30mm, 18kt pink gold Chronograph with black tachymeter dial marked "Anti-Magnetique," ref. 3055. © Sotheby's, Inc., 1991.
Circa 1930

A stainless steel gold marker day/date/calendar Oyster Chronograph with nickel movement and three subsidiary dials, 36mm. © Sotheby's, Inc., 1990.
Circa 1945

18kt gold chronograph with flexible moving lugs, two registers and oval pushers. © Sotheby's, Inc., 1990.
Circa 1940

Rare small stainless steel single button Zerograph with hour marker bezel, ref. 3462. 29mm. © Sotheby's, Inc., 1991.
Circa 1938

Rare small stainless single button Chronograph with register. Ref. 3055. This example only 26mm and with single button above winding crown. © Sotheby's, Inc., 1991.
Circa 1930

Gold Chronograph Daytona with tachymeter and registers, with black bezel and non-screw down pushers. © Sotheby's, Inc., 1990.
Circa 1968

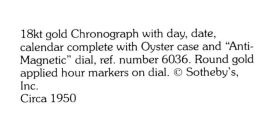

18kt gold Chronograph with day, date, calendar complete with Oyster case and "Anti-Magnetic" dial, ref. number 6036. Round gold applied hour markers on dial. © Sotheby's, Inc.
Circa 1950

The Chronograph: Stopping Time 229

Rare 14kt gold Cosmograph with "Paul Newman" or "exotic" dial. This example was retailed by Cartier and has hand-stamped Cartier numbers on the back of the lug. Ref. 6241. © Sotheby's, Inc., 1991.
Circa 1968

A pink 18kt gold non-Oyster Chronograph with flared lugs. Ref. 4062, 36mm. © Sotheby's, Inc., 1991.
Circa 1950

Rare small, 30mm, 18kt pink gold Chronograph, "Anti-Magnetic," with two-tone dial and oval pushers. Ref. 3055 with 18kt pink gold Rolex Oyster bracelet. © Sotheby's, Inc., 1992.
Circa 1935

Steel Chronograph with register, tachymeter, and telemeter. Flat bezel and flexible lugs. Gold applied markers. Ref. 2918, 33mm. © Sotheby's, Inc., 1992.
Circa 1930

Small 18kt gold Chronograph with registers, tachymeter and telemeter, Ref. 3834. © Sotheby's, Inc., 1992.
Circa 1940

18kt pink gold Chronograph register, tachymeter, and telemeter, "Anti-Magnetic." Pink matte dial with Arabic and baton numerals. Ref. 3484. © Sotheby's, 1992.
Circa 1945

Pink gold and stainless steel Chronograph with registers, calendar and tachymeter, "Anti-Magnetic." Silvered matte dial with Arabic and square numerals, three subsidiary dials, windows for day and month, outer ring calibrated for date and tachymeter, lobed lugs. Ref. 4768. © Sotheby's, 1992.
Circa 1955

Stainless steel back & pink gold top Chronograph with tachymeter, telemeter, and registers. Unusual tear drop lugs, ref. 4313. © Sotheby's, Inc., 1992.
Circa 1940

230 The Chronograph: Stopping Time

Stainless steel Chronograph with registers, calendar and tachymeter, "Anti-magnetic." Silvered matte dial with Arabic and square numerals, three subsidiary dials, windows for day and month, outer ring calibrated for date and tachymeter, lobed lugs. Ref. 4768. © Sotheby's, 1992.
Circa 1955

Stainless steel Oyster "Anti-Magnetic" Chronograph, silvered dial with outer telemeter and tachymeter scales, gilt raised triangle markers, three subsidiary dials for running seconds and registers for 30 minutes and 12 hours. Case with screwed back and winder, and round pushers. © Christie's, London.
Model no. 6034
Circa 1950

Stainless steel square antimagnetic chronograph with register and red "snail shell" tachymeter, silvered dial with Arabic and Roman numerals, two flat pushers. © Christie's, London.
Model no. 3529
Circa 1942

18kt gold small Chronograph, silvered dial with Arabic numerals and subsidiary dials for running seconds and 30 minute register, outer tachymetric scale, sweep center seconds operated via two flat buttons. Movement jeweled to the center with gold alloy balance. © Christie's, London.
Model no. 3055
Case no. 164986
Circa 1942

Stainless steel current model Oyster Perpetual Cosmograph Daytona, white dial with black register inserts and steel baton numerals.
Model no. 16520
Circa 1996

The Chronograph: Stopping Time 231

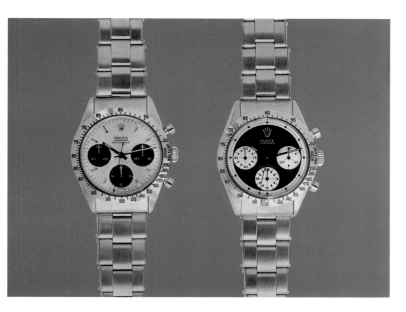

English brochure from 1964, the only known published picture showing both the standard and "Paul Newman" dials together.

U.S. market brochure for the Daytona Chronograph printed in December, 1966. This is the first brochure to use the word Daytona on the front cover, which also bears a photograph of NASCAR stock cars at the Daytona oval. It is also worth noting that the watch is described as being "protected from dirt, dust and perspiration by the....twinlock crown." Nowhere in this brochure is their any mention of the watch being waterproof!

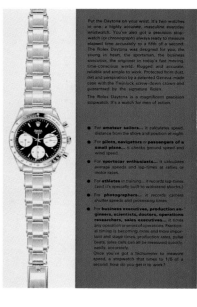

Chapter 10
A DATE(JUST) WITH DESTINY

In 1943, during the depths of World War II, Rolex made a move that was probably the most significant one in its history. It started with the simple decision to introduce a larger version of the bubbleback. This new watch was powered by a perpetual version of the classic 10 -1/2''' Hunter. The new calibre, called the A295 (or calibre 720), was initially used in a simple large bubbleback with the reference number 5026. But three years after the introduction of the new movement Rolex made two additions to it: a sweep seconds hand and a date disc visible through an aperture in the dial at 3. Renamed the calibre 740, it required only 13 additional parts over the calibre 720, but changed the fortunes of Rolex as dramatically as the Oyster and the Perpetual had in their own days.

Launched in 1945 as the Jubilee Datejust, model 4467, it was initially available only in 18kt gold and became the flagship model of the Rolex line, just as the Prince went into a sales decline. The first Datejust was a seminal watch in the history of Rolex, launching a model and a look that would last for over 50 years. It identified the market that Rolex would attain and then totally conquer.

This first model is, arguably, the most beautiful of all the Datejusts. It features a finely machined coin edge bezel, a gold-edged date window, alternating red/black date figures, a tiny bubbleback winder, and a deeply domed back. This deeply domed back was needed because the addition of the date disc and its mechanism, to say nothing of the indirect drive to the center seconds, made an already thick watch even more so.

Rolex was unsure about the name of the new model and registered the name Ritedate in Geneva on the 1st of October, 1945. But by the time the watch was introduced at the Hotel des Bergues in Geneva on November 24, 1945 on the occasion of the celebration of the 40th jubilee of the founding of Rolex,[41] it was called "Datejust" and the model was "Jubilee," although this name came later to be applied only to the bracelet of the new watch. Even the name "Jubilee" was in doubt until just prior to its introduction. Wilsdorf was set on calling his new flagship model the "Victory," after the recent allied victory over the forces of darkness. However his Swiss colleagues were adamant that this was not exactly the Swiss way of doing things and Wilsdorf finally relented.

When Rolex introduced the watch at the Jubilee in 1945, it had not been finalized for production and the picture shown here, which was the one used in the literature given out at the Jubilee party shows a bracelet which differs from the production version in one major function: it was not a deployant bracelet, but an opening two-piece one.

Today's version of the Datejust with its sleek lines is barely discernable from the current Rolex President model. The subtle changes in design over the years have come so gradually that one is shocked when directly comparing the original three-piece "Ovettone" (big egg) model with today's version.

The Ovettone gave way to a similarly shaped watch that was made from only two case parts but still retained the protruding back, commonly referred to as the "bubble," and the finely milled bezel. The date function also retained its alternating red and black numbers, presumably to let the wearer know that the date had indeed changed. The only obvious change was that the winding crown, previously the small bubbleback crown, became larger when Rolex increased the underwater depth to which the watch could be worn from 60 feet to 165. Before this Rolex had simply claimed that their Oyster line was waterproof. Now they could claim that all Oyster watches were waterproof to 50 meters.

The most dramatic change in the Datejust happened in 1954 when Rolex introduced the cyclops date magnifier at Basel. This little bump in the crystal enabled one to see the date more clearly and has since become a signature for the Rolex line. The company soon introduced the new slimmer movement which of course allowed the company to fit a flatter back, not unlike today's model. The new movement, the calibre 1065 (introduced in 1957 at the Swiss Watch Fair at Basel) was much lighter in weight than its predecessor and much slimmer, making the older rounded back obsolete.

To complete the "new" look, a new milled bezel was affixed which was similar to bezels used today. This new "fluted" or "diamond cut" bezel replaced the "coin edge" and plain bezels seen on the Ovettone models. It complemented the slightly slimmer new design to produce a much sleeker look that the company seems to favor, even to this day.

These newer model Datejusts were manufactured primarily in yellow gold, though they were also available in pink gold, mostly for the South American market. A scant few have been seen in white gold, making them much sought after and prized by collectors and dealers alike. Around 1957 they became available in stainless steel for the first time. We have it on the highest authority that a very small number of platinum Datejusts were produced, but we have not had the opportunity to study one and cannot discuss them in greater detail.

Not quite so rare, but still highly desirable, are the so called "left-handed" Datejusts also from this period. These are normal looking Datejusts with the winder on the left side of the watch. We use the word "so called" when describing them because they were never intended for use by left-handed people. They were in fact designed for people who wore their watches low on the wrist and were always complaining about the button sticking into their wrists. Whatever the reason they were made, they are still a rare and desirable watch.

When the Day-Date model (the "President") was introduced around 1956, the company for the very first time allowed the Datejust to be ordered with something other than the Jubilee bracelet. After all, image was very important to Rolex, so why would they allow their flagship model, produced during their 40th jubilee, to have anything but a Jubilee bracelet? When the Day-Date became their flagship, some of the barriers came down and one could now order the Datejust with a Jubilee bracelet, an Oyster bracelet, or a leather strap.

The stainless and gold Datejust was introduced around 1962 and has proven to be one of the most popular models Rolex has ever produced. It was around this time period that the powers that be at Rolex realized, to their horror, that the stainless Datejusts were far outselling the gold Datejusts, whether in 18kt or 14kt gold. There was a large gap between the two models, which Rolex interpreted as a need for an intermediately priced watch. The stainless and gold (sometimes called the "steel and gold" or the "two-tone" and, in the United Kingdom, the "bi-color") seemed to fill this need rather handily.

Slowly the world seemed to change as we moved toward the "Swinging 60's." In England, Harold Macmillan proclaimed to the British public that "You've never had it so good." This was in tune with the mood in the U.S., where John F. Kennedy's "Camelot" was in full swing. Suddenly ostentation was back "in." The new "two-tone" look not only fit the times, but enabled Rolex to increase the margins on a very popular watch. They were now able to sell the new watch for close to twice the price fetched by the stainless model. And the good part was that only a few francs worth of gold were needed to accomplish this amazing feat of marketing! Now "everyman" could wear a "gold" watch and the fashion-conscious needn't worry about the white gold/yellow gold controversy. Once again Rolex had created the right watch for the right time.

The Datejust stainless and gold bands have gone through many changes and incarnations over the years. The bracelets were primarily manufactured in Switzerland and the United States. The early United States versions featured individual links with an "oval" design, while the Swiss versions were primarily of a half-oval variety, virtually the same style seen today. Bracelets were made in the United States to conform to the strict import laws that the United States Congress had imposed earlier in an attempt help the floundering U.S. watch industry. Watch bracelets and straps were levied with a heavy duty to encourage the purchase of American products. The tax levied on imports made them too expensive for the average American to afford. These laws still exist today in the United States even though the watch manufacturing industry essentially died with the demise of Hamilton in the early 1960s and the closing of the last remaining case company, the Star Watch Case Company of Ludington, Michigan in 1981.

Both the Swiss and the American versions of the two-tone band were manufactured exclusively in 14kt gold until 1977. Around this time the public began clamoring for gold. Gold was beginning a rise in value that would take it near 800 dollars (around £400) per ounce, creating an atmosphere of a modern day gold rush. What better time for Rolex to upgrade the gold content on the Datejust to 18kt gold?

With this change came another important milestone in the metamorphosis of the Datejust. The old curved-edged dial (sometimes referred to as the "sloped-edged" dial or the "pie pan" dial) gave way to the new flat dial which in turn was facilitated by the new "Rapid Date Change" or "quickset" feature. This now familiar feature was much needed and became very desirable. The new calibre 3085 movement not only enabled the wearer to change the date in the window to the proper date much faster, it also meant that there was not so much wear on the hand setting mechanism. Most importantly, it allowed the wearer to change the date at the end of the month from 28 or 30 to 1 without changing the time, thereby upsetting the accuracy of the watch.

In 1956 a new model of the Datejust was launched combining the advantages of the standard Datejust with those of the now discontinued "Turn-O-Graph." Launched as the model 1625, it featured a gold (14kt or 18kt depending on the market) rotating bezel in which was cast with the minute divisions and very fine dividing lines. In one of the first catalogs to feature this watch, the U.S. catalog for 1960, the watch is shown with a dial bearing the insignia of the U.S. Air Force aerobatic flying team "The Thunderbirds" (who took their name from the F105 Thunderchief they flew). From this association this is known as a "Thunderbird" bezel in the U.S., though, confusingly, it is known as the "Turn-O-Graph" in Europe. Initially only available in 18kt gold, the watch has never proved massively popular[42] but it is still in the catalog and has benefitted from subsequent mechanical and cosmetic updates.

When the manual "Oysterdate" was introduced in the early 1950s it came in two sizes, one about the size of the modern "Air-King" (ref 6494) and one a little larger than a bubbleback (ref 6466). By the late 1960s the smaller one was dropped leaving the company with no watch in a size between the current Datejust and the lady's models. This problem was solved in 1970 with the introduction of the 2030 movement. This was a scaled down version of the 1030 movement and was immediately fitted to a new series of watches universally known as the "mid-size." These watches were perfect 80% replicas of the Oyster Perpetual and the Datejust and were available in all the metal and dial combinations that their big brothers were. They have proved to be one of the company's major successes, being worn by many men in Europe, South America, and Asia. They also appealed to the woman for whom the normal lady's size was just too small.

Since the end of the 1980s, all Datejusts have been equipped with a "sapphire" crystal. This crystal is sliced out of a lab-grown sapphire with the cyclops date window cut from a smaller crystal and then applied. The purpose of the sapphire crystal is two-fold. Aesthetically, it is very appealing as it further slims the look of the Datejust. Practically, it protects the dial better in case of severe trauma and virtually eliminates the common scratches that inevitably occurred by the hundreds on the older, plastic crystals during the course of normal wear.

The success of the "two-tone" Datejust can only be called a phenomenon. This is proven by the fact that it is the most commonly copied watch in the world[43]. Everyone from the street corner hustlers of cheap low grade look-alikes to some of the most widely known watch manufacturers have attempted to sell their own little version of the stainless and gold Datejust. While Rolex can do little to prevent other watch manufacturers from making watches with a similar look now that the patent on the watch design has expired, the company is understandably aggressive in its pursuit of those who use the Rolex trademark on any watch which is not manufactured by them.

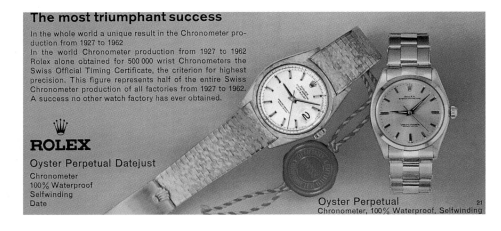

Datejust from a 1964 Bucherer catalogue showing the unusual "bark" finish flexible bracelet and bezel.

[41] Actually it was the 40th anniversary of the founding of Wilsdorf & Davis.
[42] Mainly because it has always been the most expensive of all the Datejust models. In fact for quite a time, prior to the introduction of the Day-date, it was the most expensive Rolex cataloged.
[43] For more information on "Fakes" see Chapter 24..

1966 British brochure showing variations on the Datejust.

Rolex 18kt gold three-piece case Oyster Perpetual Datejust with coin-edge style milled bezel original dial with applied triangle markers and aperture at three for the date. 10-1/2''' Hunter style movement, bubbleback numbered G23159.
Model no. 4467
Circa 1955

Another early example of an 18kt Oyster Perpetual Datejust, with "coin edge" bezel, silvered dial with triangle and baton markers. 10-1/2''' bubbleback chronometer movement numbered G23154. © Christie's, London
Model no. 4467
Circa 1955

Steel and yellow gold Datejust with steel/gold Jubilee band. This watch has a plastic glass, with hack feature, and is non-quickset.
Model 1603.
Circa late 1960s

236 A Date(just) with Destiny

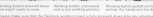
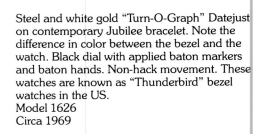

Brochure from 1967, showing both the gent's and lady's model Datejusts, ref. 1601 and 6517 respectively.

Steel and white gold "Turn-O-Graph" Datejust on contemporary Jubilee bracelet. Note the difference in color between the bezel and the watch. Black dial with applied baton markers and baton hands. Non-hack movement. These watches are known as "Thunderbird" bezel watches in the US.
Model 1626
Circa 1969

Stainless steel Rolex Oyster Perpetual Date, new old stock with original riveted Oyster band and original paperwork. The watch was purchased at the NAAFI (UK equivalent of PX) in the far east. These Date models are some 5% smaller than a Datejust and therefore better for some people of smaller build.
Model 1500
Circa 1964

Stainless steel Datejust with unusual steel crenelated bezel and double stepped silvered dial with applied baton numerals. It sits on its original chronometer papers.
Model no. 6605
Case no. 1728479
Circa 1969

Stainless steel Datejust on original Jubilee bracelet. Notice the difference in this standard steel bezel and the crenelated bezel shown above. Silvered dial with sweep seconds and applied baton markers.
Model no. 6605
Circa 1965

18kt pink gold Datejust with rare purple enamel dial, applied faceted numerals, leaf hands, and sweep seconds. The color of the dial is so dark that it is almost impossible to read the Rolex information on the dial, which is printed in black.
Model no. 1601
Circa late 1960s

Stainless steel "Datejust." Note the much wider divisions between the milling on the steel bezel than on the gold one.
Model no. 1522
Case no. 7187653
Circa late 1970s

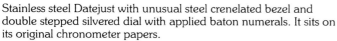

18kt white gold Datejust with rare lapis lazuli mineral dial.
Model no. 1601
Case no. 3924613
Circa 1975

Rare Oysterdate Perpetual in stainless steel, silvered dial with applied diamond cut markers, triangle hands, sweep seconds hand and raised gilt metal frame around the date window. How this watch differed from the normal Datejust is difficult to know.
Circa 1958

Rare Air-King-Date model in stainless steel. These are the predecessors of the "Date" model, but unlike them these watches are not chronometers.
Model no. 5700
Circa 1968

Steel and 18kt yellow gold Datejust with Jubilee band. Watch has black dial and baton markers and hands. Dating from the early 1980s, this watch still has the plastic crystal, but now has both the hack and quickset date features.
Model no. 16013
Circa 1978

14kt yellow gold Bombé Oyster Perpetual presentation watch bearing the logo "Eaton" and "1/4 Century Club" on the dial. It is interesting to note that the watch was presented in 1971, by which time this model watch had been out of production for over 10 years, reinforcing the theory that Eaton would buy the watches in bulk (because of the dial complexity) and warehouse them until needed.
Model no. 6092
Circa 1960

Oyster Perpetual Datejust in steel and gold with Turn-O-Graph or "Thunderbird" bezel. Note the unusual dial which has the "Datejust" in red and below it is the depth capability of the watch, "50 meters=165 feet." The dial still retains the gilt frame around the date window and the date disc has red numerals for the even numbers and black for the odd dates.
Model no. 6309
Case no. 141318, dated 1.56
Movement no. 98336

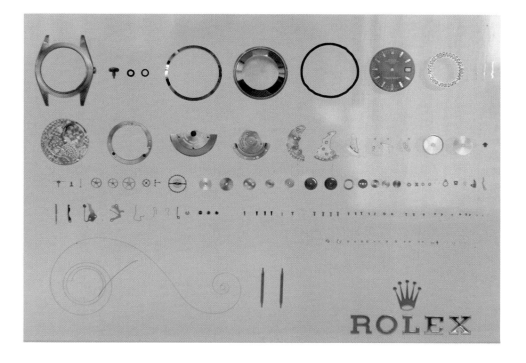

Oyster Perpetual Date in disassembled state showing all constituent parts.

CHAPTER 11
THE EXPLORER

*We shall not cease from exploration
And the end of all our exploring
Will be to arrive where we started
And know the place for the first time.*
T.S. Eliot

After the two-tone Datejust, the Explorer is one of the most easily recognizable of all Rolex models. With its black dial, large luminous triangle marker at 12, and luminous arabic numerals for the other quarters, it is the perfect mixture of a sport and a dress watch. It seems to have been around as long as there have been Rolexes, but that is not exactly true. However, as with most legends, finding the truth is never simple.

Before we get into the history of the Explorer it is perhaps first worth defining the watch itself. The generally accepted definition of an Explorer is that it is any watch with the dial described above. Unfortunately, one often encounters watches with the word "Explorer" proudly printed on their dial, which bear no resemblance to this description. More rarely, one sees an Explorer dial on other Rolex models. In this discussion of the Explorer we are exercising our writers' prerogative and choosing to include all of the above as Explorers.

The generally accepted origin of the Explorer is that it was first designed and made in honor of Edmund Hilary and Tenzing Norgay, who, on May 29, 1953, were the first to reach the summit of Everest and who did so wearing Rolex wristwatches. The only problem with this hypothesis is that it can not possibly be true. The climbers on Everest were, in fact, wearing Explorers, so the watch had to have been introduced before the climb and not after. One of the watches worn on that expedition was auctioned by Sotheby's London on July 19, 1988 as lot 117, (see photograph on page 243). As you can see the watch was a classic early Explorer down to the "Mercedes" hands, except for the absence of the word "Explorer" on the dial. The shape of the watch (and the description by Sotheby's as a "Bubble Back" Explorer) leads us to believe the watch is in fact a model 6350. This hypothesis is strengthened by the photograph on page 246 which shows another model 6350 with an almost identical dial. The main difference between the two is that the Sotheby watch lacks "Explorer" but has the word "Precision" above the 6, whereas the other watch has the word "Explorer" but lacks either "Precision" or "Officially Certified Chronometer" above the 6. Instead it has a British military marking in their place, as well as on the case back. In the early 1950s, the period these watches were made, Rolex often stamped the inside of the case back with the date of manufacture. The military 6350, marked Explorer, shows the manufacture date as IV 53, meaning the 4th month of 1953. As stated above, Everest was conquered on the May 29, 1953. Using these facts, it seems

This 1966 British catalog features a Submariner with an "Explorer" style dial.

♛ *Rolex watches are not "styled"—they are designed. With purpose. Consider these facts:*

Rolex invented the waterproof wrist-watch. The Oyster case, entirely Geneva-made, is still the only one of its kind in the world. It takes 162 separate precision operations to create the Oyster case from a solid block of fine stainless steel or gold. It is not moulded, pressed from sheet metal or soldered; it has immense strength. The Oyster case alone costs as much to make as most other complete watches.

Rolex made the first rotor self-winding wrist-watch—so successful that all other automatic watch manufacturers just had to adopt the Rolex principle. Rolex, with constant research and development, keeps ahead.

Rolex Oyster watches have been official equipment on rugged Himalayan expeditions, and have been down to the bottom of the deepest ocean chasm. They have never been affected by heat and cold, or water, or sand, or dust.

Rolex make but a very small percentage of the 50 million watches made in Switzerland every year. Yet Rolex have been awarded nearly half of the Swiss Official Chronometer Certificates during the last 40 years. A chronometer takes nearly a year to make. It must undergo a series of stringent tests and pass with the mention "Especially Good Results."

When a man has a world in his hands, you find a Rolex on his wrist.

Rolex Oyster Perpetual Submariner
Created originally for deep-sea divers, the Submariner is nearly half an inch thick and weighs a hefty 3 ounces. It's guaranteed waterproof down to 660 feet underwater and is self-winding. The calibrated auto-lock turning bezel permits easy and precise calculation of decompression times and also the length of telephone conversations, photo development, etc.
The Submariner has a wealth of other features (and recently, a number of imitators). It has all the character of an original. Officially supplied to the Admiralty for use by Royal Navy divers.
1029
Stainless steel with "Flush-fit" bracelet £70 8s.

Rolex Oyster Perpetual Explorer
Has been used in the Arctic, the tropics and the Himalayas. Supplied to the first successful Everest expedition. In the words of Sir John Hunt, leader of the Expedition, who is seen below: "We have come to look upon Rolex watches as an important part of high-climbing equipment."
Officially Certified Chronometer with the special mention "Especially Good Results." Self-winding.
1025
Stainless steel with "Flush-fit" bracelet £72 18s.

likely that Explorers were in production prior to the conquest of Everest[44]. It is also worth noting that the name "Explorer" was registered in Geneva on the January 26, 1953, obviously well before the conquest of the world's highest mountain.

While it is true that many of the members of the successful Everest expedition were issued with Rolex watches (*see the advertisement on page 243*), the embarrassing fact for Rolex was that only one of the two climbers at the top was wearing a Rolex. This watch, worn by Tenzing Norgay, is now in the Rolex Museum in Geneva. Although Rolex was an official supplier to the Everest expedition, so was the English watch company Smith's and Edmund Hilary chose to wear a Smith's watch (*see the advertisement below*). In the end it was the Rolex publicity machine that triumphed. Interestingly, due to a pact made by Tenzing and Hilary, we will never know which watch was first at the summit; both climbers have always said that it mattered nothing who was first.[45]

The real origins of the Explorer are revealed by its name. It was designed for explorers and so it had a high visibility dial, an extra strong case, and, on request, it could even be lubricated with a special oil which could withstand temperatures of between -20°C and +40° without changes in its viscosity. As such it was used by many expeditions both before and after the successful Everest expedition.

The look of the Explorer is all about the dial, which is a mixture of a number of previously used styles. The large triangle at "12" was first used by the company on the mixed Roman and Arabic dial of the early 1940s. The large arabic numerals for the quarters and bars for the remainder are seen on many of the very first model cushion Oysters. Despite this somewhat mixed parentage, the dial has taken on an identity of its own and can never be confused with any other.

These first Explorers (6350 models) used the "big bubbleback" 10-1/2''' A296 movement. Most of the other 6350 models to have surfaced do not have the classic "Mercedes"[46] hands. Instead they have heavily luminized versions of the standard parallel hands of the period. While the sweep seconds hand is very strange, it looks similar to the current hand having a large circular luminous insert. Closer examination reveals that this circular insert is at the tip of the hand, not 4 mm from the tip as now. Most of the early 6350 dials are also unusual in that they are "honeycomb" textured (how the company managed to print on this surface is a mystery) and are signed as "Officially Certified Chronometer." It is difficult to know how successful this model was, for it is not exactly a common piece and seems to have been replaced by the 6150 model within a year or so. Powered by the same movement, the 6150 was distinguishable from the 6350 by being 2mm. larger and was only available as a Precision model. Not all the 6150s were classic Explorers. The one shown in the photograph on page 249 is unusual in that, while it bears the normal reference number, the dial is previously unknown. The 6150 was made until 1959 when it was replaced by the 6610, which appears identical, but in fact can be identified by its flatter back, a consequence of using the newer 1030 calibre movement[47]. The dial of the 6610 is signed "Chronometer." The simplest method of recognizing any of the early Explorers is by looking at the dial. Although they are all steel watches, all of the printing on the dials (the minute track, the Rolex logo and the model name) is in gold

During these early days of the Explorer, Rolex was unsure of the model's potential. As a result the name was affixed to a number of watches not immediately recognizable as Explorers. Today the name and the look are so intimately entwined it seems ridiculous to apply the name to watches which were so obviously not what we would call Explorers. But it happened, and the results are some of the rarest Explorer models known. There were two distinct variations on the theme and they seem to have been aimed at two distinct markets. The first was the so-called "Air-King" Explorer. This was an Explorer bearing the

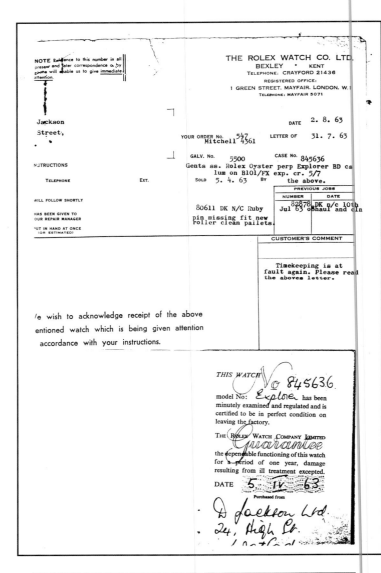

5500 model number usually applied to the Air-King, but with an Explorer dial that is marked "Precision" rather than the "Superlative Chronometer Officially Certified" we would expect to see. There has been some doubt that these watches are real, for, if we examine the watch closely, we can see that the dial is smaller than a normal Explorer dial. But the receipt from Rolex, shown on the previous page, lists a 5500 Explorer. With this information we think we can validate these as genuine Rolex pieces. A closer examination of all of these models which have turned up with original paperwork reveals one further interesting fact. All of them were purchased from N.A.A.F.I. (the British equivalent of the PX) in the middle or far east. Whether or not the model was made as a comparatively inexpensive military style watch for officers to purchase with their own money is just one more question waiting to be resolved.

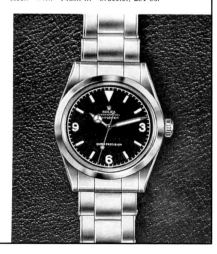

The only known advertisement showing the 5500 Super Precision "Explorer." Note the similar price to the Air-King shown to its left.

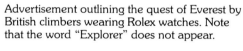

Advertisement outlining the quest of Everest by British climbers wearing Rolex watches. Note that the word "Explorer" does not appear.

The earliest known Explorer, although, in fact, it does not have the word Explorer on the dial. This is one of the watches worn on the John Hunt Expedition which succeeded in conquering Everest at the end of May, 1953.

The second variation on the theme are "dress Explorers." These are standard Oyster Perpetuals in steel or gold with white (or more rarely, black) non-Explorer dials featuring gold markers and hands, but signed "Explorer" on the dials. Seen in both date and non-date forms, these watches all seem to have been sold in the North American market. The Explorer Date, shown bears a model number 5700, previously unseen on any other Explorer. The non-date model is a model 5504, which interestingly is more often seen on Explorers with the standard dial. We have even seen a gold capped Explorer with a Tiffany dial with the reference no 5510. The earliest one of these to turn up has been a model 6298, having the manufacture date of 111 53, and bearing the phrase "Self Winding" more often seen on Tudor watches. It seems that these watches were made in the 1950s or 1960s when Rolex was unsure if the Explorer would succeed or not, and attempted to increase the popularity of the watch by broadening the line. When the mainstream Explorer began to sell it seems that these "piggyback" models were withdrawn. This is one more example of the bizarre fact that it is a company's failures that become the most desirable and valuable items in any collection.

Perhaps the strangest "Explorer" model is not really an Explorer, but a Submariner. It is a model 5513 non-date, non-chronometer watch with a classic "Explorer" style dial. The watch seems only to have been sold in the U.K. and only for a short period in the 1960s. The catalog shown bears the date 1966 and was a catalog only for the British market.[48] One of these watches has surfaced with a hacking seconds movement (which would date it after 1972), also in the U.K.

The introduction of the 6610 model in 1959 gave us the Explorer in its most recognizable form. Only in production for five years, the 6610 differed from the 1016 that replaced it in calibre; the 6610's calibre 1030 was replaced with the more modern calibre 1560[49]. While the cases of the two watches looked almost identical, the new model was now guaranteed waterproof to a pressure of 10atm rather than 5atm of the 6610.

The 1016 Explorer was the longest running of all the models, being in production from 1963 right through to 1989. During this period it did not, however, remain unchanged. The first model (in production from the start to around 1975) used the basic calibre 1560, but, as is the general Rolex policy, there were certain updates and modifications. While there are some watches signed Explorer that do not look like regular Explorers, the 1016 also exists with another name on the dial: "Space Dweller." The model was first introduced into the Japanese market in 1963, just after a visit to Japan by the Mercury astronauts. A trial run of Space Dwellers was made to honor these men, who were (at that time) seen as the ultimate explorers. The watch was not a major seller, either in Japan or elsewhere, and very few of the watches so signed have ever surfaced.

The second version of the 1016 was really the second version of the 1560 calibre (now renamed 1570), because the major recognizable difference is in the movement. The "hack" feature, stops the second hand when the winding crown is pulled out to the hand setting position. By stopping the hand at the "12" position, it is possible to synchronize your time with that of a known source, a radio or telephone time signal for example.

Simultaneously with the movement change, Rolex introduced the new Oyster bracelet. It had links machined from solid steel, instead of the folded sheet steel of earlier bracelets. In this revised form the Explorer continued through to 1989, when, to the astonishment of the Rolex retailers, it was removed from the new catalog. It was only six months before a new very heavily revised Explorer arose from the ashes of the 1016. The new model, bearing the model designation 14270, sported a new case, dial, movement, and glass. It seemed as if the hands and the name were the only things carried over from the 1016. Almost thirteen years after Rolex had first introduced the sap-

Stainless steel Oyster Perpetual Explorer Precision, non-chronometer, silver dial, gilt markers and hands.
Model no. 5504
Circa 1955

Rolex Oyster Perpetual Explorer Officially Certified Chronometer. This watch has the honeycomb dial and the very early parallel gilt hands, however, with very heavy luminous inserts.
Model no. 6350
Case no. 955901
Movement no. H81697
Circa 1954

phire crystal, the Explorer was finally fitted with one. Under this new crystal, the dial featured white gold skeleton markers with luminous tritium fillings; these replaced the previous painted dial markers. Under the dial was the very latest fast beat calibre 3000. These modifications brought the Explorer in line with all other Rolex models and because the cosmetics of the new watch were so different, they sent the prices of the older models in the collectors market on an upward spiral. The result is that, at this writing (early 1996), collectors are paying more for the older models than the retail price of the new one.

The other Explorer is the Explorer II. Introduced in the early 1970s as model 1655, it is essentially a GMT-Master with a fixed bezel. Using the same calibre 1575 movement as a GMT-Master, it also had a fourth hand which rotated once every 24 hours, however on the Explorer II the hour was read from this from a fixed, engraved 24 hour steel bezel. The watch was introduced as being especially useful for the speleologist (or cave explorer), who, Rolex claimed "soon loses all notion of time: morning, afternoon, day, or night." For these intrepid souls Rolex developed the watch which would tell them whether the "2" on the dial was 2:00 a.m. or 2:00 p.m. (14:00 hrs). This may well be true, and perhaps cave dwellers are more susceptible to losing track of time than others. We would suggest, however, that the demand for a watch specifically targeted at speleologists would find a tiny market, and that even its limited popularity was due to its acceptance by others who work in civilian and 24 hour time systems, such as pilots and air traffic controllers. The watch went through two styles. The first, made for only three years, used an orange 24 hour hand, and the following model, made until 1985, used a red one. The 1655 Explorer II and the 1019 Milgauss are the only Rolex models which use hands that are used by no other model.

We used the phrase "limited popularity" above intentionally for this watch always was one that was never very widely distributed. It was not really popular until 1991, five years after it was relaunched with a sapphire glass and the same movement as the GMT-Master II. This allowed the owner of the watch to set the hour hand backwards or forwards in one hour jumps without disturbing the second and minute hands. This facility allowed the owner to change time zones without losing a preset accurate time.

The new watches, like the first model Explorer IIs had the bezel numbers enamelled in black, but the very first of the new model Explorer IIs had very unusual red bezel enamelling. These early white dial models had the tritium dots and the hands outlined in white gold, while the following models had them outlined in black.

The name, Explorer, and its whole history of promotional material have made much of the watch's ability to resist hazards and we are sure that this reflected glory is the reason for the watch's continued popularity. In the end wouldn't we all like to be thought of as "Explorers"?

---

[44] The conquest of Everest was important to Rolex because they had identified themselves with Everest for over 20 years prior to the conquest. See the accompanying advertisement.

[45] Interestingly, the watch sold at Sotheby's should have been the first watch up Everest. It was issued to Tom Bourdillon, the climber who was chosen to make the initial assault, but was forced back by oxygen problems only 800 feet below the summit.

[46] At Rolex these hands are known as "skelette" or skeleton hands never as "Mercedes," but for convenience we will use the term more commonly used by collectors.

[47] The 1030 calibre was a major breakthrough for Rolex. It was their first automatic movement in which the rotor wound the movement while moving in either direction.

[48] The catalog shown bears the date 1966 and was only for the British market.

[49] The calibre 1560 movement introduced the micro-stella screws for adjustment of the balance. This was the first calibre Rolex had produced which was free-sprung from its introduction. They had, of course, used free-sprung balances in many of their Kew A watches, but these were essentially hand made watches.

Rolex Oyster Perpetual Explorer Super Precision (non-chronometer model). On this model the writing and the chapter ring are still in gilt finish.
Case no. indistinct
Model no. 6150
Circa 1955

Rolex Oyster Perpetual Explorer Chronometer. Although the writing is still in gilt the model is now an Officially Certified Chronometer.
Model no. 6610
Case no. 268255
Circa 1955

246 The Explorer

Stainless steel Rolex Oyster Perpetual Explorer. Original black dial, British military issue. Notice the T on the Dial for Tritium, the material used to luminize the watch, and the military engravings on the back which stand for Hydrographic Survey, which is the department of the Royal Navy which makes maps and charts.
Model no. 6350 or 6150 (The watch actually bears both numbers)
Case no. 930934
Movement no. 96884
Circa 1954

Rolex Oyster Perpetual Explorer Precision. This model is slightly smaller than the standard Explorer and takes a 19mm strap rather than the conventional 20mm, and is overall about 2mm smaller in diameter. It is, of course, a precision watch rather than a chronometer.
Model no. 5500
Case no. 1112592
Circa 1960

Rolex Oyster Perpetual Explorer Officially Certified Chronometer. Once again the honeycomb finished dial, but this time with slightly later Mercedes style hands. Notice the bubbleback style fully enclosed movement. Inside of case dated 4/53 meaning last quarter of 1953.
Model no. 6350
Case no. indistinct
Circa 1953

Stainless steel Rolex Oyster Perpetual Explorer Officially Certified Chronometer. Flat dial (non-honeycomb) differentiates this watch from the previous model.
Model no. 6350
Case no. indistinct
Circa 1960

The Explorer

Rolex Oyster Explorer with honeycomb finished white dial. Signed "Shock Resistant" on the dial, with silvered metal and Arabic applied numerals. Signed inside the case 3/55 meaning of course the third quarter of 1955.
Model no. 6298
Case no. 282632
Signed 1955

Rolex Oyster Perpetual. This is the penultimate model Explorer before the launch of the new crystal glass, displaying all the signs of the current Explorer. Writing is now in white and the hands are also in white metal and there is no outlining to any of the numerals on the dial.
Model no. 1016
Case no. 2173480
Movement no. 78360
Circa 1967

The Explorer 249

Rolex Oyster Perpetual Precision Explorer also signed on the dial "Rotor Self-Winding." Applied gilt metal hour markers and gilt hands. This is a very strange watch. The "rotor self winding" logo is normally only seen on Tudors, but this watch is quite definably original and in fact is a 6150 which is a normal Explorer reference number.
Case no. 941699
Case marked 1/53

Rolex 1016 Explorer with hacking seconds. Externally identical to the watch on the previous page, but with the stop seconds operated by the winding crown.
Model 1016
Case no. 6151764
Movement no. 78360
Circa 1980

Perhaps the rarest of all Explorers is the 1016, signed "Space-Dweller." A very small number of these watches were made and were released on the Japanese market after the visit of the Mercury astronauts. This model was produced for only a period of 18 months and apart from the writing on the dial is identical in every way to a normal 1016.
Case no. 1040738
Circa 1968

Stainless steel Oyster Perpetual Explorer, black "honeycomb" dial with gilt writing and minute track. Notice the early heavily luminous parallel gilt hands and unusual sweep seconds hand with large luminous dot at the very tip, rather than 4mm from the end as is used currently.
Model no. 6350
Case no. 955466
Movement no. 8422896
Case stamped "IV 53"

250 The Explorer

Previous watch in an early Explorer box, on an official Swiss chronometer timing certificate along with Rolex explanation and receipt. It is very rare to find a watch with all of its original papers and packing.

Explorer II, final model of the 1655 with hacking seconds and heavyweight bracelet.
Model no. 1655
Case no. 3097615
Circa 1971

Stainless steel Oyster Perpetual Explorer Date Precision. Very rare 1960s "dress" model of the Explorer with white dial, gold triangle and baton markers. Explorers with date feature are very rare and seem to have been made for the U.S. market only in the early 1960s. Model no. 5700
Case no. 515218
Circa 1965

The Explorer, an Oyster Perpetual Chronometer, was created specially for the first successful Everest expedition. The specification was simple yet exacting — to design a thoroughly dependable, shockproof rugged watch to perform in the most difficult circumstances. After the expedition had successfully completed its mission, we received these comments:

BRIGADIER SIR JOHN HUNT

... As I have emphasized before, this expedition was built on the experience and achievement of others. Rolex Oyster watches have accompanied many previous pioneering expeditions. They performed splendidly and we have indeed come to look upon Rolex Oysters as an important part of high climbing equipment...

SIR EDMUND HILLARY

... I wore the watch (Rolex Oyster Perpetual) continuously night and day... In the course of the expedition it experienced considerable extremes of temperature from the great heat of India to the cold temperatures at over 22,000 feet and seemed unaffected by the knocks it received on the rock climbs or the continual jarring of long spells of step cutting in ice... Its accuracy is all one could desire and it has run continuously ever since I put it on some nine months ago...

SHERPA TENSING

...your watches behaved splendidly under any and all the worst conditions imaginable, not only this time but also on previous expeditions to Mount Everest, of which I happened to have also been a member.

The Explorer has become recognized the world over as the watch for climbers, cave explorers, skiers:

BRITISH NORTH GREENLAND EXPEDITION

...occasional time signals broadcast from England proved that my Rolex watch was maintaining a remarkable accuracy. On no occasion did it require to be wound by hand. When on the ice-cap, away from base for several weeks at a time, it was of inestimable value to have on my wrist a watch whose accuracy could be relied upon at all times...

ARGENTINE ARMY EXPEDITION TO THE ANTARCTIC

An officer says:
As a pilot of a Cessna reconnaissance plane and later at the base of Sobral (81° S latitude and 40° W longitude) I had occasions to test out the qualities of my Rolex. It was exposed to temperatures of –53° C and to very rugged conditions. Its accuracy and reliability make it of incalculable value to all who have to work in completely adverse conditions, such as are found in the Antarctic.

BRITISH SPELEOLOGICAL EXPEDITION TO THE CANTABRIAN MOUNTAINS (SPAIN)

Caving is not a gentle activity and of necessity the Rolex Explorer watches received a great deal of rough treatment including immersion in mud and water, and scratching and banging against rock, but the accurate reading of time, which is an important safety factor in caving, never gave us any worries.

A. C. Huntington

ROYAL SOCIETY ANTARCTIC EXPEDITION

(My Rolex) ... became part of me – an unquestioned, reliable part – doing no more than gain a matter of seconds per week, although exposed to such extremes of temperature as a hot cooking stove, or digging out stores in minus 60° Fahrenheit.
Surgeon Lt-Cdr Dalgliesh, Leader Advance Party

The Explorer has a specially strengthened **pressure-proof** Oyster case which is waterproof and swimproof to 330 feet. It has been specially designed, with its officially certified chronometer movement, to be fully shockproof, to withstand the jars and knocks of mountain climbing, of exploring. It can safely be worn while operating a pneumatic drill or cutting rock-steps. It is selfwound by the Rolex Perpetual mechanism and the mainspring cannot overwind. Like all Rolex Oysters, the Explorer has the exclusive Twinlock screw-down winding button. The crown becomes an integral part of the case and is protected against accidental damage.

Not everyone is an explorer but there are many who are forced to treat their watch roughly. With the Explorer you can at last own a really accurate watch that is tough.

November, 1966 brochure for the first of the 1016 model Explorers. The inside of the brochure bears many testimonials to the strength and endurance of the watch. It is interesting to note that the watch claimed to be waterproof to 330 feet, whilst the Datejust and GMT-Masters of the same period were said to be safe only to 165 feet.

252 The Explorer

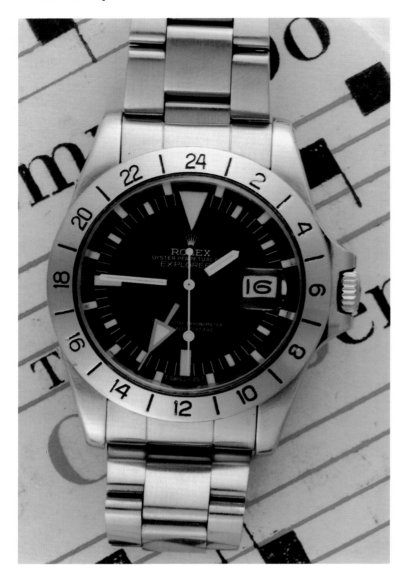

Rolex Oyster Perpetual Explorer II, fixed 24 hour bezel, dial with 24 luminous hour markers and large orange or red 24 hour hand. Never one of the more popular models, the Explorer II has, by virtue of its subsequent rarity, become one of the more collectible 1970s/1980s Rolex watches.
Circa 1978

Stainless steel and gold Rolex Oyster Perpetual Explorer. Note the textured dial and the unusual gold bezel and winder. The watch uses the A295 movement. Ref 6299. © Sotheby's, 1994.
Circa 1958

Unusual "boy's size" Explorer. Based on the Air-King models these Super Precision (or non-chronometer) Explorers are around 2mm smaller than a 1016 Explorer and the dials are not interchangeable. Note on this model the luminous insert for the seconds hand is missing.
Model no. 5500
Circa 1975

Probably one of the first examples of the new model sapphire crystal Explorer II. This was evidently provided by Rolex for the Steger International Polar Expedition of 1986. It is assumed that the accompanying cloth strap was used over the bulky garments needed for such temperatures.
Case no. 8597482
Model no. 16550

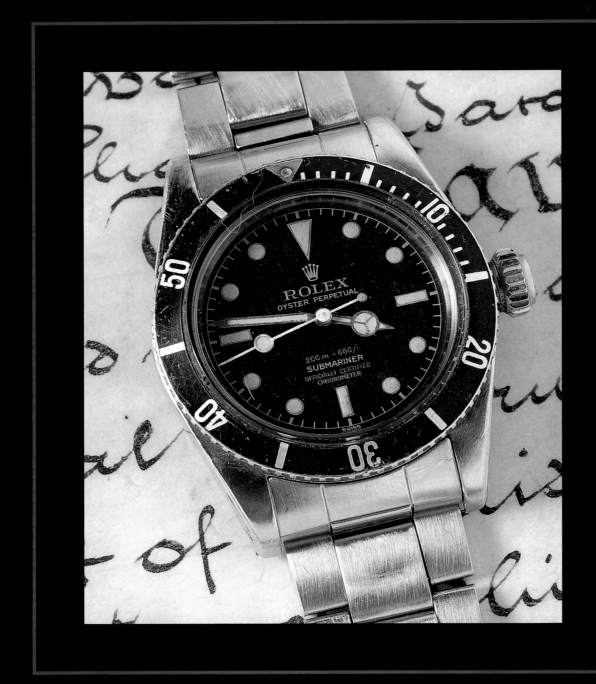

The Oyster Perpetual Submariner still does not have the crown guard, the writing is still in gilt, but the watch is now an Officially Certified Chronometer as seen on the dial. Notice the unusual red triangle at twelve on this model.
Model no. 6238
Case no. 385893
Case stamped "11 58" which stands for the second quarter of 1958

CHAPTER 12
THE SUBMARINER AND THE SEA-DWELLER
The Stock Car, The Diver, and The Gas Escape Valve

The Submariner: 660 feet and counting
The Rolex stand at the 1954 Basel Spring fair was one of their largest and most colorful ever. The five large windows were gaily decorated in pastel shades, and held some of the company's most important new products since the Datejust. In fact the Datejust itself was new and improved, for it was at this show that the company chose to launch the "Cyclops" date lens. Still, the brand new products stole the show.

There was a lady's Perpetual with chronometer certification, the Explorer, and the Turn-O-Graph. But the honor of place was given to a bizarre, oversized steel watch with a hemispherical crystal, proudly bearing the name "Submariner".

The Submariner was Rolex's pride and joy. Less than six months earlier, on September 30, 1953, Professor Auguste Piccard[50] and his son Jacques had piloted their new bathyscaphe to a world depth record of 10,335 feet below the ocean's surface. The watch had accompanied the two on their quest, although it had not been of much use to the good Professor and Jacques. It had been fixed to the OUTSIDE of the craft. At a depth of 3,150 meters, or 10,335 feet, the watch had been subject to a pressure of over 600 atmospheres or two tons per square inch. It had survived this gruelling test and had returned to the surface, after a two and a quarter hour venture, in the same condition as when it had entered.

The skills needed[51] to produce this special watch had been used in the case, crown and glass design. It carried the model number 6204 and, as a production watch, was waterproof to a more realistic 200 meters (or 660 feet). The watch was sold as "the diver's friend."

One of the reasons for the immediate success of the watch were that the time was simply right for it. For the first time movie and television viewers were being entranced by the work of such undersea pioneers as Hans and Lotte Haas and Jacques Cousteau, to say nothing of the weekly exploits of Lloyd Bridges in "Sea Hunt." The development of SCUBA (Self Contained Underwater Breathing Apparatus) diving as a sport really began to take off toward the end of the fifties. A diver needs a reliable watch as much as he needs air, for without knowledge of how long he has been submerged he could suffer serious decompression sickness (the bends). The rotating bezel on the Submariner enabled a diver to keep accurate track of his elapsed time under water.

Rolex was fortunate that one of its directors, René-P. Jeanneret[52], was an enthusiastic amateur in the skin-diving world and was therefore able to give input from both the underwater and horological viewpoints. This advantage was not shared by any of the other companies developing competing products.

The first Submariner model, the 6204, differed considerably from the watch we now know, lacking nearly all of the classic Submariner features. The very first model had simple parallel hands, not the "Mercedes" hands we now recognize. It had no protective shoulders by the winding crown and the rotating bezel had only five minute markers. Apart from the bezel calibrations, the watch looked identical to the "Turn-O-Graph" 6202. (Intriguingly the "professional" Submariner managed to get by with only 5 minute divisions while the "civilian" Turn-O-Graph had markings for every minute.) The original model was rated as being waterproof to 600 feet, rather than the 660 feet we are more used to seeing.

The first Submariner was in production for less than two years before being replaced by the newer model 6538. The 6538 looked similar in many way, but was the first watch to be rated at 660 feet.

As is the case today two Submariner models existed early on, though the difference then was in the waterproof depth rather than the presence of a date feature. While the 6538 was guaranteed to 660 feet, the less expensive model 6536 looked identical but was only guaranteed waterproof to 100 meters or 330 feet. These two models are the ones generally known amongst collectors as the "James Bond" models, although we think that the very first model (6204) could also well be included in this group. The first change to both models was made within months with the introduction of the "Mercedes" style hands.

The watch continued in production in this format for two years, until it was supplanted by a complete new model, which, interestingly, kept the same model number 6538. This new model was much more like the current Submariner. It introduced the more robust (some might say, bulky) case we now know, with the bezel markings for the first 15 minutes and a red triangle at the "12" position. It also had the larger "Triplock" style crown, although at this point the name had not been registered. The crown, rather than having the three dots under the crown, had the word "brevette" or patented around the circumference. The dial still bore its markings in gilt but now carried the words "Officially Certified Chronometer" under the word "Submariner." It was able to carry this statement because the watch now was powered by the new 1030 movement. Simultaneous with the launch of the new 6238, the company also launched the 6200, which was cosmetically identical to the 6238, but still used the old style A296/775 movement and the old ungraduated bezel.

It was obvious that the Submariner had started something, because the year after its introduction the 1955 Basel show featured a large number of watches from Rolex's competitors, proving once again that imitation is the sincerest form of flattery. Blancpain brought out the "Fifty Fathoms" and Eterna came out with the "Kon Tiki." In the years that followed every watch company of note seemed to bring out a diver's watch, almost all of them being near clones of the Submariner.

In 1959 a new model arrived, the 5512. This introduced the new case featuring the protective shoulders which were an integral part of the case body and rose from the sides of the case to almost completely surround the Triplock winding crown. They protected the weakest part of the watch from inadvertant knocks and damage whilst underwater, where of course such damage could prove fatal both to the watch and the wearer. All of the dial printing was in white rather than gold, and it had a bezel which still featured minute markers for the numbers between "0" and "15". The triangle at "12" was now silver. The watch

Submariner brochure for 1961.

was also available as a non-chronometer or Precision model, as the model 5513, but was still waterproof to 660 feet. It had completely acquired the Submariner "look" that we know today. The previous model, the 6536, was kept in production with a new number, 5508. It was now stated to be waterproof to only 330 feet, although the two models appear to be identical.

This new Submariner was launched on the back of Jacques Piccard's latest venture. On January 23, 1960 using the bathyscaphe "Trieste," Piccard[53] dove to the bottom of the Challenger Deep of the Marianas Trench, located off the Philippines in the deepest part of any ocean in the world. Once again a Rolex was affixed to the outside of the Trieste and after the 8 hour 41 minute dive to a new record depth of 35,798 feet (or 10,916 meters), where the watch was exposed to pressures of over seven tons per square inch, it was found that it was still in perfect working order with no evidence of moisture inside the case.

As evidence of the company's serious involvement with SCUBA diving, in 1961 Rolex produced a Skin Diver handbook which was given with each model 5512 Submariner. This booklet gave a full decompression table, courtesy of the U.S. Navy and a dictionary of diver's hand signals.

The next major breakthrough came in 1965 when the 1565 calibre movement was fitted to the Submariner. This brought the combined benefits of both a chronometer movement and a date function to the watch. This new model, the 1680, looked otherwise identical to the previous model, the 5512. It was the

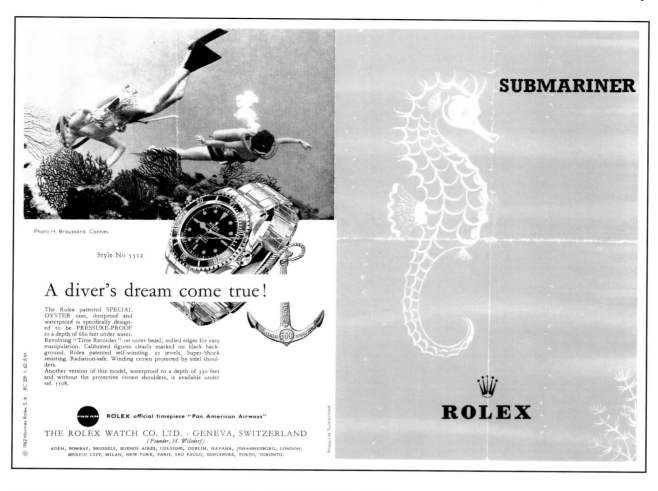
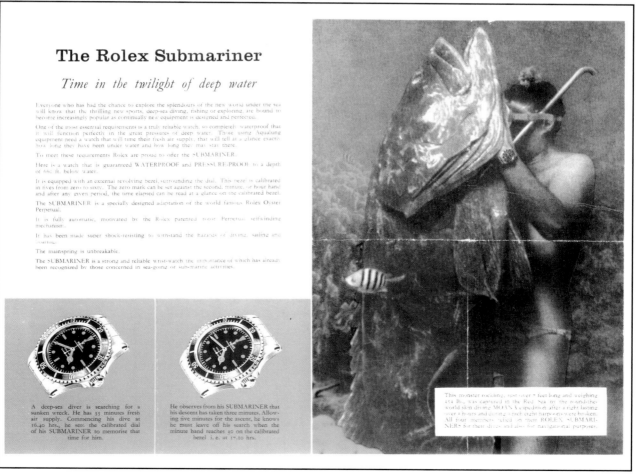

Brochure for the Rolex Submariner, 1962.

first Submariner to be available in 18kt gold and in steel with gold bezel. With the new model becoming the premium Submariner the old 5508 model was discontinued and the 5513, the non-chronometer model with shoulders became the "entry" level Submariner.

The 5513 became the first Rolex for many years to become an official British military issue watch, when it was issued to the divers of the Royal Navy in 1965. As is usually the situation, the military wanted certain modifications, in this case the modifications were so major as to almost be a new watch. First they wanted a new dial featuring a large "T"[54] in a circle under the center post. The bezel was to have no minute markers all the way around its circumference, not even for the first fifteen minutes. The high-visibility hands were much larger and of a completely different shapes. The hour hand was diamond shaped, while the minute hand was much wider than on a conventional Submariner and the seconds hand had a unique diamond luminous indicator at its tip.

Even the case was different from a standard Submariner. On these military models the spring bars were replaced with solid steel bars welded into the spring bar holes. The case was then polished so the hole was hardly visible. The rear of the case was then engraved with all the military specification and issue numbers which enabled the quartermaster to identify the watch throughout its life[55].

Most unusually, the Admiralty authorized Rolex to issue a press release when the watch became official issue. It makes interesting reading as it states (in part) "This official Royal Navy divers' watch was put into production only after two years of exhaustive Admiralty exposure and timing tests in England and the Mediterranean. It has proved itself waterproof even down to 400 feet below the surface of the water."

Later, at the end of the 1960s, another branch of the British military; the Royal Marines ordered Submariners for their frogmen. These were even more immediately identifiable than the more prevalent Royal Navy models because they had even more alterations. The bezel was a completely new one with minute markers all the way around its circumference, rather than just for the first fifteen minutes. More interestingly, the movement was also special. These watches were still non-chronometer, plastic glass 5513 models, yet they had the hacking seconds feature, something that was only introduced on the chronometer models three years later. The one shown here is one of these Royal Marine models, but is even more interesting in that it was never issued by the quartermaster's department and was disposed of to the public as surplus without ever being used by a Marine.

The next major update to the Submariner did not come directly from Rolex, but through one of its customers, the diving contractor Comex. This was of course the watch that began life as a converted Submariner, metamorphosed through the model 1665 Submariner Sea Dweller, and became the Sea Dweller of today.

The Sea Dweller was introduced in 1971 at the same time as the next major update of the Submariner. The glass was much thicker and and the word "Submariner" was now printed in red. This red printing lasted for only four years, making the Submariner from this period one of the most easily identifiable models, and, due to the short production run, one of the most collectible.

Toward the end of the 1970s the Submariner, along with most other Rolex date models, was fitted with the quick set feature. Then, in 1986, the Submariner received its most significant update yet. The new model was numbered 16800, to fit in with Rolex's new numbering system. It featured a sapphire crystal and was rated as being capable of operating at a depth of 1,000 feet (or 330 meters). The watch was identifiable by the new dial which featured white gold settings in which the luminous indices were placed.

But only three years later, in 1989, it was all changed again. This time it was the replacement of the 16800 model by the new 16610 model which looked externally identical, but was now powered by the new fast beat calibre 3135. The bezel now was fitted with a safety feature for divers, which allowed the bezel to be turned only in an anti-clockwise rotation and only when a strong vertical pressure was applied to it. This meant that even if the bezel were inadvertently moved while a diver was submerged, it could only move in such a way as to indicate a shorter, and therefore safer, diving time.

The biggest compliment ever paid to the Submariner came in 1993, when it was copied, believe it or not, by Rolex. The model 1662 Yachtmaster was, to all intents and purposes, an 18kt clone of the Submariner, differing only in the cosmetics. The bezel was solid gold rather than a screen-printed colored insert and the dial was now cream instead of blue or black, although further dial colors were subsequently introduced in 1995. Also introduced at this time was a lady's version of the watch. Though Tudor diving watches for ladies had been produced for some years, this was the first time Rolex had produced a lady's version of any of their diving watches. Interestingly despite the crown guard and the "Triplock" crown, the watch was waterproof to only 330 feet (100 meters) rather than the 1,000 feet (300 meters) of its stablemate the Submariner, which, incidentally, was around £300.00 (or $500.00) less expensive.

In the forty-plus years that it has been in production the Submariner has, by dint of Rolex's detailed progressive improvements, moved from being a niche model to one of the company's most popular and distinctive models. Despite numerous imitators it is still the standard by which all diver's watches are judged.

The Sea Dweller
In 1951 Chrysler, the U.S. auto maker, introduced their new V8 engine. Called the "Hemi" because of its hemispherical combustion chamber, it was the first modern design V8 after General Motors' short stroke 1949 offering, first seen in the Cadillac. Its initial output of 180 h.p. was 20 more than Cadillac could offer and Chrysler made the most of it. Stock car racing, in which production based vehicles compete over a closed oval circuit, was just coming into favor and Chrysler chose to become part of it. Cadillac was not going to take it lying down and upped their engine to 190 h.p. for the 1952 line. By 1954 Chrysler was rating their engines at 245 h.p. and only four years later 400 h.p. was offered.

While this horsepower race was in full swing President Eisenhower put into motion the Highway Trust Fund Act of 1956 and the U.S. began to build an interstate highway network covering the whole country.

The effect of these two developments was that a lot of people began to travel these newly opened interstates, sometimes at speeds in excess of 100 m.p.h. A huge amount of gasoline went down the throats of those four-barrel carburetors. The U.S. was on the way to its first energy crisis. Domestic oil production, which had sustained the U.S. through World War II, had been seriously depleted and the major oil companies began to search the world for additional sources.

The oil companies had experience in drilling for oil offshore in places like the Gulf of Mexico but these were all comparatively shallow areas, with water depths of less than 200 feet. Initial testing had discovered probable reserves in much deeper waters, but new technology was required to reach them.

The first hurdle to be overcome was to find a way in which people could work at these depths for periods of two or three hours at a time; anything less than this made it very difficult for complex tasks to be completed. When a diver ascends from deep water too quickly, the sudden reduction of pressure releases the nitrogen compressed in the air he breathes and forces

Model 5513 Submariner with "Explorer" style dial. Note the bottom of the advertisement mentioning that the watch was also used by many other navies.

it out of suspension in the blood as bubbles. In simple terms, working deep underwater without taking precautions turns his blood into Perrier water with the obvious deleterious effects.

The problem was solved, first by switching the diver's air from a natural mixture of oxygen and nitrogen to an artificial mixture of oxygen and helium. Second, to avoid the time consuming task of decompressing after every dive, special living chambers were constructed. A diver would enter the chamber at the beginning of a work period of many days. The air in the chamber would gradually be changed to the oxygen-helium mixture and the air pressure would be increased to match that of the sea floor. This would be a gradual change, taking a few days to complete. When the correct conditions were met and the adjustments made by the diver, he would commute to and from the work site in a diving bell designed to maintain the same conditions. After his tour of duty he would return to the living quarters and the process would be reversed, gradually returning to a normal air mixture and pressure.

As ingenious as the system was it created unforeseen problems for the Submariner watch. While it was capable of working at the depths the divers did, the long exposure to high pressures had not been anticipated by its designers. Ironically it was the superior construction of the watch that led to the problems. The Submariner was designed to be waterproof and airtight and to survive the pressures of the deep water. It had no problem withstanding the increase in pressure, and over a short dive it maintained sea level pressure inside its works. Over the long period of these high pressure dives, the helium gas gradually permeated the crystal of the watch, equalizing its internal pressure with its surroundings.

This was no problem as long as the diver was in a high pressure environment. But when the chamber was decompressed, returning to sea level atmospheric pressure, it was not possible for the high pressure helium to escape from the watch as rapidly as the pressure in the chamber was dropping. The internal air pressure of the watch would simply blow the crystal straight out of the watch! This really is not a fun thing to happen to you at the best of times, but when you are living in a chamber the size of a couple of phone booths with two or three other guys for more than ten days at a time it is really not to be recommended.

The problem occurred with sufficient frequency for Comex (COmpagnie Maritime d'EXpertise), the French company who had developed the pressure chambers and employed all the divers, to decide to do something about it. They approached Rolex and the companies began working together on a solution. The first thing they did was to take one of the normal Submariners and modify it with a one way gas escape valve on the side of the case immediately opposite the winder. The modification was patented in Switzerland under number 492,246 and proved successful enough for Rolex to decide to make a production version of the Submariner fitted with this gas escape valve and a strengthened case, it was launched in 1971 as the model 1665 Sea Dweller.

First model 6204 Submariner. Note the parallel hands and simple bezel.

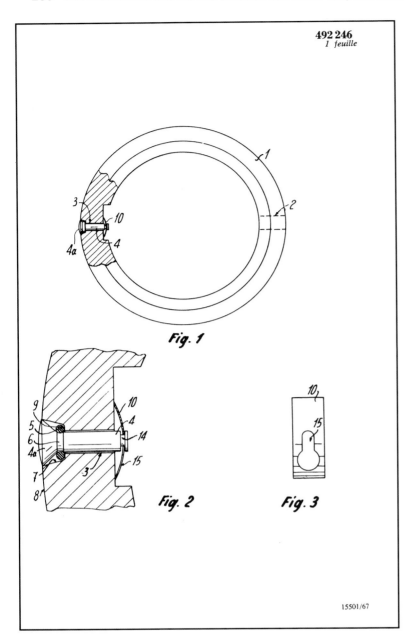

The red printing on the dial and the "Submariner 2000" disappeared in 1973 and in this revised form the watch continued until 1980, when it received its major overhaul and was promoted from model 1665 to 16660. The new watch was equipped with a rapid date changing mechanism, an artificial sapphire crystal, and a revised gas escape valve, and was now rated at 4,000 feet (or 1220 meters). This watch has recently been replaced by the latest model Sea Dweller, the 16600 which is fitted with the "fast-beat" 3135 calibre.

The Sea Dweller differs from a regular Submariner by being much thicker and heavier. It has no "cyclops" date lens because the glass on a Sea Dweller is thicker than on a Submariner. This would place the lens at a greater distance from the date disc, thereby being unable to focus correctly. A "cyclops" lens for a Sea Dweller would need to be almost twice the size of a regular one.

The other major difference is the bracelet, where both the end piece and the expansion link differ from the "Submariner" style bracelet. The end piece or "Flushfit" on all other Rolex watches is made from pressed stainless steel sheet metal, while on the Sea Dweller the piece is machined from solid metal, as are all the other links on all current Rolex bracelets. The expansion link (which allows the use of the watch over a diving suit) on a Submariner is made from two pieces of sheet steel; on the Sea-Dweller the second link from the clasp is machined from solid steel with a cut out for the expansion link spring bar.

Some of these prototype Sea Dwellers have surfaced and they are very interesting watches. They are model 5513, non-chronometer, plastic glass Submariners with the dial marked 660 feet/200 meters. There is no other marking on the dial, but the rear of the watch bears the full Comex markings and the side of the case reveals the gas escape valve in its original form. Comex divers were equipped with a variety of Rolex models, the Sea Dwellers were used only by the "saturation" divers, while the other "atmospheric" divers used standard Submariners (both date and non-date models). Due to the arduous conditions under which they operated, the Comex watches were withdrawn from service every six months and returned to the factory for a meticulous overhaul. Because of the relationship between Rolex and Comex, these watches were not just serviced, but upgraded. It is not unusual to see a 1970s watch with the later "hacking" movement and a 1990s dial. This was done because it was realized that the divers required the best equipment that was available and also because it would seem "penny-pinching" to deny the latest and best equipment to men for whom the watch literally meant the difference between life and death.

Guaranteed safe to depth of 2,000[56] feet (or 660 meters), the first Sea Dwellers were actually labelled as "Submariner 2000 Sea Dweller" on the dial printed in red. These early models are the most desirable of all the regular production Sea Dwellers.

Stainless steel Rolex Oyster Manual Precision, two-piece case, silver metal dial with triangular markers and luminous hands. Dial signed "Oyster 50 meters = 165 feet." It has always been assumed that the Turn-O-Graph was the first step on the road to the Submariner. This is false. They were in fact introduced at the same time in the 1954 Basel Show. The first antecedent of the Submariner was the new improved case introduced in 1952, which for the first time guaranteed that the Rolex Oysters would be waterproofed to at least 165 feet. With this new case it was not difficult to improve on the water resistance. The company doubled it and from there it was a small step to the first Submariner.
Circa 1954

The Sea Dweller is the direct lineal descendant of the octagon Oyster which Mercedes Gleitz wore when swimming the English Channel. It is the quintessential Rolex Oyster. Anything you may ask of a watch, the Sea Dweller provides, but at the cost of being the chunkiest watch Rolex has ever produced. It has what may be called "wrist presence," which is merely another way of saying it is a very big watch[57]. The watch, which has only been available in stainless steel[58], has never been out of the Rolex catalog in over 20 years, and yet its popularity can only be described as bizarre. If you think about it with any logic, the watch is around 10% more expensive than a Submariner, which to 99.99% of the world looks identical. The watch is thicker on the wrist than a Submariner and therefore less wearable. Its only advantage over a Submariner is its ability to undergo repeated compression/decompression cycles in a helium-oxygen atmosphere, not really something that most people have a regular use for. However, as a professional tool for the people for whom it was developed, the Sea Dweller is without peer; no diver working for Comex is allowed to dive without one on his wrist and as long ago as 1976 Comex divers had amassed over 250,000 hours of diving time wearing these watches.

Rolex Oyster Submariner Perpetual. Although this one is not signed Submariner on the dial, it has the much larger 8mm Twinlock style crown. It does not yet have a crown guard and still has no minute markings on the bezel, but the dial is the Explorer style dial which was used on all Submariners sold in Britain for the first four or five years.
Model no. 6200
Case no. 32120
Circa 1957

Stainless steel Rolex Oyster Perpetual Submariner. This is the second model 6536 Submariner recognizable by the small winder with no crown guard, the lack of any minute markings to the bezel, and the gilt writing on the dial for the markers and all other writing.
Circa 1956

---

[50] Piccard was an unusual character, more like a Swiss Indiana Jones (now there's a concept!) than a conventional scientist. In August, 1932 he took a balloon to the unknown heights of 10-1/2 miles. This made him, after his bathyscaphe journey, the holder of the world's records for both height and depth, a feat never achieved before or since.

[51] The skills used were not only from Rolex. Many other Swiss companies provided input, particularly in the form of ultra high pressure testing equipment.

[52] Jeanneret was much more than the man behind the Submariner. After the death of Wilsdorf he became the guiding light of the company and was instrumental in pushing it to develop the GMT-Master and the Explorer. He referred to these watches (and the Submariner) as "tool" watches, specifically designed to be used by practitioners of a single profession or hobby.

[53] This time the Trieste flew the U.S. flag. A month after the Soviet Union's launch of Sputnik, the U.S. Navy purchased the bathyscaphe and hired Piccard as a scientific consultant, although the dive was nominally under the command of Lt. Don Walsh USN.

[54] The "T" stands for tritium and is the material used to illuminate the hands and markers. The symbol is used on all modern Rolex watches, but is normally seen next to the "Swiss Made" at the very bottom of the dial in microscopic type.

[55] Although these watches all seem to have been made in the late 1960s and early 1970s, their serial numbers suggest that many were not issued until as long as 10 years later.

[56] This depth indicates just how over-engineered Rolex watches are. Six divers working for Comex, wearing Sea Dwellers of course, achieved the world diving record in 1975 at a depth of 1,070 feet or 325 meters. That is a little over half the capability of the first model Sea Dweller!

[57] Despite being the largest watch the company makes, it still looks more or less like a conventional Submariner. The Omega professional diver's watch, introduced at around the same time, is so large and cumbersome that it cannot be worn as a normal everyday watch.

[58] This fact marks the Sea-Dweller as one of the four "professional" watches Rolex makes. These are watches, which are not available in either gold or in two-tone are: the Sea-Dweller (although the Submariner is available in a "dress" version), the GMT-Master (although not the GMT-Master II), and both of the Explorer models. These are the watches Rene Jeanneret used to call the "tool" watches.

The Submariner now has acquired the minute markings on the first 15 minutes of the bezel, although the writing on the dial is still in gilt and the watch does not, as yet, have the crown guard.
Model no. 6538
Case no. 448951
Circa 1964

With this watch the Submariner acquires the characteristics which make it more like the current model. It now has the crown guard and the writing on the dial is in silver.
Model no. 6205
Case no. 21529
Circa 1964

When the new model Submariner acquired the crown guards, the less expensive model did not and it was only guaranteed to 100 meters or 330 feet as noted on the dial.
Model no. 6508 (Sometimes known as the James Bond Submariner).
Case no. 764019
Circa 1965

The later plastic glass Submariner now is beginning to look very similar to the watch we know today. It is still a 660 feet depth watch, but now has the crown guard, the marked bezel and the look and size of the current watch. However this Model 5513 still bears the Explorer style dial sold in Britain.
Case no. 6045010
Circa 1980

In the late 1960s the British Government acquired a number of Rolex Submariners especially configured for the Royal Navy Frogman branch. These watches differ considerable from the standard model. Note the solid bars and the unusual hour, minute and second hands, and of course they were engraved with full military markings on the rear.
Model no. 5513
Case no. 2669290
Circa 1967

A little later the British Government changed the specification on the watch and it now acquired a bezel which was marked at every minute. However, when these watches were declared surplus and sold on the market, many of them were sent back to Rolex for refurbishment and when Rolex refurbished the watches they automatically changed the hands to the standard model. This has happened with this watch although every other part of the watch is standard military specification. It is still a 5513.
Circa 1975

The Submariner acquired a Date disk and the new name Oyster Perpetual Date Submariner. On some of the first of these models the name Submariner was in red and these of course are much more collectible than the standard model.
Model no. 1680
Circa 1974

Another Rolex Military Submariner Model 5513 with the earlier standard style bezel and the normal hands.

This is another 5513 Military specification Submariner, but one that is absolutely unissued and therefore in pristine condition. It does not even bear any engraving on the back as it was never issued.
Circa 1970

This is the current Oyster Perpetual Date Submariner. Now waterproofed to 1,000 feet with sapphire glass and quick set date.
Circa 1980

The Submariner and the Sea-Dweller: The Stock Car, The Diver, and The Gas Escape Valve

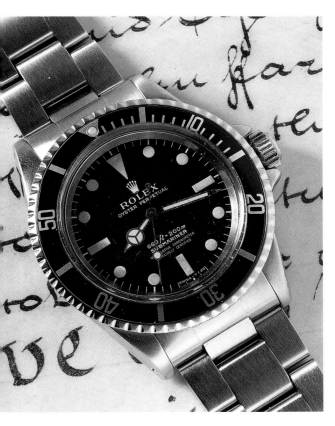

Stainless steel non-date Submariner, however note the dial is signed "Superlative Chronometer Officially Certified." These chronometer versions are quite rare and have their own model number 5512, whilst the non chronometer version is, of course, numbered 5513.
Model no. 5512
Case no. 2509528
Circa 1969

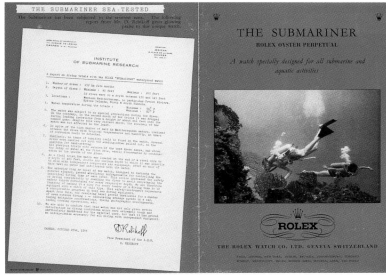

The earliest known Submariner brochure from 1954, showing the first model 6204 with parallel hands and simple bezel. Note the similarity of this watch to the model 6202 "Turn-O-Graph" shown elsewhere.

Late November, 1966 Submariner brochure showing the 5512 (chronometer) and 5513 (non-chronometer) models. It is interesting that the diver on the front cover is wearing a much earlier model 6538 which has no crown guards, on a NATO strap on his right arm. The watch he is wearing also has the rare "Explorer" style dial.

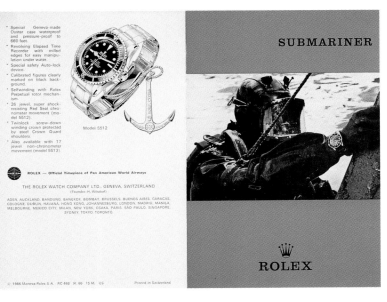

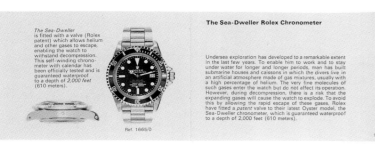
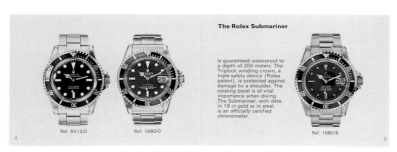

May, 1975 brochure showing the Submariner 1680 (with dial nomenclature in red) and the first Sea-Dweller 1665. The gold watches shown already have the luminous indices in gold settings, while on the steel watches the luminosity is merely painted on the dial.

Current steel and gold Submariner, silvered dial with applied sapphire and diamond markers (known as the "Sultan" dial) and blue bezel.
Model no. 161133
Current Model

Steel and 18kt yellow gold current model Oyster Perpetual Submariner, hack feature, quickset date feature and sapphire glass.
Model no. 16803
Current Model

Stainless steel British Royal Marines Submariner, with different bezel, dial, hands and the solid lugs. Interestingly, although the cases on these watches are marked as 5513 (non-chronometer) watches, they are fitted with hacking seconds chronometer movements, the only plastic glass Submariners to be so fitted.
Model no. 5513
Circa 1970

Model 5513 Submariners in stainless steel, early model bottom and late model top. The difference in thickness between these watches shows most notably in the profile of the crown guards. Also, the early crown does not have the three dots signifying a triplock crown, although it is the same size.

Rolex Oyster Perpetual Submariner with "Explorer" style dial. This watch is one of the earliest of the 5513 models and also is fitted with an early riveted style expanding bracelet.
Model no. 5513
Circa 1970

Stainless steel Submariner issued to a Comex diver. Although the watch has a plastic crystal, it has the much later dial with the indices in white metal settings. This is because Comex watches are always returned to the factory for servicing and, while there, will always be fitted with the latest updates. The case back is also marked Comex, this time it is engraved. Strangely no issue number usually is found on Comex watches.
Model no. 5513
Circa 1974

The very first prototypes of the Sea-Dweller were in fact Submariners with a built-in gas escape valve. They were built for the diving company Comex and looked exactly like a 5513 and in fact bear that number. They are identical except for the small, almost hand made, gas escape valve at the nine position. Circa 1968

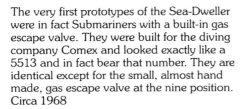

The Sea-Dweller and the gas escape valve converted watches did not completely supplant the normal Submariners within Comex and a number of standard plastic glass Submariners were still made. This is one of them. Circa 1968

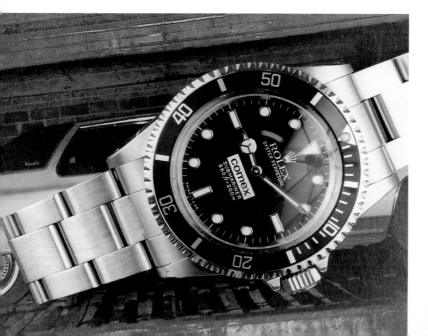

Oyster Perpetual Date Sea-Dweller Submariner 2000. As is evident from the name of this watch the very fist Sea-Dwellers were in fact versions of the Submariner. With the gas escape valve and the heavy construction they were now warranted to 2000 feet. Notice the lack of a cyclops magnifier on the glass.
Model no. 1665
Case no. 3078339
Circa 1971

Second model Sea-Dweller. The dial is no longer signed "Submariner 2,000."
Model no. 1655
Case no. 5619287
Circa 1980

Even now, Comex equips its long term saturation divers with the Rolex Sea-Dweller. These models are standard production watches with the only variations being the dial and the rear of the case, which are customized for the company.
Circa 1988

NATIONAL PHYSICAL LABORATORY
Teddington   Middlesex   TW11 0LW   England

Report

TEST OF A ROLEX WRIST WATCH

A wrist-watch, movement number 4272, was submitted to the A test at the National Physical Laboratory, Teddington, in 1949, by the Rolex Watch Co and as a result of the test a Class A especially good Kew Certificate was issued for the watch.

The following information is abstracted from the records of the test:-

The mean rate, in the pendant* up position was + 3.1 seconds per day.
The mean rate, in the pendant* right position was + 2.5 seconds per day.
The mean rate, in the pendant* left position was + 3.1 seconds per day.
The mean rate, in the dial up position was + 1.6 seconds per day.
The mean rate, in the dial down position was - 1.3 seconds per day.
The mean variation of rate (average for all periods) was 0.49 seconds per day.
The mean change of rate per 1° Fahr. was 0.061 seconds per day.
The maximum difference between any two individual rates during the test was 6.9 seconds. per day.

(The + sign denotes a gaining rate.)

* In the case of wrist watches, 6 (vi) o'clock uppermost corresponds to pendant up.

<u>Marks awarded</u>

In respect of consistency of rate	30.2
In respect of consistency of rate with change of position	34.9
In respect of temperature compensation	16.0
Total marks	81.1

Note: It should be understood that the results given in this Report do not necessarily represent the present performance of the watch.

DATE: Issued 6 February 1992

SIGNED   B. R. Stoubey for Director

This Report may not be published except in full, unless permission for the publication of an approved extract has been obtained in writing from the Director.

CHAPTER 13
THE KEW "A"
The Rarest Rolex

In the early 1940s, during the war, Rolex gave up signing the inside of their watches with the number of world records they had achieved and simply signed them with the new company name, "Rolex SA." This did not mean that Rolex had given up on trying to obtain worlds records. As the forties slipped into the fifties Rolex undertook a challenge that had never been attempted before (or since). They produced a series of 140+ watches and submitted them to the Royal Observatory at Kew for chronometer testing. Almost all of them were model 5056 boy's size Oysters with simple dials signed "Oyster Speedking." These watches were fitted with 10-1/2''' movements with hand made balance staffs, wheels, and cocks. Amazingly, practically every one of the watches passed and received the coveted Kew "A" certificates, with 16 of the batch also obtaining the distinction "especially good results," meaning that they produced ratings in the top 20% of the accuracy needed to obtain a Kew "A" certificate. A certificate for one of these watches is shown and it is amazing to see that the mean average daily variation was less than half a second. While there is no proof, it appears that the watches were made to celebrate the return of Rolex watches to Britain.

The construction of the watches was very unusual, in that it utilized the basic components from two different models, with the base plate from classic 10-1/2''' Hunter and the gear train, bridges, and cocks from the 720 perpetual. The movements were instantly identifiable by virtue of their two color finish, with the main plates being colored rose gold with the balance cock only in stainless steel. The movement was 18 jewels with a fine seconds train of 18,000 beats per hour. The pallets were very fine bevelled and cross cut steel with a very delicate nickel/brass bi-metallic cut balance wheel which had 20 mass screws and 2 regulating timing screws with 3 spare screw holes each side for compensation. The balance staff was an extremely fine one with 7/100 pivots. The double roller was rectangular with a fine ruby pin. The balance spring was a free sprung Breguet type made from Nivarox, and the balance jewels were not shock protected. The top endstone is mounted on to a very distinguishable polished steel plate screwed onto the top surface of the cock with two screws.

The parts share one intriguing detail: they all tell a lie! Because Rolex was using production parts wherever possible, the main winding wheel is inscribed, as all contemporary 10-1/2''' Hunter movements were, "Patented Superbalance." While, of course, the balances were the crowning achievement of these movements, they were not superbalances. They were one of the few Rolex watches where a glass exhibition back would be an asset. The craftsmanship evident in them is exquisite to behold.

These watches were all adjusted and timed by Jean Matile, the master timer of Rolex. He had timed all of Rolex's world record watches from 1925 onwards, including the 500 Princes consecutively numbered from 501 to 1,000. Produced for the Silver Jubilee of George V in 1935, all of them received the accolade of "Especially Good Results" from the Bienne Rating office. The feat was even more amazing when you consider that these watches were produced in the period of 146 days. It would seem that Matile had a group of "régleurs", the athletes of chronometry, as David Landes called them, working with him to achieve the output he did. This hypothesis can be confirmed by the fact that each of the watches was marked by the "régleur" before he assembled the watch. He would mark the hidden underside of the balance cock with the serial number of the movement. So far the author has seen three different handwriting styles on the seven movements he has had the privilege of examining.

The first boy's Oyster arrived in Teddington, where the "Kew" tests had been done since 1912, on January 14, 1947, in a group of four watches. Two weeks later another four arrived; each group of four averaged 50% with two certificates and two failures per group.

The watches which failed these stringent timekeeping tests often did so because their extremely delicate balance staffs were displaced in transit from Geneva to London. Despite the most painstaking packing, it was inevitable that some of the watches did not arrive in the condition they were in when they left. The pieces that failed were returned to Geneva where the "régleurs" went back to work on them and soon they were on their way back to the testing benches at Teddington. While some watches took as little as eleven days to go from Teddington to Geneva and back, others took almost a year. The watches normally spent between 10 and 15 weeks back in Geneva. Many watches made the journey so many times they could probably have found their own way! One watch (4187) travelled nine times, it was one of the second batch sent on January 28, 1947, and it finally received its certificate on July 1, 1950. Two watches made eight journeys. One of them (4216,) when it finally received its certificate, proved to have obtained the second highest rating of the entire production, 85.3. While four watches made seven journeys and three made six, most watches obtained their certificates on their first or second time on the bench.

Out of the total number of 145 watches submitted only nine did not eventually gain a certificate and these probably only because Teddington ceased to perform the Kew "A" test in early 1951, before these last few pieces could be retested. Rather than concentrating on these few rare failures, we should congratulate the "régleurs" of Rolex and their master, Jean Matile, for their success in producing 136 watches that all obtained that coveted piece of paper and the sixteen that gained their rightful places in that "pantheon" of timekeeping, places given solely to the bearers of Kew "A" Especially Good certificates.

After obtaining their certificates, the movements were returned to Switzerland and cased prior to sale. Because some movements required multiple testing it was quite common for watches with sequential movement numbers to have case numbers all over the place. For example movement number 4261 is only 4 digits later than 4257 yet it has a case serial number almost 700 earlier. This is because 4261 needed two attempts to pass, over a seven month period, while 4257 sailed straight through on its first attempt.

Kew Tests and Their Results

DATE SUBMITTED	MVMT NO.	RESULT	DATE	NO. OF ATTEMPTS	DATE SUBMITTED	MVMT NO.	RESULT	DATE	NO. OF ATTEMPTS
01-Apr-48	4148	Fail	02-Jun-48	1	23-Nov-49	4182	Fail	25-Jan-50	4
29-Sep-47	4151	A	27-Nov-47	1	25-Mar-50	4182	Fail	02-Jun-50	5
29-Oct-47	4152	A	30-Dec-47	1	30-Jan-51	4182	A e/g 80.5	06-Apr-51	6
26-Feb-48	4153	A	26-Apr-48	1	26-Jul-48	4183	A	30-Sep-48	1
01-Apr-48	4154	Fail	02-Jun-48	1	12-May-48	4184	A	13-Jul-48	1
26-Jul-48	4154	Fail	30-Sep-48	2	06-Jul-49	4185	Fail	13-Sep-49	1
01-Mar-49	4154	A	02-May-49	3	13-Jun-48	4186	Fail	31-Jul-48	2
16-Oct-47	4155	A	05-Dec-47	1	16-Dec-48	4186	A	25-Feb-49	3
14-Sep-48	4156	Fail	17-Nov-48	1	28-Jan-47	4187	Fail	31-Mar-47	1
03-Sep-49	4156	Fail	02-Nov-49	2	04-Jul-47	4187	Fail	12-Sep-47	2
14-Jun-50	4156	Fail	23-Aug-50	3	13-Nov-47	4187	Fail	21-Jan-48	3
30-Jan-51	4156	A	06-Apr-51	4	11-Mar-48	4187	Fail	20-May-48	4
25-Nov-47	4157	Fail	27-Jan-48	1	30-Aug-48	4187	Fail	30-Nov-48	5
28-Apr-48	4157	Fail	08-Aug-48	2	30-Dec-48	4187	Fail	07-Mar-49	6
08-Nov-48	4157	Fail	20-Jan-49	3	06-Jul-49	4187	Fail	13-Sep-49	7
12-May-49	4157	Fail	20-Jul-49	4	02-Jan-50	4187	Fail	10-Mar-50	8
01-Dec-49	4157	A	01-Feb-50	5	28-Apr-50	4187	A	01-Jul-50	9
01-Jan-48	4158	Fail	02-Mar-48	1	14-Jan-47	4188	A	05-Mar-47	1
12-May-48	4158	A	13-Jul-48	2	25-Feb-47	4189	Fail	13-May-47	1
01-Jan-48	4159	Fail	02-Mar-48	1	30-Jun-47	4189	Fail	29-Aug-47	1
28-Apr-48	4159	Fail	08-Aug-48	2	13-Jan-48	4189	Fail	15-Mar-48	1
14-Sep-48	4159	Fail	17-Nov-48	3	13-Apr-48	4189	Fail	08-Jun-48	2
06-Jul-49	4159	A	13-Sep-49	4	26-Jul-48	4189	Fail	30-Sep-48	3
12-May-48	4160	Fail	13-Jul-48	1	30-Dec-48	4189	Fail	07-Mar-49	4
30-Dec-48	4160	Fail	07-Mar-49	2	12-May-49	4189	Fail	20-Jul-49	5
06-Jul-49	4160	Fail	13-Sep-49	3	13-Sep-49	4189	Fail	29-Nov-49	6
28-Apr-50	4160	A	01-Jul-50	4	28-Apr-50	4189	A	01-Jul-50	7
25-Nov-47	4161	Fail	27-Jan-48	1	14-Jan-47	4190	Fail	05-Mar-47	1
11-Mar-48	4161	A	30-Apr-48	2	14-Jun-47	4190	A	08-Aug-47	2
01-Jan-48	4162	Fail	02-Mar-48	1	14-May-47	4191	Fail	29-Jun-47	1
26-Jul-48	4162	Fail	30-Sep-48	2	08-Sep-47	4191	Fail	31-Oct-47	2
18-Jan-49	4162	A	02-Apr-49	3	26-Feb-48	4191	A e/g 81	26-Apr-48	3
13-Jun-48	4163	Fail	31-Jul-48	1	14-Jan-47	4192	A	05-Mar-47	1
30-Sep-48	4163	Fail	10-Dec-48	2	25-Feb-47	4193	A	28-Apr-47	1
14-Apr-49	4163	A e/g 82.7	05-Jul-49	3	28-Jan-47	4194	Fail	31-Mar-47	1
27-May-48	4164	Fail	20-Jul-48	1	04-Jul-47	4194	A e/g 80.5	12-Sep-47	2
14-Sep-48	4164	Fail	17-Nov-48	2	14-Jan-47	4195	Fail	05-Mar-47	1
06-Jul-49	4164	A	13-Sep-49	3	08-Sep-47	4195	Fail	31-Oct-47	2
26-Feb-48	4165	Fail	26-Apr-48	1	11-Mar-48	4195	A	20-Jul-48	3
27-May-48	4165	A	20-Jul-48	2	25-Feb-47	4196	Fail	28-Apr-47	1
28-Jan-48	4166	A	21-Mar-48	1	30-May-47	4196	Fail	25-Jul-47	2
10-Feb-48	4167	A	08-Apr-48	1	29-Oct-47	4196	A	30-Dec-47	3
13-Nov-47	4168	Fail	21-Jan-48	1	28-Jan-47	4197	A	16-Mar-47	1
26-Feb-48	4168	A	26-Apr-48	2	08-Sep-47	4198	Fail	31-Oct-47	1
01-Jan-48	4169	Fail	02-Mar-48	1	01-Apr-48	4198	A	02-Jun-48	1
27-May-48	4169	A	20-Jul-48	2	27-May-48	4199	Fail	20-Jul-48	1
11-Mar-48	4170	Fail	20-May-48	1	16-Dec-48	4199	Fail	25-Feb-49	2
14-Sep-48	4170	Fail	17-Nov-48	2	03-Sep-49	4199	A	02-Nov-49	3
03-Sep-49	4170	Fail	02-Nov-49	3	13-Jun-48	4200	Fail	31-Jul-48	1
13-Feb-50	4170	A	26-Apr-50	4	08-Nov-48	4200	Fail	20-Jan-49	2
28-Apr-48	4171	Fail	08-Aug-48	1	13-Sep-49	4200	Fail	29-Nov-49	3
12-Oct-48	4171	Fail	30-Dec-48	2	29-Jun-48	4201	Fail	01-Sep-48	1
06-Jul-49	4171	Fail	13-Sep-49	3	16-Dec-48	4201	A	25-Feb-49	2
28-Apr-50	4171	Fail	01-Jul-50	3	13-Jun-48	4202	A	31-Jul-48	1
30-Jan-51	4171	Fail	06-Apr-51	4	26-Jul-48	4203	Fail	30-Sep-48	1
10-Feb-48	4172	Fail	08-Apr-48	1	01-Mar-49	4203	Fail	02-May-49	2
12-May-48	4172	A	13-Jul-48	2	23-Nov-49	4203	Fail	25-Jan-50	3
11-Mar-48	4173	A	30-Apr-48	1	25-Mar-50	4203	A	02-Jun-50	4
28-Jan-48	4174	Fail	01-Apr-48	1	26-Jul-48	4204	Fail	30-Sep-48	1
27-May-48	4174	A	20-Jul-48	1	18-Jan-49	4204	Fail	02-Apr-49	2
29-Jun-48	4175	Fail	01-Sep-48	2	12-May-49	4204	Fail	20-Jul-49	3
12-Oct-48	4175	A	30-Dec-48	3	23-Nov-49	4204	Fail	25-Jan-50	4
12-Oct-48	4176	Fail	30-Dec-48	2	25-Mar-50	4204	Fail	02-Jun-50	5
03-Sep-49	4176	Fail	02-Nov-49	3	29-Jun-48	4205	A	01-Sep-48	1
14-Jun-50	4176	A	23-Aug-50	4	13-Jun-48	4206	A	31-Jul-48	1
13-Jun-48	4177	Fail	31-Jul-48	1	13-Jun-48	4207	A	31-Jul-48	1
16-Dec-48	4177	Fail	25-Feb-49	2	27-May-48	4208	A	20-Jul-48	1
12-May-49	4177	Fail	20-Jul-49	3	30-Aug-48	4209	A	30-Nov-48	1
01-Dec-49	4177	Fail	01/Feb-50	4	13-Jun-48	4210	Fail	31-Jul-48	1
15-Sep-50	4177	Fail	23-Nov-50	5	12-Oct-48	4210	A	30-Dec-48	2
30-Jan-51	4177	Fail	06-Apr-51	6	25-Feb-47	4211	A e/g 81.1	28-Apr-47	1
26-Jul-48	4178	A e/g 82.3	30-Sep-48	1	14-May-47	4212	A	29-Jun-47	1
27-May-48	4179	Fail	20-Jul-48	1	30-May-47	4213	Fail	25-Jul-47	1
06-Jul-49	4179	Fail	13-Sep-49	2	11-Sep-47	4213	Fail	07-Nov-47	2
02-Jan-50	4179	A	10-Mar-50	3	26-Jul-48	4213	A	30-Sep-48	3
27-May-48	4180	A	20-Jul-48	1	25-Feb-47	4214	Fail	13-May-47	1
28-Apr-48	4181	A	08-Aug-48	1	12-Aug-47	4214	Fail	08-Oct-47	2
28-Apr-48	4182	Fail	08-Aug-48	1	26-Jul-48	4214	Fail	30-Sep-48	3
30-Sep-48	4182	Fail	10-Dec-48	2	01-Mar-49	4214	A	02-May-49	4
01-Mar-49	4182	Fail	02-May-49	3	25-Apr-47	4215	A e/g 81.2	26-Jun-47	1

DATE SUBMITTED	MVMT NO.	RESULT	DATE	NO. OF ATTEMPTS
25-Feb-47	4216	Fail	13-May-47	1
25-Nov-47	4216	Fail	27-Jan-48	2
13-Jun-48	4216	Fail	31-Jul-48	3
12-Oct-48	4216	Fail	30-Dec-48	4
14-Apr-49	4216	Fail	05-Jul-49	5
13-Sep-49	4216	Fail	29-Nov-49	6
13-Feb-50	4216	Fail	26-Apr-50	7
15-Sep-50	4216	A e/g 85.3	23-Nov-50	8
28-Feb-47	4217	A	20-May-47	1
28-Feb-47	4218	A	20-May-47	1
28-Feb-47	4219	Fail	20-May-47	1
29-Sep-47	4219	A	27-Nov-47	2
25-Feb-47	4220	Fail	28-Apr-47	1
30-Jun-47	4220	Fail	29-Aug-47	2
30-Aug-48	4220	A	30-Nov-48	3
25-Feb-47	4221	Fail	13-May-47	1
30-Jun-47	4221	Fail	29-Aug-47	2
13-Nov-47	4221	A	21-Jan-48	3
25-Feb-47	4222	A e/g 81.2	28-Apr-47	1
25-Feb-47	4223	A e/g 84.9	03-May-47	1
14-May-47	4224	A	29-Jun-47	1
12-Aug-47	4225	Fail	08-Oct-47	1
28-Jan-48	4225	Fail	01-Apr-48	2
26-Jul-48	4225	A	30-Sep-48	3
30-May-47	4226	A	25-Jul-47	1
25-Apr-47	4227	A	26-Jun-47	1
30-Jun-47	4228	Fail	29-Aug-47	1
25-Nov-47	4228	Fail	27-Jan-48	2
28-Apr-48	4228	A	08-Aug-48	3
25-Feb-47	4229	A	03-May-47	1
04-Jul-47	4230	Fail	12-Sep-47	1
29-Oct-47	4230	A	30-Dec-47	2
14-Jun-47	4231	A	08-Aug-47	1
25-Feb-47	4232	Fail	13-May-47	1
04-Jul-47	4232	Fail	12-Sep-47	2
13-Nov-47	4232	Fail	21-Jan-48	3
01-Apr-48	4232	Untested	02-Jun-48	4
06-Jul-49	4232	Fail	13-Sep-49	5
01-Dec-49	4232	Fail	01-Feb-50	6
15-Sep-50	4232	Fail	23-Nov-50	7
30-Jan-51	4232	Fail	06-Apr-51	8
24-Jul-47	4233	A	26-Sep-47	1
25-Feb-47	4234	A	03-May-45	1
13-Nov-47	4235	A	21-Jan-48	1
12-Oct-48	4236	A	30-Dec-48	1
08-Sep-47	4237	Fail	31-Oct-47	1
11-Mar-48	4237	A	20-Jul-48	2
08-Sep-47	4238	Fail	31-Oct-47	1
11-Mar-48	4238	Fail	20-May-48	2
12-Oct-48	4238	Fail	30-Dec-48	3
06-Jul-49	4238	Fail	13-Sep-49	4
02-Jan-50	4238	A	10-Mar-50	5
29-Sep-47	4239	A	27-Nov-47	1
16-Oct-47	4240	Fail	05-Dec-47	1
13-Apr-48	4240	A	08-Jun-48	2
08-Sep-47	4241	Fail	31-Oct-47	1
13-Jan-48	4241	Fail	15-Mar-48	1
27-May-48	4241	Fail	20-Jul-48	2
30-Aug-48	4241	Fail	30-Nov-48	3
01-Mar-49	4241	Fail	02-May-49	4
13-Sep-49	4241	Fail	29-Nov-49	
30-Jan-50	4241	Fail	04-Apr-50	5
15-Sep-50	4241	A e/g 80.2	23-Nov-50	6
13-Jan-48	4242	A	15-Mar-48	1
28-Jan-48	4243	A	21-Mar-48	1
13-Nov-47	4244	Fail	21-Jan-48	1
01-Apr-48	4244	A	02-Jun-48	1
13-Nov-47	4245	A	21-Jan-48	1
01-Jan-48	4246	Fail	02-Mar-48	1
13-Jun-48	4246	A	31-Jul-48	2
30-Jun-47	4247	A	21-Aug-47	1
29-Oct-47	4248	Fail	30-Dec-47	1
14-Sep-48	4248	Fail	17-Nov-48	2
14-Apr-49	4248	A	05-Jul-49	3
08-Sep-47	4249	Fail	31-Oct-47	1
26-Feb-48	4249	Fail	26-Apr-48	2
27-May-48	4249	Fail	20-Jul-48	3
30-Aug-48	4249	Fail	30-Nov-48	4
18-Jan-49	4249	Fail	02-Apr-49	5
06-Jul-49	4249	Fail	13-Sep-49	6
23-Nov-49	4249	A	25-Jan-50	7
30-Jun-47	4250	A	21-Aug-47	1
24-Jul-47	4251	Fail	26-Sep-47	1
01-Jan-48	4251	A	02-Mar-48	2
26-Jul-48	4252	A	30-Sep-48	1
14-Jun-47	4253	Fail	08-Aug-47	1
16-Oct-47	4253	Fail	05-Dec-47	2
29-Jun-48	4253	Fail	01-Sep-48	3
12-Oct-48	4253	A	30-Dec-48	4
04-Jul-47	4254	Fail	12-Sep-47	1
13-Nov-47	4254	A	21-Jan-48	2
29-Sep-47	4255	Fail	27-Nov-47	1
10-Feb-48	4255	Fail	08-Apr-48	2
29-Jun-48	4255	Fail	01-Sep-48	3
08-Nov-48	4255	A	20-Jan-49	4
11-Sep-47	4256	Fail	07-Nov-47	1
30-Jan-50	4256	Fail	04-Apr-50	2
24-Jul-47	4257	A	26-Sep-47	1
24-Jul-47	4258	Fail	26-Sep-47	1
25-Nov-47	4258	A	27-Jan-48	2
14-May-47	4259	A	29-Jun-47	1
26-Feb-48	4259	Fail	26-Apr-48	1
12-May-49	4259	Fail	20-Jul-49	1
25-Feb-47	4260	A	28-Apr-47	1
30-Jun-47	4261	Fail	29-Aug-47	1
25-Nov-47	4261	A	27-Jan-48	2
04-Jul-47	4262	A	12-Sep-47	1
25-Feb-47	4263	Fail	28-Apr-47	1
29-Sep-47	4263	Fail	27-Nov-47	2
16-Oct-47	4263	A e/g 81.3	05-Dec-47	3
07-Feb-47	4264	A	14-Apr-47	1
28-Feb-47	4265	A	03-May-47	1
25-Feb-47	4266	Fail	28-Apr-47	1
14-Jun-47	4266	A e/g 87.2	08-Aug-47	2
07-Feb-47	4267	A	14-Apr-47	1
28-Jan-47	4268	A	16-Mar-47	1
28-Feb-47	4269	Fail	20-May-47	1
12-Aug-47	4269	A	08-Oct-47	2
25-Feb-47	4270	A	03-May-47	1
26-Jul-48	4271	Fail	30-Sep-48	1
01-Mar-49	4271	A	02-May-49	2
30-Aug-48	4272	Fail	30-Nov-48	1
18-Jan-49	4272	Fail	02-Apr-49	2
06-Jul-49	4272	A e/g 81.1	13-Sep-49	3
26-Jul-48	4273	Fail	30-Sep-48	1
01-Mar-49	4273	A	02-May-49	2
26-Jul-48	4274	A e/g 80.0	30-Sep-48	1
30-Sep-48	4275	Fail	10-Dec-48	1
06-Jul-49	4275	A	13-Sep-49	2
30-Sep-48	4276	A	10-Dec-48	1
30-Aug-48	4277	A	30-Nov-48	1
30-Aug-48	4278	Fail	30-Nov-48	1
18-Jan-49	4278	A	02-Apr-49	2
14-Sep-48	4279	Fail	17-Nov-48	1
14-Apr-49	4279	A e/g 80.8	05-Jul-49	2
30-Aug-48	4280	Fail	30-Nov-48	1
30-Dec-48	4280	Fail	07-Mar-49	2
06-Jul-49	4280	Fail	13-Sep-49	3
30-Jan-50	4280	Fail	04-Apr-50	4
14-Jun-50	4280	Fail	23-Aug-50	5
30-Jan-51	4280	Fail	06-Apr-51	6
30-Sep-48	4281	Fail	10-Dec-48	1
06-Jul-49	4281	A	13-Sep-49	2
30-Sep-48	4282	A	10-Dec-48	1
24-Jul-47	4283	Fail	26-Sep-47	1
13-Apr-48	4283	A	08-Jun-48	2
30-May-47	4284	Fail	25-Jul-47	1
11-Sep-47	4284	Fail	07-Nov-47	2
13-Apr-48	4284	A	08-Jun-48	3
25-Apr-47	4285	Fail	26-Jun-47	1
24-Jul-47	4285	A	26-Jul-47	2
08-Sep-47	4286	Fail	31-Oct-47	1
11-Mar-48	4286	A	30-Apr-48	2
30-May-47	4287	A	25-Jul-47	1
13-Nov-47	4288	Fail	21-Jan-48	1
12-May-48	4288	Fail	13-Jul-48	
30-Sep-48	4288	A	10-Dec-48	2
30-Jun-47	4289	A	21-Aug-47	1
14-Jun-47	4290	A	08-Aug-47	1
14-May-47	4291	A	29-Jun-47	1
14-Jun-47	4292	Fail	08-Aug-47	1
13-Nov-47	4292	Fail	21-Jan-48	2
28-Apr-48	4292	Fail	08-Aug-48	3
30-Sep-48	4292	A	10-Dec-48	4
11-Sep-47	4293	Fail	07-Nov-47	1
10-Feb-48	4293	A	08-Apr-48	2
04-Jul-47	4294	A	12-Sep-47	1

274 The Kew, "A": The Rarest Rolex

The wonder watch that defies the elements

HERE IS the Rolex Oyster, first and most famous *waterproof* wrist-watch in the world.

How was such a watch made a reality? It was the result of *years* of experiment by Rolex technicians. Imagine their excitement when, in 1927, Miss Mercedes Gleitze, a London stenographer, startled the world by swimming the English Channel wearing . . . a Rolex Oyster!

Rolex had perfected their unique waterproofing method— it *permanently* protects the movement's accuracy against dirt and moisture. No wonder the Rolex Oyster is world-famous!

A NEW ROLEX TRIUMPH! *The coveted Kew 'A' Certificate has been awarded to 116 Rolex Oyster wrist-watches during recent tests.*

ROLEX
Leaders in fashion and precision
THE ROLEX WATCH COMPANY LIMITED (H. WILSDORF, *Governing Director*)

While the theme of the advertisement is Mercedes Gleitze's 1927 swim of the English Channel, the award of a Kew A Certificate to 116 Rolex wristwatches earns a sidebar.

Movement case and dial of two Kew A certificated watches, having passed the stringent rules of the toughest timekeeping trials in the world. The watch on the left is no. 4272. Right is no. 4237. The close-up is of watch 4272.
1949

Photographs showing Kew A certificate watch no. 4237 in detail notice the pink main bridges and the steel free sprung balance cock. These watches were completely handmade by only the best watchmakers at the Rolex factory.
1948

The Kew, "A": The Rarest Rolex

Kew A certificated watch no. 4257. It is interesting to note the simplicity of the case and dials of these watches, in contrast to the exquisite finish of the movement.
1947

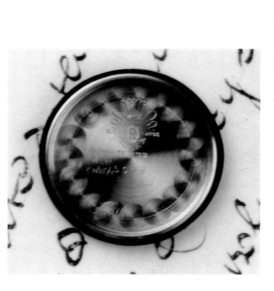

Internal and external details of Kew A certificate no. 4272.
This watch was one of only 16 to achieve the rare distinction of an "Especially Good" certificate. An "Especially Good" rating from Kew was the highest rating available in the world at the time. The mean variation of rate was only 0.49 seconds per day, an extraordinary feat in 1949.
1949

ROLEX AND NASA

CHAPTER 14
THE GMT-MASTER
The Correct Time Times Two

On the 15th of July, 1954 a new Boeing airplane took off from the company's strip at Renton, Washington. Built to the U.S. Air Force type request number KC 135, it was a four jet engine, air-to-air tanker designed to extend the range of the new B47 jet bomber to intercontinental range. A year later the government, who had funded the development of the KC 135, allowed Boeing to begin development of a commercial passenger version. This new version, renamed the 707, became the world's first successful intercontinental jet passenger aircraft. The first airline to fly the new plane was Pan American Airways in 1956. The world's largest airline, Pan Am soon instituted transcontinental services and shortly afterwards began to fly the Atlantic from New York's Idlewild to London and Paris.

Travelers loved the fact that trans-Atlantic journey times were now halved, from 13 hours to seven, but they suffered from a new phenomenon: "Jet Lag." Simply put, the body's internal rhythms were knocked off balance by the sudden difference between time zones 5 or 6 hours apart. Pan-American was worried enough about the effect on its pilots to commission research, and the result was that they were advised to keep their pilots on "home" time during the period they were away from base. At the same time, they needed to be aware of the local time at their various venues.

To meet both of these needs Pan Am asked Rolex to develop a watch. A joint task force from both companies was assigned the problem. Pan Am's team was lead by the inimitable Captain Frederick Libby, one of the original "Skygods" (the name other Pan Am pilots gave to the original cadre of pilots who had flown for Pan Am before World War II), while the Rolex team was led by the redoubtable René-P. Jeanneret. Jeanneret was called the company's Head of Public Relations, but in fact was the company's Jack-of-All-Trades (and master of all of them)[59]. In remarkably short order the task force produced the "G.M.T. Master" (model no. 6542), named after Greenwich Mean Time, the world's standard time, the GMT-Master was a heavily revised version of the "Turn-O-Graph" (model 6202) with the calibre 1035 movement upgraded with the addition of a supplementary 24 hour driving wheel. The new movement, now called the calibre 1065, powered a watch that featured four hands, having a 24 hour hand in addition to the normal hour, minute and seconds hands. This new hand was complemented by a rotatable 24 hour bezel. Together they enabled the wearer to read the time in any two time zones. Produced only until 1959, the model 6542 is the rarest of all GMT-Masters. It is immediately recognizable by the absence of protective shoulders at the crown and by its plastic (rather than painted) 24 hour bezel. The 6542 is also slightly smaller, by 2mm., than the 1675 which replaced it in 1960.

The new watch was powered by a new movement too. The calibre 1565 was based on the new 1530 calibre, which had been introduced a year or so previously. The new watch was distinguished by the much brighter, screen printed 24 hour bezel insert and by the protected crown which made the watch look similar to the Submariner.

The untold story of the GMT-Master is of its role in the United States' space missions of the late 1960s and early 1970s. Everyone knows that the NASA astronauts were issued with Omega Speedmasters, but anyone who has read Tom Wolfe's definitive book *The Right Stuff* or who has seen Philip Kauffman's movie made from the book will know that the astronauts were not men who accepted authority easily. They were mostly ex-fighter or test pilots (or both) and most definitely not conformists. Instead they were the last of the rugged individualists with a long tradition of obeying orders only when they agreed with them. Following the tradition started by RAF pilots during World War II, U.S. pilots also wore Rolex watches, and those who later became test pilots continued the tradition. Chuck Yeager, a fighter pilot during the war, became the most famous pilot in the world when his flight of October 14, 1947 became public knowledge. The first man to exceed Mach 1 (the speed of sound) did so with his own Rolex Oyster on his wrist.

Almost all of the astronauts owned their own GMT-Masters, which had become the standard aviator's timepiece, and continued to wear them at all times including during space flights. Those from the initial NASA group who had flown the "X" planes would have been issued GMT-Masters[60]. The Speedmasters were relegated to occasions when they had to be worn, such as during space walks when the special extended bracelet allowed them to be worn outside the space suits.

It was a GMT-Master on the wrist of Jack Swigert that helped the crew of Apollo 13 to make it back to earth safely after their on board oxygen tank ruptured. While the recorded history of

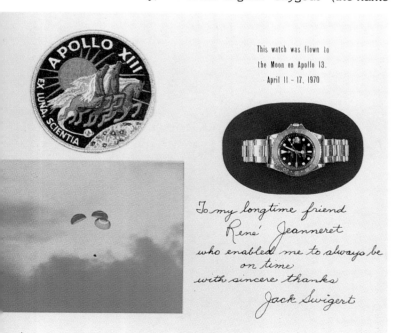

Above and left:
Jack Swigerts' Rolex worn on the Apollo XIII mission. It hangs on a wall honoring the long relationship between NASA and Rolex.

278 The GMT Master: The Correct Time Times Two

this mission makes much of how mission commander James Lovell used his Speedmaster to time the short engine firings that saved the craft, within a month of their successful return to earth Jack Swigert had his GMT-Master mounted on a board, along with a photograph of the "splashdown," a copy of the shoulder patch worn by the crew, and a hand-written note which said "To my long-time friend René Jeanneret who enabled me to always be on time. With sincere thanks. Jack Swigert." Above the watch was the legend "This watch was flown to the Moon on Apollo 13. April 11-17, 1970." This plaque was proudly placed on a wall in Rolex's Geneva head office with the signed photographs of nine other astronauts, all Rolex wearers.

It could be argued that the only reason the astronauts wore Omegas in the first place was due to a foul-up in the New York City office of Rolex. Noticing that most of the astronauts wore Rolex GMT-Master watches and recalling the X plane project where Rolex had supplied GMT-Masters, a purchasing official at NASA called Rolex and asked for fifty watches to be made available for the Space Program. He was curtly informed that the company only had a dozen or so in stock and was not expecting a further delivery for a few weeks. Understandably NASA went elsewhere and, shortly afterwards, when knowledge of the situation reached Geneva very serious reprimands were forthcoming.

The astronauts themselves discovered an unforeseen problem with their Rolex watches while on missions longer than 48 hours: the watches stopped. Putting it simply, the self-winding mechanism had ceased to function in the zero gravity of space. The perpetual mechanism relies on gravity to cause the inherently unbalanced rotor to swing and thereby wind the watch. In space this did not happen. Obviously the cure was simply to wind the watches manually and this was what they then did.

The 1675 went through all the usual Rolex changes between its 1959 introduction and its withdrawal at the end of the 1980s. In the mid-1960s it became available in 18kt gold and in steel with a gold bezel and crown. Around 1976 it received the hacking seconds movement and a few years later the "quickset" feature was also added. The sapphire glass was added just prior to the model being withdrawn.

The withdrawal was temporary and the 1675 was replaced with two new models. The first of these was the 16700 GMT-Master which could be differentiated from the earlier models by the white gold circular settings for its luminous hour markers, though the gold and two-tone models had been fitted with yellow gold settings since their inception. The second and newest model was the 16710 GMT-Master II. This watch used the new calibre 3085 movement which allowed the hour hand to be moved forward or backwards in precise one hour jumps, this feat could be performed without losing the precise accuracy which was usually the reason the watch was bought in the first place. The GMT-Master remains the premier timepiece for aviators and all those whose lifestyle requires knowledge of multiple time zones, whether for international telephone calls or intercontinental travel.

---

[59] The launch of the GMT-Master was the start of a long relationship between Pan Am and Rolex. From 1958 Rolex became the official watch of Pan Am and in 1973 the chairman of Pan Am, General William T. Sewell, appeared in one of the Rolex advertisements.

[60] The GMT-Masters seem to have been given to NASA. An unsolicited testimonial from an X-15 pilot, published by Rolex in the early 1970s begins "The GMT-Masters that you have sent...."

First model GMT-Master in stainless steel on later Jubilee bracelet. This model is slightly smaller than the 1675 which followed it, note the lack of a crown guard and the clear plastic bezel which is painted with the 24 hours underneath.
Model no. 6542
Circa 1956

Stainless steel GMT-Master, black dial, red/blue bezel and "Mercedes" style hands. This watch still has a plastic crystal, but does have both the hack and quickset date features.
Model no. 16750
Circa 1987

The GMT Master: The Correct Time Times Two 279

Stainless steel Rolex Oyster Perpetual GMT-Master with sweep seconds and calendar. Ref. 1675. © Sotheby's, 1994. Circa 1970

December, 1966, brochure for the model 1675 GMT-Master, giving details of uses of the watch, including (on the back page) a method of using the watch as a compass.

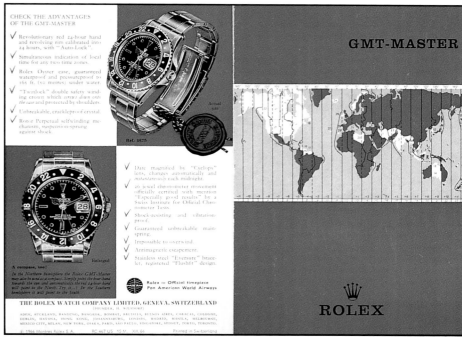

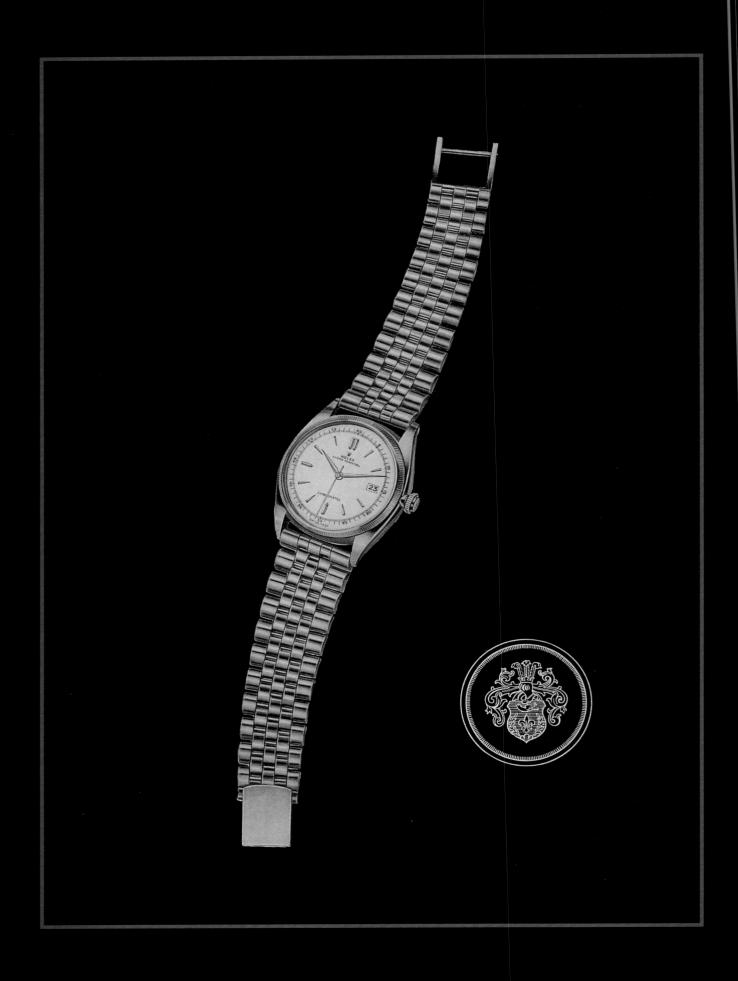

Chapter 15
THE PRESIDENT OR DAY-DATE
Attaining a New Summit

By the time it introduced the President (more correctly called the Day-Date) model in 1956, Rolex had just about given up on manufacturing complicated timepieces. The moonphase watches introduced a few years earlier (which were their first Day-Date watches) were still sitting on jeweler's shelves and attracting little interest. Still, at the Basel fair of 1956, Rolex introduced a new model, the "Day-Date" or model 6511. With a window at 12 spelling the day out in full, allied to the Cyclops date window at 3, this watch moved Rolex into a new position in the market. Like their previous flagship model, the Datejust, it was available only in 18kt gold or platinum, and it introduced a new matching "President" bracelet and therefore was significantly more expensive than any previous model. Now Rolex was competing at price levels previously only occupied by the likes of Patek Philippe and Vacheron et Constantin, but they now had something unique to offer. Due to the complexity of the watch, with its automatic movement, additional day and date discs and screwed back, it was a very large and thick watch which gave it great presence on the wrist. Introduced as a niche model it soon became one of the company's more popular models and certainly one of its most distinctive. The watch became even more famous when the concealed clasp bracelet was offered, giving birth to the "President" as we know it today. Little did they know they had opened a gold mine that the company would still be mining forty years later.

The product that became their flagship and savior at a time when many other watch companies were going out of business has gone through many changes over the years. The first version, introduced in 1956, came to market quickly after having been only recently patented on July 23, 1955 by Marc Huguenin of Rolex Geneva. The patent, 323982, showed a seven toothed "star" wheel which sat on top of the 31 tooth calendar wheel, both wheels would be ratcheted once every twenty-four hours at midnight. This first model 6511 lasted only a year before it was replaced by the externally identical 6611. This model featured the new 1055 calibre, which, with its free sprung Micro-Stella balance, enabled the new watch to become the first Rolex model to bear the legend "Superlative Chronometer Officially Certified" on the dial. This new standard of accuracy is exemplified by a certificate we have seen, which states that the average (or mean) daily rate was + 3.0 seconds per day, and that was in 1965.

There was always some doubt as to whether a watch with both day and date indication would be able to keep accurate time due to the amount of mainspring energy required to move both the discs simultaneously[61]. The initial models most definitely suffered from this problem. In the final hour before the discs changed, all the energy was being taken to drive the cam which triggered the discs. Rolex solved this problem by reprofiling the cam so that energy began to be stored within minutes of the change and was then subsequently stored over the whole day. Any further inaccuracies caused by the friction of the day disc on the surface of the movement were cured by fitting "table" jewels in the space between the two. Once again we see the classic Rolex methods at work: gradually modify, improve and simplify.

This popular Day-Date model was nicknamed the "President"[62] when Rolex reportedly gave one to the then-president Dwight D. Eisenhower, complete with the newly designed concealed clasp bracelet. Mr. Eisenhower was not the first American president to sport a Rolex on his wrist. Franklin Delano Roosevelt preceded him by wearing one for several years and Kennedy, Johnson, Nixon and Ford followed "Ike" in this proud tradition.

These early models did not have the hack feature and most definitely had a rather bulky look. In their never-ending quest to modernize the basic configuration of the Day-Date, Rolex designers have changed from the lumpy "bubbleback" look of early models to the slender shape (complete with sapphire crystal and slimmer case) employed today.

With only mild case design changes over the years, the first modification in the Day-Date was the addition of the "Hack" feature in 1972. This enabled the seconds hand to stop dead, or "hack," so the wearer might synchronize the time with a radio or telephone time signal. This was hailed as a major advancement in technology. By this time, the case design had stabilized into one which is essentially the same style as today's examples. White gold models and platinum models were introduced later. The now rare and collectible rose gold version was phased out in the early 1970s. Later pink gold examples were manufactured only to special order.

The "Quick Set" feature solved the major difficulty involved in setting the Day-Date function when one had neglected to use the watch for some time. Introduced in the late 1970s, Quick Set was added to all Rolex models by 1983. Instead of turning the crown round and round ad infinitum to put the day and date back into their proper spot, the wearer had to simply pull out the crown about halfway, give it a short turn or two, and the display read the correct date.

This new development coincided with another advancement in the President model: the inclusion of the new sapphire crystal, first used on the early quartz model. This not only added to the sleek new look, but also greatly added to durability. The sapphire crystal fitted more tightly than the plastic crystal, making it much less likely to pop out during rough activity, and also making it more waterproof and scratch resistant. The early Day-Date models were waterproof to only 165 feet (or 50 meters) while the current model is safe to 330 feet (or 100 meters). At the same time the latest model is significantly slimmer than the early models. When discussing depths it is worth remembering that pressure increases by the square of the depth rather than in direct proportion; by going twice as deep you will experience four times the pressure!

To complete these major improvements in the Day-Date, Rolex introduced a new case reference number system that is still in use today. They added a fifth digit to the model number. This digit, at first thought to signify a new product line, actually delineated the materials in the watch as it left the factory. This enabled potential buyers to determine that their timepiece was not a "conversion."

282 The President or Day-Date: Attaining a New Summit

0 = Stainless steel (not stainless and gold)
1 = Yellow gold-filled
2 = White gold-filled
3 = Steel and yellow gold
4 = Steel and white gold
5 = Gold shell
6 = Platinum
7 = 14-carat gold
8 = 18-carat yellow gold
9 = 18-carat white gold

This number was invaluable for inventory control, both at the factory and for the individual retailer.

The most recent advancement in the Day-Date model, introduced at the end of 1990, is the "Double Quick Set" feature. This made it even easier to align the day and date into their proper positions.

Stylistically the latest renditions of the Day-Date are standard models decorated with an ever expanding variety of precious stones. Called the "Crown Collection" by Rolex U.S.A, these are by far the most expensive Rolex watches in current production. Many of the watches feature precious stones not only on the bezel and dial but on every exposed surface of the case as well as all around the bracelet. While most people think this trend for embellishing the Day-Date is a recent phenomenon, the watch was available with diamonds on the bezel from the very start.

The Day-Date Presidential is synonymous in America with style and success and remains as one of the most prestigious additions to a gentleman's wardrobe. To own a Day-Date Presidential is to claim a rare distinction.

---

[61] Rolex likened the speed at which the discs changed to "the speed of a camera shutter."
[62] Rolex themselves apply the name "President" only to the bracelet, not to the whole watch. Interestingly Rolex call all the bracelets by this name, making no difference between the basic and the concealed clasp models.

18kt yellow gold Day-Date chronometer with silvered dial, applied baton markers, sector, and aperture for the day and date. Luminous baton hands. Rolex President bracelet with concealed clasp. © Christie's, London.
Model 1803
Hallmark 1972

The first Day-Date model both in gold and in platinum. Note the early style leaf hands.

The President or Day-Date: Attaining a New Summit

Day-date

The Rolex Oyster Perpetual Chronometer Day-Date is the crowning achievement of Rolex and, possibly, the most brilliant timepiece in the world today. It combines all the features invented and perfected by Rolex... the waterproof Oyster case, the Perpetual rotor self-winding movement—an Officially Certified Chronometer with the special mention "especially good results." The Day-Date was the first wrist-watch in the world to show the date and the day of the week spelt out in full. Both the date, which is magnified by the "Cyclops" lens, and the day of the week change automatically at midnight. It is available only in 18ct. gold or platinum, and has its own exclusive matching bracelet.

(left)
3242/7286 18ct. luxurious pink gold, with matching 18ct. gold bracelet.

(centre)
3035/7286 18ct. yellow gold, with matching 18ct. gold bracelet or with leather strap and gold buckle.

(right)
3208 Solid platinum with matching solid platinum bracelet. The dial is surrounded, and the hours marked, with carefully chosen diamonds. Also available without diamond-set bezel.

1966 brochure showing Day-date variations.

Platinum and gold Day-Dates from 1960 catalogue, note the early triangle hands and that both watches are on non-concealed clasps.

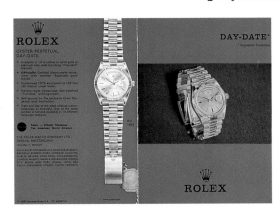

Brochure dated October, 1967, showing reference 1803 Day-Dates in both yellow and white gold. Neither of the watches shown have concealed clasp bracelets.

Day-Date from a 1964 Bucherer catalogue showing the watch and its accompanying timing certificate.

Chapter 16
THE QUARTZ REVOLUTION

It may seem strange that a company such as Rolex, which emphasized accuracy almost from the start, should produce and sell so few quartz watches. They are so much more accurate than mechanical movements that even such venerable companies as Patek Philippe use quartz movements in many of their watches and in almost all their lady's pieces. But not Rolex. While everyone from Longines to Audemars Piguet seemed to be switching wholesale to quartz movements, Rolex downplayed its involvement in them.

However, as with most things to do with Rolex, all is not always as it seems. In fact Rolex was one of the first Swiss companies involved in the development of the quartz movement, through their membership of the consortium of watch manufacturers known as "Centre Électronique Horologer" (Electronic Watch Center) or C.E.H. at Neuchâtel. This group was made up of Patek, Favre-Leuba, Zodiac, Eberhard, Ebel, Juvenia, Doxa, Cyma, Borel, Le Coultre, Elgin, Movado, Zenith, Omega, Bulova, I.W.C., Longines, Enicar, Rolex and Rado, plus three of the movement companies, Ebauches SA, Fabriques des Balanciers Réunies, and Fabriques d'Assortements Réunies SA. It was founded in 1960 and produced its first working prototype in 1966. This was the Beta 1. The following year their next prototype, the Beta 2, was entered in the annual Geneva chronometer tests (the first year that quartz watches were allowed to enter[63]). It conclusively destroyed the previous record with a score of 0.152[64], while the best mechanical timepiece scored 1.73. The new contender was twelve times more accurate than the finest product of the greatest watchmakers in the world. Its accuracy was three thousandths of a second per day. This is, more or less a second a year!

The accuracy of the new Beta 2 had two immediate results. First the investors and technicians at C.E.H. redoubled their efforts to develop a production model and, second, the Council of State announced on April 26, 1968 that the wristwatch chronometer competition was suspended. Not long afterwards the suspension was made permanent.

C.E.H. produced their first production movement in 1968 and called it the "Beta 21." This was in fact a very interesting movement combining, as it did, the features of both quartz and tuning fork movements in one. The movement used a quartz bar, maintained in a vacuum capsule, electrically excited to produce a frequency of 8,192 hertz (or vibrations per second). This vibration was divided by 32, making a frequency of 256 hertz. This was then transferred to a resonant reed which activated a ratchet wheel and pawl system to drive the hands. Despite being involved in its development, Rolex was not prepared to stake the name of the company on this new movement, without extensive testing. In mid-1968 one of the first movements was cased and worn as a watch on a daily basis for fourteen months by a number of Rolex staff. Only after these tests was it decided to go into production. By deciding to test the watch this way, Rolex was not one of the eighteen companies who were involved in the main public launch of the Beta 21 on April 10, 1970 at the Basel fair.[65] Rolex had one of their prototypes at the fair but chose not to display the watch. Instead they discreetly showed it to interested parties. They were uninterested in immediate sales, rather looking upon the watch as "testing the water." They were shocked when it proved to be an immediate success. The new Quartz Rolex was launched on June 5, 1970 as model 5100, although it was not generally available until later that year. The initial pilot series production, released in October, 1970 was completely sold out prior to delivery.

The new watch was the first production Rolex to have a sapphire crystal and it featured all of the facilities now taken for granted, hacking seconds and quick date change. It also introduced a new and different look for the case style, one in which the lugs no longer flowed smoothly into the bracelet. The new model was much more angular in shape and, despite its impressive bulk (the less complimentary might call it "clunky"), it was not an Oyster. Available only in 18kt gold, both yellow and, more rarely, white, with a matching concealed clasp bracelet it was obviously an expensive piece. Because it lacked either the day display of the "Day-Date" or the impermeability of the Oyster, it was never destined to be a major seller. This watch is the exception to the rule that new models of Rolex that do not sell initially find great demand in the post-production resale market.

Its marked lack of popularity, both in the new and used markets, is undoubtedly due to its styling. It is basically an ugly watch and was the only watch produced by Rolex which followed the other Swiss makers down the 1970s path toward "bigger is better." Most of the watches made by Rolex's competitors during the 1970s, such as Zenith's 1975 version of the "El Primero," the 1974 Certina "Argonaut" or Omega's 1970 "Memomatic" were huge and their styling was such that they dated within a few years.

The watch is also the only model that Rolex ever signed their name upon, where the company was not involved in the construction of the movement[66]. The Beta 21 itself suffered major problems and in the two years following its introduction in 1970, the factory (Ebauches Electronique, Marin or EEM) was only able to produce around 6,000 movements. This was not enough for the twenty plus companies involved and gradually many of the original investors pulled out. Even before the launch Omega, Longines, and Bulova had pulled out and Rolex was not far behind. The company had been involved in its own research into electronic timekeeping for over twenty years. In 1952, W. Emil Borer (for Rolex) obtained the first patent, number 296,731, for a Rolex electro-mechanical watch. A decade and a half later, in 1969, Rolex was one of the first important watch companies to patent a quartz watch. This Swiss patent, number 519,747 was by F. Gasser for the Rolex firm on June 13. It was followed very quickly by a patent on a Quartz digital timepiece, number 530,667 on October 20, 1970 and patent number 569,320 by P. Girard for a "Multi-display digital quartz watch."

The "Oysterquartz" movement (calibre 5055) was developed from 1972 and introduced in late 1977 to much fanfare. It featured a new Rolex developed and produced movement with a combination of quartz and mechanical technology. The quartz frequency was 32,768 hertz or four times faster than the Beta 21, (although, by then, Omega were making a quartz watch with a frequency of 2,400,00[67]). The movement featured eleven jewels and a Rolex-designed and produced stepper motor which drove the hands, which, in the case of the seconds hand, were locked between each beat by a lever and escapement wheel.

Two new watches, model 17000 Datejust and 19018 Day-Date, which finally featured a screwed crown and back, were much slimmer than the earlier Beta 21 model and even slimmer than the current mechanical models. They introduced a new flatter, angular case which Rolex even fitted with a self-winding movement for some time (model 1630).

Despite their innovation, to a large extent the Oysterquartz was doomed to long-term commercial failure. It was doomed first by what was happening in another department at Rolex. While the Oysterquartz movements were being readied for production, other Rolex technicians were preparing to launch the fast beat 3030 movement. This new movement, whose balance rotated at 28,800 revolutions per hour (or 8 times per second), was a great improvement of the previous Rolex movements which rotated at 19,800 revolutions per hour (or 5.5 per second). It was able, at a stroke, to increase the accuracy of all watches by a factor of almost 100%, from around 5 seconds per day to around 2-1/2[68]. The accuracy of the new calibre was also helped by its new escapement wheel which now had 20 teeth rather than the previous 15.

What the introduction of the new calibre proved was the company's continuing commitment to the mechanical self-winding wristwatch. While the whole of the Swiss watch industry was turning to quartz (and also toward oblivion) Rolex introduced a complete range of new mechanical calibres[69].

Meanwhile, the company continued its research into possible solid state watches and went as far as constructing prototypes of a watch that used LEDs in place of the hands. This potential movement was even given a calibre number, 7065, and featured a two digit LED readout toward the center of the watch which displayed the date. In fact in the 30 years from 1960 to 1990 21 of the 50 patents awarded to Rolex have been for watch electronics. While the company may not produce many quartz watches, it keeps itself fully aware of all the possible developments that might affect it.

Then, just as the Oysterquartz was being launched came the final blow: the arrival of large quantities of fake "Datejusts" onto the world market, all of them quartz powered. And, it seemed, that everyone in the world, from potential muggers to potential purchasers quickly learned one thing; "if the second hand "steps" it ain't the real thing!", although in this case, it was the real thing, and people still wouldn't touch them.

Finally, after the fakes, came the free gift digital quartz watch with a gallon of oil and the arrival of the Swatch. These consolidated the idea in most people's minds that quartz equals inexpensive. Since Rolex had made the decision that their quartz watches would not be significantly cheaper than their equivalent mechanical versions, most of their customers chose the mechanical version. These various factors have resulted in the current Rolex quartz production that is limited to around 10% of total production.

This book, like most books nowadays, is being produced on a computer. Almost no-one uses a typewriter today. And, as this book is being completed, news reaches us of the closure of the Remington typewriter factory, the last one in the United States and also the factory where the typewriter was born. In the end, new technology will always drive out old. Today no-one uses a sextant to navigate, when a GPS receiver will pinpoint your position to within 5 meters, anywhere in the world. Similarly, no TV news crew would use 16mm film when videotape is available. Yet Rolex has managed not only to survive using baroque technology, but has prospered. Unlike the small companies making valve amplifiers for a tiny coterie of purists unable to accept the transistor, Rolex produces and sells around half a million mechanical timepieces every year and in so doing proves, once again, that it is a law unto itself.

Today, Rolex has confined the production of quartz watches primarily to the Gent's President and Datejust models as well as some limited production in the Cellini and Tudor lines. Why these watches have not caught on with the general public is a mystery, especially since the use of the quartz movement enables the Rolex designers to make an incredibly sleek and exceedingly attractive watch. But it would seem that in the ownership of a Rolex, the romance of traditional styling and the legacy of mechanical craftsmanship has triumphed over the accuracy of science.

---

[63] It strikes us as an amazing coincidence that the first year that the Swiss produced a quartz timepiece was the same year that such watches first could be entered in the competition.

[64] A perfect score in the chronometer trials would be 0, as distinct from the Kew A tests, where a perfect score would be 100.

[65] Unfortunately, this launch was four months after Seiko launched the Quartz Astron using their new calibre 3500 as the world's first production quartz watch. While Seiko made very loud noises about being first with a quartz wristwatch, it seems very few were ever sold to the public. Most went to members of the company as part of a trial program. Seiko's triumph however was to be short lived; most of the watches were recalled due to problems.

[66] While the Rolex Chronographs always used movements from either Valjoux or Zenith, these movements were always finished by Rolex.

[67] These ultra accurate quartz movements, both Omega's 2.4mhz movement (which was used in the world's only wristwatch marine chronometer) and Seiko's Twin Quartz, which were made in the late 1970s to the early 1980s are still the most precise wristwatches the world has ever offered. However, both were commercial failures and 15 years later no-one has ever thought of making watches to challenge them, let alone surpass them. It would seem inconceivable to the "régleurs" of the chronometric societies of Switzerland, but no-one wants to beat a 15 year old record! Watches are now as accurate as anyone needs them to be.

[68] The "fast beat" movements offered other advantages. There was less change in balance amplitude as the mainspring unwound and when the watch was moved from the vertical to the horizontal (or vice-versa). Any watch so fitted was less affected by shocks and sudden changes in attitude. These beneficial effects were a result of the so called "gyroscope" effect. There was some resistance to these new movements, however, because the gear train was subjected to much greater forces and therefore much greater wear.

[69] The introduction of this new calibre which beat at 28,000 beats per hour (or 8 per second) also illustrated the innate conservatism of Rolex. Over a period of 20 years it increased the frequency of its balance rotation from 18,000 in 1957 to 19,800 in 1964, and finally to 28,000 in 1977. Most other companies went directly from 18,000 to 36,000 and ran into many problems.

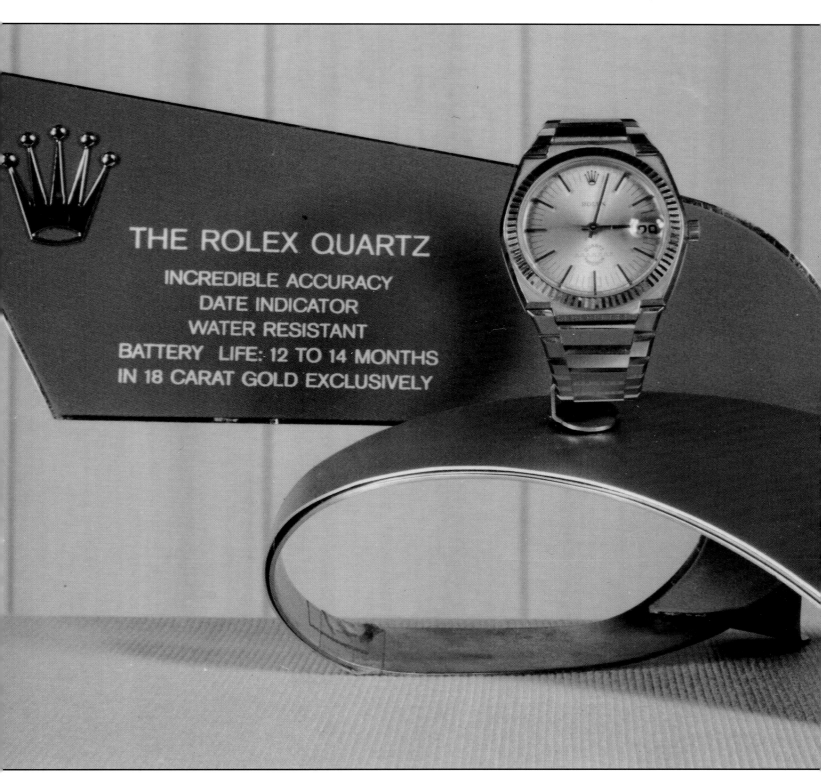

The first model Rolex quartz watch on its display unit. Note that the watch is described as "water resistant" rather than waterproof. Circa 1971

CHAPTER 17
THE LADY'S ROLEX

When Wilsdorf and Davis began their original company, their wristwatches were solely for ladies. Despite fashion breakthroughs of the First World War, it was a long time before men's Rolex watches outsold the ladies'. The famous "Daily Mail" channel swim advertisement of November 27, 1927, showed only eight man's wristwatches, two pocket watches, and fifteen lady's watches. The 1930 catalog featured 79 lady's watches and only 37 mens' models (plus 23 pocket watches). Even as late as 1934, the catalog showed 31 man's and 64 lady's. In the earliest of days, prior to the invention of the word "Rolex" on July 2, 1908, Wilsdorf & Davis signed all of their watches W&D, on the case and movement, leaving the dials unsigned. These are some of the rarest of all the watches and, interestingly, so far all of the watches so marked have been lady's models (see the picture on page 319). As is well known, all of these watches used Aegler movements signed "Rebberg" below the dial. Prior to the incorporation of Wilsdorf & Davis, Aegler himself had tried to sell complete watches and, as the blue enamelled example on page 318 shows, these watches were signed on the case and movement as Rebberg. These pieces are the earliest known antecedents of Rolex, and both of them are lady's watches. In these early days the company offered whatever it believed would sell, so in addition to wristwatches there were small fob watches and a few tiny "ball" watches which were often beautifully enamelled.

The simple early lady's watches are classic "transitional" pieces, identifying the direct route between pocket and fob (which were merely shrunken pocket) watches and the true wristwatch. The first true wristwatches were also lady's models. These were tonneau, baguette, and rectangular models introduced around 1920. They were the first watches to have a movement made by Aegler exclusively for Rolex use. Until the arrival of the cushion and octagon Oysters they were, by far, Rolex's best selling watches.

After the Prince was introduced in 1928, Rolex capitalised on its success and brought out a companion, high end lady's model called the Princess. Interestingly this was the first lady's model that Rolex sold on the basis of accuracy, the accuracy being authenticated by the issuance of a Kew A certificate to the movement. This certificate, for 86.5 marks, was the highest that Rolex had yet achieved and was the first they had ever obtained with an "especially good"[70] rating.

The Princess was, for a long time, the most expensive watch that the company sold. Even ignoring the diamond set models, the Princess was a very expensive watch to manufacture, probably more expensive than the Prince. The diamond set platinum Princess sold for 100 guineas (or $420.00) at a time when the gentleman's flared Prince "Brancard" model in platinum sold for only 65 guineas (or $286). A simple steel and gold Princess sold for 14 guineas (or $59.00) at the same time a man's steel and gold classic Prince was only 11 guineas (or $46.20).

The original Oyster was introduced in 1926 in both man's and lady's models, but seems to have been mainly a man's watch. The proportion of lady's models that have surfaced seems to have been less than 20%. As the watch did not live up to Rolex's initial expectations it was decided to try and make it more feminine by replacing the "coin edge" milling on the bezel with a more attractive engraved finish. It does not appear that this attempt was very successful for the resulting model is, in fact, very rare.

The market for lady's watches in the thirties consisted mainly of delicate baguette style miniature watches, at a time when Rolex was moving toward the two-piece Oyster, which was of course a larger watch. Rolex was able to introduce their first styled lady's Oyster, the model 2295, known as the Corona in the catalogs of the time.

In 1936, Rolex introduced six new lady's Oysters, numbered 2324, 2515, 2517, 2524, 2528 and 2532. They were all essentially variants on the Corona but had lugs that ranged from the simple to the outrageous. None of the above proved really popular and were withdrawn within four years.

During the war years Rolex focused their energies on the production of man's steel Oysters and the lady's watches were to a large extent ignored. However on the occasion of the Rolex Jubilee in November, 1945, Rolex introduced two Jublilee models. Everyone knows about the Datejust, but they also introduced a lady's version of the Perpetual, the model 4487. This first model was not an Oyster, but a lady's dress watch with central loops to take a looped strap.

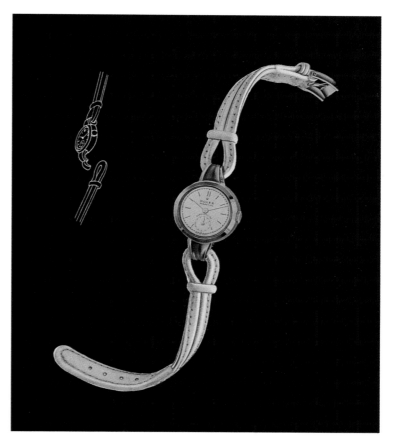

Drawings of model 4487.

290 The Lady's Rolex

It was almost another nine years before Rolex produced a lady's version of the Oyster Perpetual Chronometer, which was introduced at the 1954 Basel Spring Fair, the same fair that saw the introduction of the Submariner, the Explorer, and the Turn-O-Graph. Obviously up against this sort of competition, the lady's watch was pretty much overlooked, but closer examination would have revealed the new 7-3/4''' automatic movement, initially known as the PA movement (later called calibre 420). The new movement was the smallest automatic that Rolex had produced to date and the new watch it powered was in many ways a perfect miniature of the current man's watch. With its fluted bezel, "dagger" shaped batons, screw down crown and expanding oyster bracelet, it took all the styling points of the man's watch and recycled them for a lady's watch. The main differences were the subsidiary seconds hand, rather than the sweep on the man's, and the lack of a date window. It would be another three years before these two would arrive.

We mentioned much earlier how the lady's Princess had been the most expensive Rolex in the 1930s. This situation returned toward the end of the 1950s as Rolex began to reintroduce small, jeweled lady's watches. Made mostly in white gold or platinum they were also first introduced at the 1954 Basel spring fair. As the economies of Europe began to drag themselves out of the ruins left by World War II, there arose a demand for objects which not only spoke of prosperity and wealth but screamed of them. It was the time of South American playboys, like Porfiro Rubirosa, and of heirs to fabled wealth, like Aly Kahn or Barbara Hutton. It was for people like these that Rolex began to produce fabled creations, like the "Fleur d'Été" illustrated here, every visible surface of which is covered in diamonds. It is worth noting that the price of the watch 6,500 guineas ($18,200 at the time) was exactly five times as expensive as a platinum and diamond Day-Date. These watches seem to have been custom-made, as a catalog of 1959 showed the two watches below and states "Rolex watch creations such as these are *personally exclusive*- only one of each design is made. Your Rolex jeweler will obtain full details for you." This obviously explains why we see such exquisite pieces only as single items, never as series pieces. As far as we are aware, these lady's watches of the 1950s are the only custom pieces that Rolex has ever shown in its catalog. It seems certain that, while the movements for these watches were produced by Rolex, the case and jewel work was contracted out to a specialist. This hypothesis is reinforced by the fact that all of these pieces we have had a chance to examine all bear the casemakers mark CF.

The "Ice Flower" probably the most expensive Rolex ever made. Completely covered in white and yellow diamonds, it was awarded the Prix de la Ville de Geneve when it was first shown. Note that the accompanying description talks about the fact that such watches are completely custom made.
Circa 1955

1966 catalog

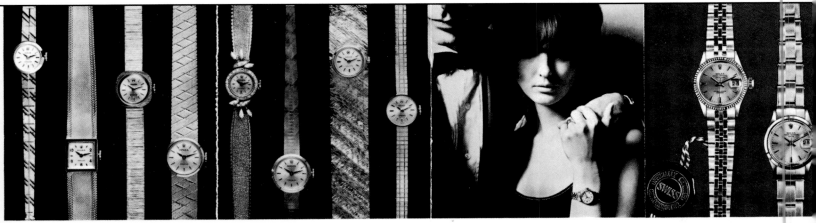

The Lady's Rolex 291

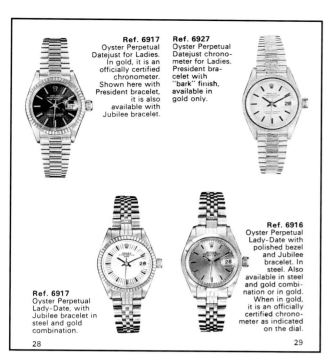

Lady's Datejust variations from 1966.

Lady's 9kt gold Egyptian style "Princess" with hinged lugs and fancy dial.
Model no. 1238
Case no. 5999
Circa 1934

More 1966 lady's models.

– the one with substantial character

ial timepiece: Pan American Airways.

These watches are examples of pieces that could only have been made in the period they were. It was a time when the machinery needed to produce reliable, accurate movements of this size was available and the craftsmen who could produce such exquisite design and setting were still around. In many ways these watches are the lost masterpieces of Rolex. Along with almost all of the lady's watches, they have been consigned to collector oblivion. In fact, the skills needed to produce these watches were much greater than those needed to produce a President. They are, in fact, the last hand made Rolex watches.

---

[70] An "especially good" rating is only issued if the watch gains a rating of over 80 points.

292 The Lady's Rolex

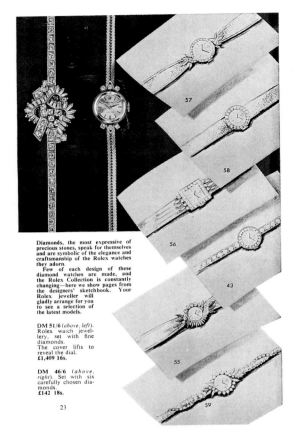

Custom-Made Lady's Rolexes.

9kt hinged case with hinged lugs. 9-3/4'" Rebberg movement.
Circa 1920

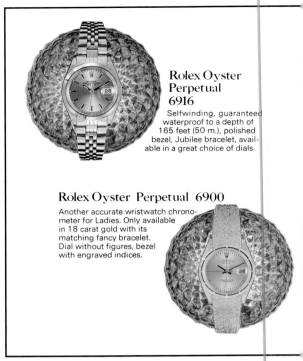

Lady's watches form 1966, including an unusual lady's "Zephyr."

Unusual oversized 15kt lady's watch with interesting original 15kt expanding bracelet. White porcelain dial with gilt minute markers, Arabic numerals including red 12. The watch is unusual in that it uses an 11'", 15 jewel movement, rather than the more common 10-1/2'" version. It is also the only 15kt Rolex to surface to date. Case made by Dennison. Case no. 91455.
Hallmarked 1914

9kt gold lady's cushion watch with round double sunken porcelain dial, hinged back case, and hinged lugs. 10-1/2'" Rebberg movement.
Circa 1915

Rolex Oyster Perpetual 6718
Selfwinding, waterproof to a depth of 165 feet (50 m.), available in steel and steel and gold combination. 18 ct yellow or white gold, engraved bezel and matching Oyster bracelet.

Rolex Oyster Perpetual 6719
This model has been specially created for the sportswoman who needed a rather simple, yet robust wristwatch. Solid gold dial figures, easily legible to facilitate life to the woman of action.

Sterling silver, enameled "ball" watch. It has a completely engine-turned case, which is then turquoise enameled with added details in gold. The periphery is studded with natural seed pearls. The tiny dial (less than 20mm) is porcelain, with black Arabic numerals and gilt hour markers. The movement is a 9-3/4''' Rebberg, with the very early signature on the side of the movement, rather than on the wheels. The case is sterling silver but is gilded on the inside and stamped with the W&D logo. The watch is otherwise completely devoid of any numeric identification.
Hallmark London 1913

294 The Lady's Rolex

9kt gold, hinged cased digital watch with 15 jewels.
Circa 1928

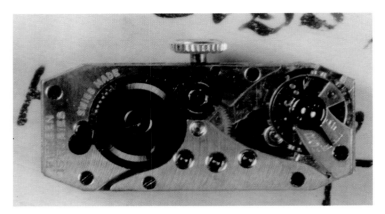
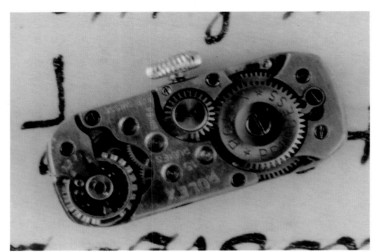

The Lady's Rolex

18kt pink gold self-winding lady's wristwatch with gold Oyster bracelet. Silvered matte dial with applied gold baton numerals, subsidiary seconds. Ref. 6504. © Sotheby's, 1994.
Circa 1957

Two-tone lady's hooded wristwatch, Rolex, gold screw-down crown, Extra-Prima lever movement, silvered matte dial, black Arabic and baton numerals, subsidiary seconds, water-resistant-type case, with Rolex steel bracelet composed of bamboo links. Diameter 22mm. © Sotheby's, Inc., 1990.
Circa 1930

Gold decorative lady's watch.
Circa 1936

Lady's 18kt bubbleback. Dial signed "Rolex Oyster Perpetual Precision." 24mm. © Sotheby's, Inc. 1991.
Circa 1955

Late Deco square 9kt gold lady's watch with circular dial and outer chapter ring with enameled numbers.
Hallmarked Glasgow 1938
Model no. 3303
Case no. 46053

296 The Lady's Rolex

Two examples of lady's platinum watches. Left: Diamond bezel model with subsidiary seconds. Right: Diamond bezel oval dress watch, ca. 1935.
Circa 1948

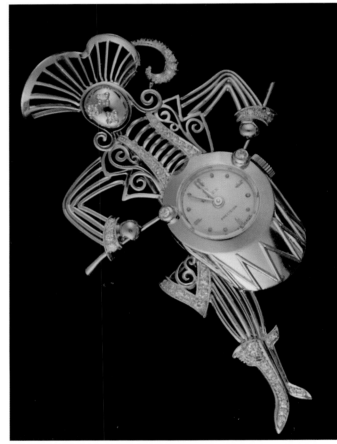

Rare and unique figural fantasy platinum brooch-watch in the shape of a Revolutionary War drummer boy, with diamonds on cuffs, boots, coattails, lapel, face, hat and drumsticks. Case, dial and movement all signed Rolex complete with Swiss platinum control marks. High grade 17 jewel movement, jewelled to the center. Ref. 765.
Believed to have been made for the US Bicentennial in 1976.

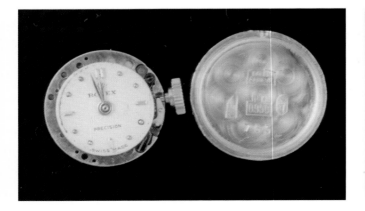
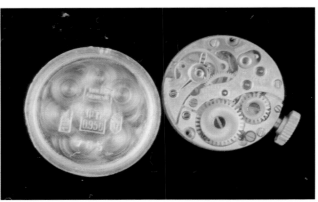

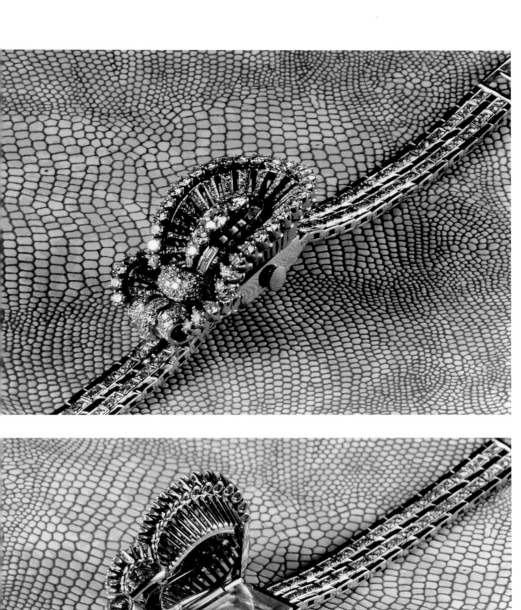

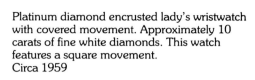

Platinum diamond encrusted lady's wristwatch with covered movement. Approximately 10 carats of fine white diamonds. This watch features a square movement.
Circa 1959

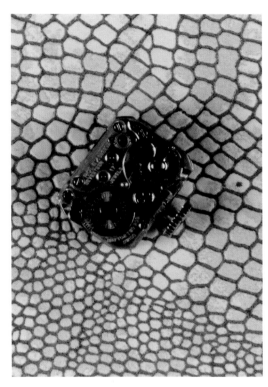

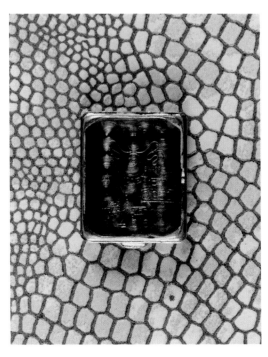

298 The Lady's Rolex

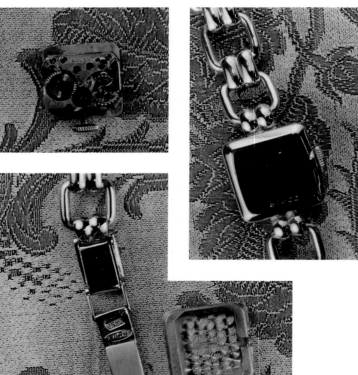

Retro Deco 18kt pink gold square lady's dress watch with mint original dial. Bracelet with square and oval links.
Circa 1946

Current model lady's 18kt Datejust with concealed clasp "President" band. Gilt dial with baton markers and hands. Quickset date feature and sapphire glass.
Model no. 69178

18kt pink gold Lady's bubbleback, with milled bezel and "triangle" dial, subsidiary seconds. Notice the dial is signed "Super Precision" around the seconds dial. A chronometer certification for this size movement was still a couple of years away. Has its original leather strap with signed buckle.
Model no. 5003
Case no. 683110.
Circa 1956

The Lady's Rolex 299

18kt "Top Hat" square watch. Black dial with triangular and dot markers, subsidiary seconds. 8-3/4''' Hunter movement, superbalance & capped endstones. This watch is a perfect example of the highly stylized watches made in the 1950s as an answer to the watches being made in the U.S. at the time.
Model no. 9
Case no. 8030
Circa 1955

Steel and 18kt yellow gold lady's Oyster perpetual (non-date) watch with milled bezel, sapphire glass, gilt dial and baton markers and hands. These non-date models are a little smaller than the Datejust models.
Model no. 67193
1991

300 The Lady's Rolex

Rare 18kt Rolex Precision ring watch with unusual channeled "hoods." Silvered dial with triangle and dot markers. Movement 4''' 17 jewel reserved calibre.
Model no. 8309
Case no. 695075.
Circa 1955

18kt yellow gold lady's Oyster Perpetual. Brushed gilt dial with raised baton numerals and luminous baton hands. Note that this watch is neither a chronometer nor a date model. © Christie's, London.
Model no. 6619
Circa 1974

Pink 18kt gold lady's Oyster Perpetual. Pink gilt dial with painted baton markers, subsidiary seconds and pink gold leaf hands. Pink gold riveted oyster bracelet with deployant clasp. © Christie's, London.
Model no. 4486
Circa 1984

18kt white gold lady's oyster perpetual Datejust, milled bezel, white dial with applied arabic numerals, sweep seconds hand, luminous baton hands and date aperture. © Christie's, London.
Model no. 6900
Current Model

The Lady's Rolex

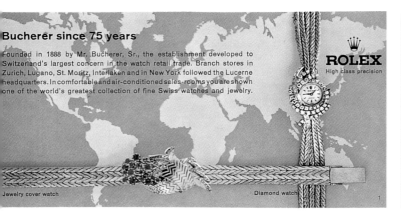
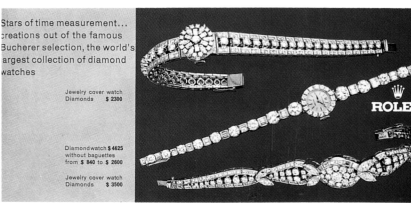

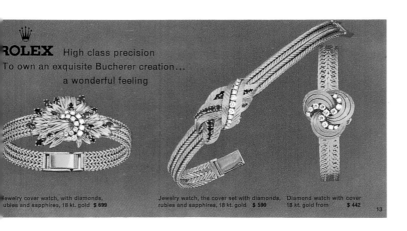

A selection of lady's jeweled watches from a Bucherer catalog dated July, 1964. These exquisite jewelry watches. The brochure describes these watches as "Bucherer creations with Rolex high precision movements," implying that the watches were made by Bucherer and that Rolex simply supplied the movements. However the vast majority of these jewelry watches are signed Rolex on their cases, so we may never know the correct story.

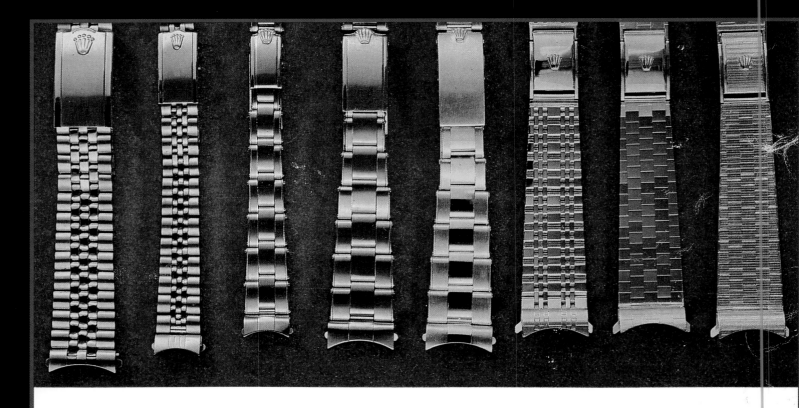

B200	B201	6634	6635	6635	2544	2557	2553
Stainless Steel	Stainless Steel	Stainless Steel	Stainless Steel	9ct. gold	9ct. gold	9ct. gold	9ct. gold

Distinctive bracelets designed specifically for Rolex Oyster watches. The "Flush-fit" ends eliminate the open spaces between the ends of the bracelet and the Oyster case. These bracelets are available separately.

Bracelets from a 1969 brochure.

Chapter 18
BRACELETS
Around the Wrists and Around the World

Part of the Rolex "look" is the immediately recognizable Rolex bracelet. Now available as the "President," "Jubilee," and "Oyster" models, they are so much a part of the watch that a modern Rolex Oyster on a leather strap looks strange. But it was not always so; the very first Oysters from 1926 managed without the advantage of factory metal bracelets.

A page from the 1930 catalog shown here features a wide variety of bracelets available from the factory, but none of them could be described as suitable for a sport watch. Despite this, many customers had the supplying retailer fit their Rolex with one of the flexible metal bracelets that had been widely available since the early 1920s. The most common of these "aftermarket" bracelets was the Bonklip model, sometimes known as the "bamboo" bracelet. This was a simple bracelet made from a series of hollow pressed steel rectangles joined together by flat pieces of steel folded over to make the links. They closed by means of a clip on the end of one side of the bracelet which was simply pushed into the gap between two links of the other side. They were much more adjustable than a leather strap, because the distance between the links was much less than the holes in a strap. They also had the advantages of being much less likely to wear than a strap, were inherently much stronger and, like the Oyster, were waterproof. These bracelets were made by a variety of suppliers but the best of them was Gay Frères of Geneva which had specialized in jewelry and particularly bracelets for many years.

When Rolex introduced the Imperial, in 1933, it was available either as a steel or as a gold watch. The following year, when the Perpetual was introduced in a mixture of steel and gold as well as the regular solid materials, the Imperial also became available in mixed metal, and for the first time a factory bracelet was listed in the catalog. It was an option, not standard, but it was a factory bracelet and it was available both in steel and in mixed metal to match the individual models. It was not an inexpensive option, the steel bracelet cost an additional £1.10s 0d (£1.50) on top of the £11.5s 0d (£11.25) for the watch, or an additional 13.33%. The two-tone bracelet cost an additional £5.5s 0d (£5.25), when the watch was £13.5s 0d (£13.25) or almost another 40% on the price of the watch. It is interesting to note that while the watch was available in both 9kt and 18kt gold, the bracelet was not manufactured in either. Obviously the initial feeling was that this was a sports watch and a metal strap was even more sporty, and of course no-one would wear a gold watch for sports.

This first bracelet was a variation on the original Bonklip, constructed in the same way but using a 3-piece deployant clip with the top section bearing seven closely spaced holes (less than 2mm apart) to allow the end of the bracelet to be moved for size adjustment. The bracelet was made by Gay Frères. (The latest Datejust uses a deployant clip which looks and operates in exactly the same way as this original, even down to the twin parallel reinforcing indentations on the two interior sections of the clip.) Although it was introduced on the Imperial, it was not long before the benefits of this bracelet were made available on other Rolex models. Oddly, other than the Imperial, the watch on which the bracelet is most commonly found is the Prince, probably the least sporty of all Rolex watches.

Bamboo sytle bracelet from Gay frères.

Rolex now found itself with a flagship product, the Perpetual (remember it was 35% more expensive than the Prince), which was the only one of their mainstream products not available with a factory bracelet. In 1940 they turned to Gay Frères once more for a design. In response, Gay Frères produced a bracelet that is one of the most flexible and distinctive ever to grace the lugs of a watch: the "Beaded" or "Grain of Rice" bracelet. Both names are very descriptive of the construction which features alternating bands of three and then four small oval links only 7mm long and 2mm wide. The groups of four were soldered at their ends to a pin running the width of the bracelet while the groups of three were free to swivel around the same pins. It was the size and number of the links which gave the bracelet its great flexibility. It is, in truth, more like a piece of chain mail than a series of segments of solid metal as are most modern bracelets.

The bracelet was available in all materials, steel, steel and 9kt gold, and all gold, although it seems to have been most popular in the materials that the bubbleback made popular, steel and gold. It was not an inexpensive bracelet. At a time when the two-tone bubbleback cost 385 SF, a matching bracelet was

Beaded or "grain of rice" bracelet by Gay frères.

an additional 165 SF. For the most expensive bubbleback of the time (the model 3372), which cost 865 SF, the matching bracelet was an astonishing 720 SF extra. One can only assume that this cost was the reason for it being such a rare bracelet today. We must also consider that the bracelet was produced primarily between 1940 and 1945, a time when most people had more things to do than worry about matching bracelets for their watches.

It is worth noting that, while both of the above bracelets are always stamped "Rolex" when fitted to factory supplied watches, they were not Rolex designs and Gay Frères produced them for many other manufacturers[71] and also sold them direct to jewelers.

The "Bamboo" bracelet was no longer in production and there remained a need for a less expensive bracelet than the "beaded" model. Again Rolex turned to an outside contractor, this time a company called Mecan. From this company Rolex purchased a new style of bracelet which was soon to become almost synonymous with the company. It was the first "Oyster" bracelet. This Mecan bracelet had 12 links (5+7), each link being made from three equally sized pieces of sheet metal bent to form a flattened loop. These loops were joined by a horizontal rod which ran through the top of the two outer loops and the bottom of the center one. This rod was then secured through a blanking piece of metal which covered the profile of the outer loop. In a new touch, the loops were sprung so as to enable a close fit, although the old three-piece deployant buckle was still used. These early bracelets are immediately recognizable because they are parallel along their length at 16mm wide, unlike the later (1947 on) bracelets which taper to a 10mm wide buckle.

In late November, 1945, only a couple of months after the end of World War II, Hans Wilsdorf summoned the heads of the Rolex distribution companies throughout the world to Geneva. They were invited to a dinner at the prestigious Hotel des Bergues[72], the purpose of the dinner was to celebrate 40 years since the foundation of Wilsdorf & Davis. At the celebration of this 40th Jubilee Rolex launched two new watches, model 4487, Rolex's first Lady's automatic, and the model 4467 "Datejust." While the Datejust was most noted for the date window and the new case style, it did not go unnoticed that the watch also featured a newly designed bracelet, called Jubilee in honor of the occasion. This new bracelet was the first one that Rolex had designed themselves and therefore which was not available to any other manufacturer.

Though most of the watches Rolex sold were steel, the Jubilee bracelet was not made in steel. This was a conscious decision on the part of Rolex. They wanted the Jubilee to be available only in the most expensive metals, as the "President" bracelet is today. To meet the need for a steel bracelet Rolex designed the first "Oyster" style bracelet. They were granted Swiss patent 257,185 on February 5, 1947 and it was in production very shortly afterwards. This bracelet was very similar to the early parallel sided "Oyster" bracelet, in that it was a sprung, expanding, externally riveted bracelet. In the new design, the bracelet was no longer parallel, tapering instead toward a 10mm wide deployant buckle. The three metal loops were no longer similarly sized, the center link being much wider than the outer ones. In Rolex advertisements the Oyster bracelet was not shown until 1948, when it was available only in the straight end (or bubbleback) style. Catalogs of the period show the majority of watches still on leather straps, but bubblebacks and chronographs on "Oyster" bracelets.

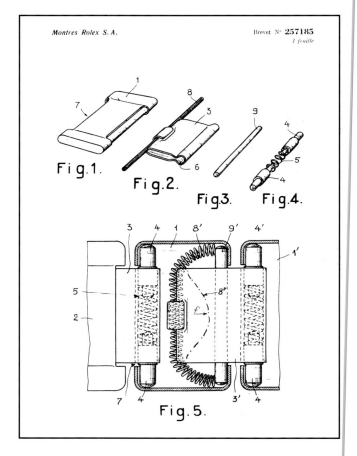

The next major development along the road to the current bracelet was described in Swiss patent 303,005 issued on November 25, 1952 and launched the following year. The subject of the patent was a small curved piece of material which would fit in the gap between the last link of the bracelet and the case. It was launched with some fanfare in 1954 as the "Flush-fit"

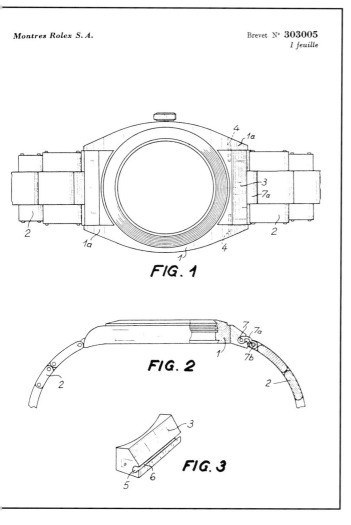

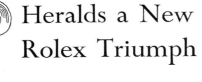

The New Elizabethan Age Heralds a New Rolex Triumph

A watch to measure the adventurous hours of the new reign

FOR this exciting age, Rolex have developed an exciting new wristwatch, in keeping with the spirit of adventure and initiative.

The watch has been called the Rolex "Explorer," and it has been designed to withstand every conceivable hazard.

It is lubricated by a special Arctic oil (tested on several recent Arctic Expeditions), which is guaranteed to retain its fluidity at temperatures as low as minus 40°F.

The famous Oyster waterproof case has been strengthened to withstand tremendous pressures. The Rolex "Explorer" will function perfectly under water to depths of at least 300 ft. and, in the air, to the fantastic height of 12 miles.

The "Explorer" is automatically wound by the unique Rolex Perpetual self-winding "rotor," which, by keeping an even tension on the mainspring, ensures the utmost accuracy, and you may still wear it should your travels take you to atomic energy plants or uranium mines, for it is equipped to withstand most powerful magnetic forces without detriment to its precision.

The numerals of the Rolex "Explorer" are clearly, neatly and elegantly set upon the dial against a jet-black background. And of course they are luminous.

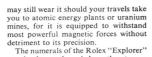

A landmark in the history of Time measurement

ROLEX WATCH COMPANY LIMITED (*H. WILSDORF, GOVERNING DIRECTOR*), 1 GREEN STREET, MAYFAIR, LONDON, W.1

---

bracelet, and the first model to feature it was the new GMT-Master. The first Explorer had missed the boat and was initially introduced with the old style bracelet (*see the accompanying advertisement*), though within a year it was available with the new style. The understanding was that the Oyster was the "sporty" bracelet, and the Jubilee was the "dress" bracelet.

This changed when the Day-Date was introduced as the new flagship watch. The Datejust and its bracelet were demoted from their premier position, shortly afterwards they became available in both steel and steel/gold[73] mixture. Of course, the introduction of the Day-Date brought forth a new bracelet for Rolex, "the President," bringing the total to three. This was in many ways the simplest of the Rolex bracelets. Made from three adjoining pieces per link, it had a massive look at a time when all other Rolex bracelets were made from pressed or folded metal. It looked like what it was. The links were machined from solid metal, either 18kt gold or platinum, and after 40 years it has never been made from any other materials.

In the decision not to make a steel or mixed metal President bracelet we see the long term thinking and intelligence of the management of Rolex. We can not think of a similar British or American company that, when producing such a popular and distinctive product, would not have attempted to "cash in" on that popularity by making the product cheaper and more widely available. Although Rolex is only 90 years old, the company has never looked at things in the short term.

With the Jubilee now available in base and mixed metals it was time for the Oyster bracelet to become available on Datejust watches for the first time. This option appeared in catalogs from around 1958, depending on the country.

Rolex had their dress watch cases made in Britain, France and Canada as well as in Switzerland, but all Oyster cases came only from the Geneva Rolex factory. While this was true, Rolex had no problems with the Oyster or Jubilee bracelets being made overseas. To save on onerous import duties bracelets were made in the U.S., Britain, Japan, and Argentina. All of these bracelets differed slightly from their Swiss originals and are shown alongside. The Japanese "President" bracelets were made in the 1960s by one man, a Mr. Hata, who worked for Isei jewelers. They can be instantly recognized, as each bears a unique number in the range J1 to J200. The American bracelets are the ones most likely to be encountered and they differed from contemporary Swiss ones in a number of aspects. Nowadays all bracelets are made in Switzerland, except for a small number of gold bracelets still made in Argentina. These are immediately recognizable by the lack of the word "Swiss" anywhere on the bracelet and the mark "K18D'A". They are, of course, rarely encountered outside of South America, but are genuine although very few people believe this.

While the Oyster bracelet remains in production to this day, there is still one unique version that differs from all the rest. This is the bracelet bearing model number 93160, which is only fitted to the Sea Dweller. This bracelet differs from all others in that the end pieces which attach the bracelet to the watch are machined from solid metal, rather than being pressed sheet metal, as they are on all other Oyster bracelets, even the Submariner.

---

[71] Which is why versions of this bracelet have turned up marked "Patek, Philippe."

[72] Now used as the site for horological auctions by Antiquorum.

[73] Due to the strict hallmarking laws in the United Kingdom, noble metals and base metals can not be sold together. So in all Rolex advertising in the U.K., mixed metal watches are described as being "stainless steel and yellow metal," although they are enhanced with the same 18kt gold as elsewhere in the world. In fact, the bracelets carry the same 18kt mark, but they can not carry a U.K. hallmark.

306 Bracelets: Around the Wrists and Around the World

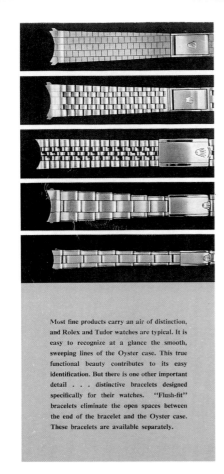

1962 brochure showing Jubilee, Oyster, & Dress Model bracelets.

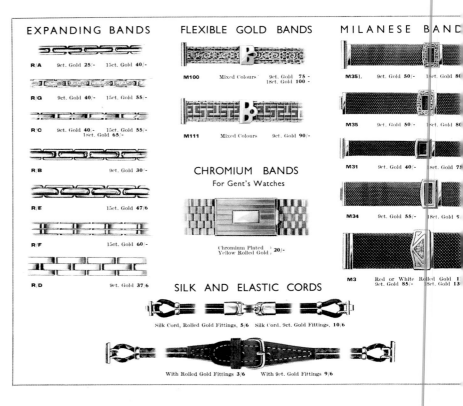

1930 British catalog showing the available bracelets. Not that there is only one band for a man's watch.

The first metal bracelet with the Rolex signature was this Bamboo or "Bonklip" style bracelet, most often found fitted to Viceroy style watches.
Circa 1935

Bracelets: Around the Wrists and Around the World 307

Steel and gold version of the Bamboo bracelet.
Circa 1935

18kt yellow gold version of the "Bamboo"
bracelet.
Circa 1937

Steel and gold Bamboo bracelet, probably not
for Rolex, with later Rolex buckle added. The
crown on the buckle was not used for another
20 years.
Circa 1937

18kt pink gold version of the "Bamboo"
bracelet. The stamps on the buckles were hand
stamped and are in differing positions.
Circa 1937

308 Bracelets: Around the Wrists and Around the World

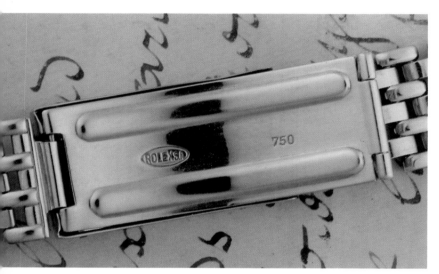

The "beaded" or "grain of rice" bracelet was the next bracelet available for the Oyster. Like the Bamboo and most subsequent Rolex bracelets, it was manufactured by Gay Freres, who made this bracelet for many other watch manufacturers and also for direct purchase.
Circa 1940

"Beaded" bracelet in 18kt pink gold.
Circa 1940

The earliest Oyster style bracelets, like this 9kt version, were parallel sided and did not taper towards the clasp. These also featured Z-spring loaded expanding links. As can be seen from this side view, they were riveted together.
Circa 1945

Bracelets: Around the Wrists and Around the World 309

18kt Oyster expanding bracelet. Notice the flat endpieces which were in use until the mid-1950s.
Circa 1948

Reverse of clasp for 18kt riveted expanding Oyster bracelet made in England for sale in England.
Circa 1948

Reverse of clasp from 9kt Oyster expanding bracelet made for sale in England by CY and Co.

Expanding riveted Oyster bracelet showing date markings 2-54 and manufacturer's name Mecan. Notice this is still a flat end bracelet.
Dated 1951

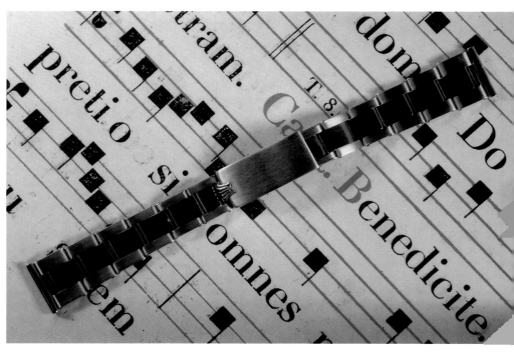

310 Bracelets: Around the Wrists and Around the World

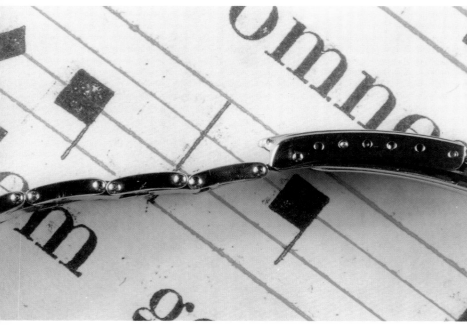

Stainless steel and gold riveted expanding Oyster bracelet. Notice the large "9" next to the date stamp indicating the fineness of the gold used, 9kt.
Dated 5/58

18kt pink gold expanding Oyster bracelet. One of the last of the straight ended bracelets.
Circa 1952

Bracelets: Around the Wrists and Around the World

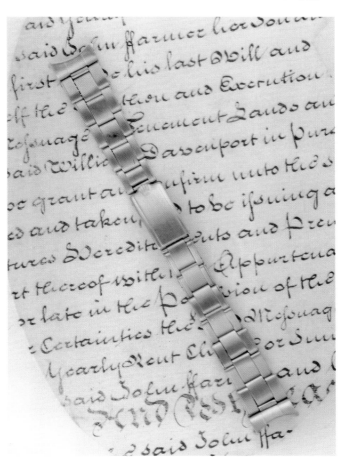

18kt yellow gold expanding Oyster bracelet. One of the first bracelets with the "flush-fit" end pieces, otherwise identical to the previous bracelets.
Circa 1954

Rolex stainless steel expanding Oyster bracelet. This construction no longer uses rivets, but merely folded strips of metal. These were not in production for very long.
Circa 1954

Stainless steel expanding folded metal bracelets.
Dated 4/64

Buckle of a stainless steel Oyster expanding bracelet signed "Rolex USA" and "C&I." These were sold only in the U.S.
Circa 1968

Stainless steel expanding and riveted Oyster bracelet in "New Old Stock" condition.
Circa 1967

Stainless steel Jubilee bracelet marked "USA" with no makers mark. Circa 1965

Reverse of concealed clasp for Oyster Perpetual Day-Date with unusual signature on the right showing that this bracelet was made for Rolex Argentina. This clasp does not bear any of the usual Swiss control or hallmarks. This and the Japanese version are the only authorized President bands to have been manufactured outside of Geneva. Production has now ceased.
Circa 1988

A comparison between the bracelets fitted to the Submariner and to the Sea-Dweller. The Submariner bracelet (top) has pressed steel "Flushfit" endpieces while the Sea-Dweller has them machined from solid. The extension links, once again, are solid in the Sea-Dweller and simply sheet metal in the Submariner.
Circa 1980

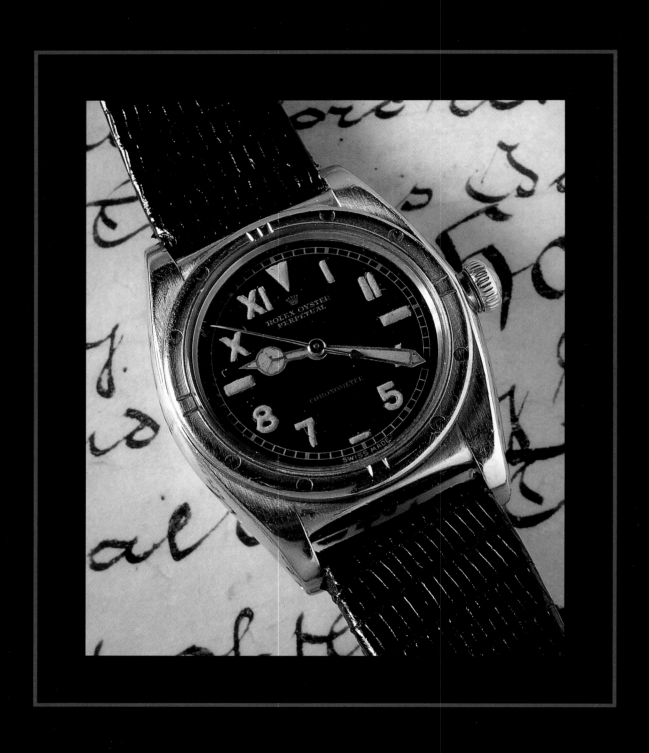

CHAPTER 19
MOVEMENT
Under the Covers

Most of the methods of identifying different Rolex movements require either an encyclopedic knowledge of the movements used or the freedom to explore under the dial. Most people have neither, so we are attempting to set down a method of identifying the main movements used by the company and the dates they were produced while only being able to examine the rear plate (usually the only part of the movement immediately available to a prospective purchaser).

Rolex have always used movements from Aegler. The very first wristwatches the company ever made were powered by jeweled lever movements from the factory of Jean Aegler, a factory started at the end of the 19th century and one which had only made small movements. The location of the factory, in the street called Rebberg, gave its name to the movement that made the company famous. The Rebberg movement was introduced in the early years of this century and was initially available with both lever and cylinder escapements. During the early years of these movements, Aegler sold raw movements to other companies who finished, cased, and sold them under their own names. They also had some movements cased themselves and sold them signed with the name "Rebberg Watch Co."

All of this happened before Aegler had ever heard of Hans Wilsdorf. But Wilsdorf had heard of Aegler during his 1902 stay in La Chaux de Fonds. Ten years later, convinced of the imminent arrival of the wristwatch, he gave Hermann Aegler (Jean's son) the largest order that either firm had ever received or given. The order was the starting point for a relationship which was to last for the rest of their lives. Before long Wilsdorf & Davis wristwatches were being sold throughout the British Empire. These watches were all powered by Rebberg movements from Aegler ranging in size from 8-3/4''' to 13'''.

The Rebberg was produced in two different grades, 15 and 7 jewels. The better version had an all over machined finish to the plates and featured a main winding wheel bearing the legend "Rolex 15 Jewels." The lower grade version had plain polished plates and the winding wheel was signed simply "Rolex." There are a very few versions of the 13''' Rebberg which were made in "Prima" quality. These are always 18 jewel movements and have a checkered finish to the main plate, capped endstones, and plain finish winding wheels with all the engraving on the main plate. These are extremely rare movements and always seem to turn up in high quality cases, often the English made "Dennison" one, which was the highest quality case ever made for the Rebberg. The Rebberg continued in use for nearly a quarter century, from the early days of this century up until around 1927. It was still produced for a few years after the introduction of the Hunter movement in 1923, and the late Rebberg movements are immediately identifiable by the plate finish which mimics that of the recently introduced Hunter.

When the Rebberg movement was finally laid to rest, Wilsdorf and Aegler could hardly have thought that its replacement, the Hunter, would in fact have a production run of almost twice as long. First introduced in 1923 in a 10-1/2''' size it finally ceased production as Calibre 700 in 1969 in the same size. In between those two dates the movement had been produced 7-3/4''', 8-3/4''', 9-3/4''', as well as the original 10-1/2''' and the latter three had also been converted into self-winding versions. The success of the Hunter, and its longevity, can in a large part be ascribed to the continual improvements Rolex made to it and to the fact that all seven versions of the movement were essentially identical, differing only in scale. This made things much easier both for production as well as for repair.

The first 10-1/2''' Hunter movement from 1923 would not be recognizable as such to most people, because the center wheel bridge is completely different from the later, more common version. It is almost "S" shaped and covers much less of the movement than the later version. This first version seems only to have been produced until 1925 and is not even used on all the watches in that period. All of the Hunter movements have a polished rhodium finish, except one or two of the very earliest versions which strangely have a finish which mimics that on the Rebberg.

The initial version of the 10-1/2''' Hunter was made in three grades, Prima, Extra Prima and Ultra Prima. All were simple 15 jewel movements. The two better grades would usually be signed "Timed 6 positions for all climates." The Hunter was the first Rolex movement that was regularly capable of being timed to chronometer precision. These movements however were often either 16, 17 or 18 jewel movements with capped escape jewels, and were always inscribed "Chronometer" on the movement and often also on the dial. The specific movements which obtained bulletins from the Swiss official timing offices are recognizable by the fact that they are numbered on the movement plate.

In 1935 Rolex patented a new balance wheel design under patent 188077 and quickly brought it to production as the "Superbalance." In this design the balance timing adjustment screws were recessed from the external surfaces of the balance wheel, all of whose exposed surfaces were then rounded to present a much more streamlined form. With the introduction of the Superbalance Rolex the three grades previously used were abandoned and all winding wheels were simply signed with the legend "Patented Superbalance."

However the abandonment of the graded and signed winding wheels did not mean that Rolex was now only producing one grade of movement. The 10-1/2''' was still mainly a 15 jewel movement but was also made in 16, 17 and 18 jewel qualities. The visible differences now were mainly in the quality of the jeweling. The basic movements having the endstones push fitted, the higher grade movements had them capped, and the highest grade versions had the endstones fitted in screwed chatons. A very few watches have been seen where the normal ruby endstones are replaced with sapphire ones. These are the highest grade of production movement made by the company at the time and are always fitted with many other high grade components such as micrometer regulation of the hairspring.

The other visible method of judging the quality of the movement was by noting the number of adjustments engraved on the main plate. The simplest movements had only two adjustments while better grade movements had five, six or seven

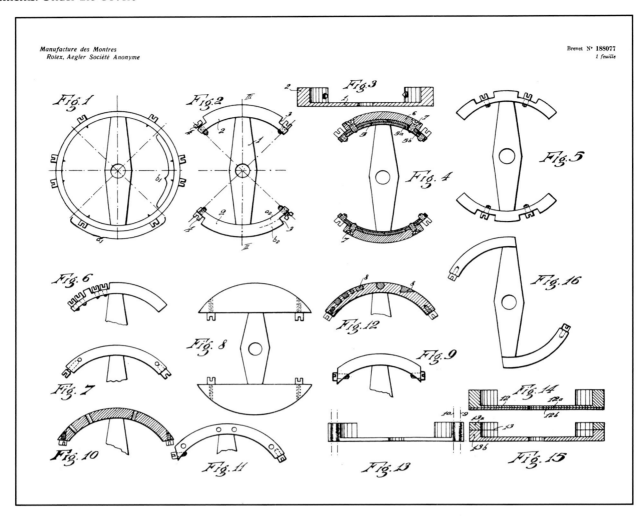

adjustments. To make a later movement appear early the winding wheels are often changed and the balance wheels sometimes are. There is a foolproof way of telling the difference between an earlier and a later movement, but it requires the ability to look at the top plate, under the dial. The early models are signed "Rolex 10-1/2"' Hunter Patent" while the later superbalance models are signed "Rolex Patent 10-1/2"' H".

Between 1935, with the introduction of the superbalance, and the demise of the 10-1/2''' Hunter in 1969, the company produced two further iterations of this same 10-1/2''' Hunter calibre. The first, which seems to have appeared around the end of World War II, has a much simplified balance wheel and the movements are usually 17 jewels and signed "Precision" on the main plate. The later one, brought in around 10 years later, also used this simplified balance wheel and is signed "Rolex 15 jewels Swiss made" on the winding wheel. These final versions are called calibre 700 by Rolex, although they are still signed "Rolex Patent 10-1/2"' H" under the dial.

All of the discussion above covers the subsidiary seconds version, which was the initial model. The sweep seconds version was introduced within a year, and came in two styles. This first model sweep was the more radically altered and even had its own name, being signed "Rolex 10-1/2"' Openface" under the dial. It is immediately recognizable because the balance wheel is now at the "12" position, rather than the normal "9" position of a 10-1/2''' Hunter. This meant, of course, that the whole winding mechanism also had to be relocated. The problems involved in producing this version caused the company to design and begin producing a sweep seconds version in 1934 which used the standard calibre. Apart from the additional driving wheel, spring, and center seconds pinion, this model can be quickly identified by a peculiar adaptation Rolex made to the normal movement. These later sweep seconds models have two much taller than normal securing screws for the crown wheel, the first and smallest of the two winding wheels. The purpose of this is to protect the center seconds wheel which sits on the surface of the movement. If, by accident, the back of the case receives a blow it will hit these raised screws before it can cause any damage to the delicate driving wheel and pinion.

While the Hunter continued almost into the 1970s, it took an important detour in the early 1930s when the 8-3/4''' version became the base calibre of the very first perpetual movement from Rolex. These very first perpetuals were simply 8-3/4''' manual movements placed inside a double stepped frame, which completely surrounded the perimeter and rear of the movement. They were still signed 8-3/4''' Hunter under the dial and the rear portion of the movement cover (around which the rotor moved) was almost completely covered with instructions (in English and French) for the unsuspecting watchmaker on all three faces of the frame.

These earliest model Perpetuals have three immediate identifying factors: the engraving discussed above; the rotor, which has a star and the words "Rolex Auto Rotor" engraved on its perfectly flat top surface; and a small slot in the top plate of the frame (at around the "7" position) to allow adjustment of the regulator without removing the rotor and its accompanying plate. These earliest versions are rare and were in production for less than three years.

They were replaced with a model which had less engraving on the frame and no longer had the star on the rotor, although the rotor is still flat. The "Swiss Made" which had been engraved around the center retaining screw is now engraved above the word "Patented."

Speaking of the center retaining screw, many people have wondered why there are two screw holes but always just one locking screw. To understand this it is necessary to look at a number of Perpetual movements. Sufficient examination of movements (or pictures) will confirm that the only screw is always on the bulge of the rotor and that no matter how many movements one looks at, there are only five positions for the locking screw. This is because the bulge of the rotor is where the matching holes are. There are five holes so that the locking screw can be fitted as close to the point at which the center retaining screw is correctly tightened. A locking screw is needed because constant rotation of the rotor may cause the center retaining screw to become undone. This center retaining screw is the single common feature linking the first Auto-Rotor movement of 1934 to its final incarnation in the calibre 640 Moonphase Oyster Perpetual.

After these first two versions of the Perpetual movement, the next iteration is recognizable by two visible features: the words "Super-Balance" are engraved on the rotor in a semi-circle; and the edge of the rotor itself now has a pronounced 45° angle. This version was produced from 1939 to 1941 and was the last movement which used the 8-3/4''' Hunter as a base. All of these previously discussed movements are only found in three-piece cases and are therefore always fitted in a movement ring.

The last of the original movements was soon replaced by the much more commonly seen version in which the chamfer on the edge of the rotor is increased to nearly 60° and the words "Super-Balance" are now engraved along the flat edge of the rotor. The movement now reaches to the edge of the case and there is no longer a movement ring (still sometimes used in 18kt three-piece cases, particularly model 3372). This version is the first one with the inscription "Chronometer," although it is often encountered without it. It retains the legend "Auto Rotor." By 1944 the movement changed again and adding the signature "Rolex Perpetual" for the first time. The rotor had a pronounced double angled slope to its edge. It was the change in the angle of the rotor edge that allowed the watch to develop its familiar "bubble" dome shape, rather than the deeper flat back of the original Perpetuals.

In 1944 the movement was changing once again. This time it grew from being based on the 9-3/4''' Hunter to the 10-1/2''' Hunter and it was from this new larger version that the Datejust and most subsequent Rolex Perpetuals grew. The smaller movement continued for another ten years but with very few visible modifications. The energies of the company were now focused on the new larger calibre. Launched as the calibre 720 it continued with almost no visible changes until 1950, when it was replaced by a completely new design, the calibre 1030. The new calibre was instantly identifiable by a new rotor design which was flat, had two angled cuts ending in circles and had the Rolex coronet engraved on the rotor with the legend "Rolex Perpetual." The rotor was no longer located by a screw, but by a circular spring which was pushed on to the central shaft. The 1030 broke with two hallowed traditions of all previous Rolex automatics: it was the first not to be based on a Hunter movement; and it was also the first Rolex movement to wind in both rotor directions.

The two cuts in the rotor were designed to solve a problem that had affected all previous Rolex automatics, the so called "rotor shake." This happened when excess movement of the heavy edges of the rotor caused the thinner top plate to twist, and it had two unfortunate effects. The top of the rotor could make contact with the inner case back and the central rotor bearings would then wear out, causing the rotor shake to become much worse and then make contact with the case back. The cuts prevented this by stopping the distortion at their outer faces. Any movement which made its way across the cuts would then be absorbed by the new four-armed spring which held the rotor in place, replacing the old style screw. The 1030 proved to be a very successful calibre for Rolex, but its reign was short lived as technology moved along and new developments forced its replacement in 1957 by the new calibre 1530.

The new calibre featured a flatter rotor with five complete annular cuts and two open cuts. This new calibre was instantly identifiable by the two click wheels which were colored red. The movement was equipped, for the first time in a production Rolex, with a "free sprung" balance. The balance, produced for Rolex by Stella, was timed by screws on the wheel rather than the traditional regulator. It was also unusual in that it featured timing screws and their singular tool, which allowed timing adjustments to be carried out without the removal of the balance wheel. The improved accuracy brought about by the introduction of the new calibre resulted in one significant change to the face of the watches so equipped: they were now signed "Superlative Chronometer Officially Certified" replacing the "Officially Certified Chronometer" of previous dials. This new designation is still in use some forty years on, although the accuracy of the watches so signed has improved dramatically.

The 1500 series of movements proved to be another in the series of extremely long-lived Rolex calibres, remaining in production from 1957 until 1977. There were, of course, a wide range of variations in the basic calibre. The first to be introduced was the 1535 with progressive calendar mechanism. This was rapidly followed by the 1555 "Day-Date" version, then by the 1565 GMT also used in the Explorer II. The first major change came in 1964 when the faster beat train was introduced, which oscillated at 19,800 per hour, 10% faster than the 18,000 per hour of the 1530. The introduction of this new balance work meant that all the calibres received new numbers. The 1530 became the 1520, the 1535 became the 1525, the 1555 became the 1536 and the 1565 GMT now operated under the designation of 1575. With the introduction of the revised calibre came a brand new version designed for the long obsolete Milgauss. The new model 1019 Milgauss sported the previously unheard of calibre 1580. All of the movements previously mentioned were chronometers, but there always were non-chronometer versions of the basic model used in watches such as the Air-King, and these were initially powered by the 1560 which was then replaced by the 1570.

It is possible to see the 1964 change to a faster beat balance as a pathfinder to the calibre 3035, a true fast beat movement whose balance rotated at 28,800 revolutions per hour. First introduced in 1977 in the "Datejust", the calibre 3035 was, at a stroke, capable of improving the accuracy of the watches fitted with it to such an extent that it destroyed the accuracy marketing edge of the newly introduced quartz movement.

The calibre 3035 and it successor calibre, 3135, are immediately recognizable without even opening the back of the case. The 3035 was the first calibre to be fitted with a "quick set" date wheel. The watches fitted with calibre 3135 can also be immediately recognized as they are only fitted to watches with a sapphire crystal. As before, the 3000 series of calibres has proved to be exceptionally long lived, having now been in production for almost twenty years.

318 Movements: Under the Covers

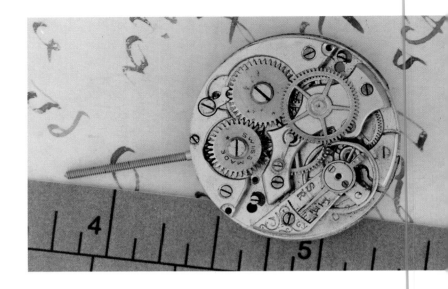

Extremely rare Rebberg 11-1/2''' cylinder escapement, sweep seconds gilt finish movement. It has always been assumed that Aegler made only lever escapement watches. The combination of a cylinder escapement and sweep seconds is rarely seen.
Circa 1900

This is the earliest known Rebberg wristwatch and is the predecessor of all Rolex watches. It shows that Aegler were making watches for direct sale prior to their association with Rolex and that they made wristwatches even in these early days. It features an 8-3/4''' movement and it is extremely unusual in that the case and movement are both signed Rebberg. Bezel shown has blue translucent enamel on top of the 14kt gold.
Case no. 517535
Circa 1900

10-1/2''' unsigned Rebberg in case signed W&D. Showing that in the early days the company used unsigned movements.
Circa 1906

Movements: Under the Covers 319

First known style of movement to come from the original W&D (Wilsdorf & Davis) company, 9-1/2''', lever escapement, 15 jewels, checkerboard engraved plates and signed W&D on the bridge. These W&D movements, which were made from 1905 to 1908 seem only to have been used in lady's watches, as production of men's watches did not begin until around 1910.
Circa 1907

13''' Rebberg movement, lower grade, 7 jewel quality. This style of movement can be recognized by the lack of any visible jewels and the fact that the winding wheel is signed simply "Rolex" and not "Rolex 15 Jewels."
Circa 1908

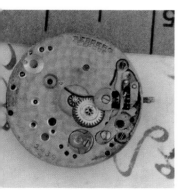 

7-1/4''' Rebberg showing earliest style of Rolex signing on the plate.
Circa 1908-1910

10-1/2''' transitional Hunter movement, notice sweeping curved center bridge, as distinct from standard model seen below. This style of movement was only made for three years between 1923 and 1925.

10-1/2''' Hunter movements as used in early cushion oyster watches, they were available in three grades: Prima, Extra Prima and Ultra Prima. As can be seen here, there is absolutely no visible difference in the quality or the finish of the movements. They were graded after timing and the correct winding wheel then fitted.
Circa 1927

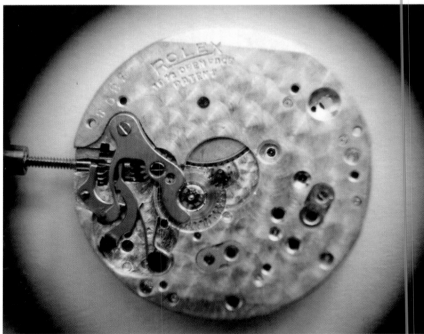

10-1/2''' Rolex open face movement. These very rare movements were made for only a few years in the late 1920s and early 1930s. Notice how the bridges are rotated by 90°. this style of movement was the earliest style of Oyster sweep seconds, but was later replaced by a conventional sweep movement shown next.
Circa 1929

Movements: Under the Covers 321

10-1/2''' Hunter movement displaying two of the later additions; the Superbalance and the indirect drive sweep seconds from a normal movement.
Circa 1936

First model Rolex automatic, which was made from a basic 8-3/4''' movement set into a frame on which the self winding assembly was fitted.
Circa 1934

First model Perpetual movement, showing watchmaker's instructions and slot for adjusting the regulator without having to remove the whole auto winding mechanism.
Circa 1934

Second model automatic, edge of rotor is now slightly angled and is now signed "Super-Balance."
Circa 1935

322 Movements: Under the Covers

Rear of movement similar to above, showing progressive disassembly. Notice full watchmaker instructions on rear of movement cover.

Movements: Under the Covers 323

Third model automatic. Notice that there are now fewer instructions on the rear of the movement cover.
Circa 1939

Fifth model automatic. No longer signed "Auto Rotor," but "Rolex Perpetual." Note that the movement is now numbered. This was normally only done for Chronometer grade watches.
Circa 1942

Fourth model automatic. Notice the much more angled edge to the rotor and that it is now signed "Chronometer."
Circa 1938

Movements: Under the Covers 325

The first two styles of Oyster winding crown. The upper style is the "Onion" shape, first used on the cushion up to 1929. The lower is the "Drum" shape which replaced the Onion, and continued until the early 1940s.

Bubbleback crowns. The crown in the lower right, signed only "Rolex Oyster," was used in the early 1940s for about two years. It was replaced with the ones above, marked with a box, which continued until the late 1940s. They were replaced with the crown at the lower left, marked with the Swiss cross.

Bubbleback crowns. The lower pair with the finer milling were used from 1936 to about 1938, before being replaced with the upper ones which continued for another two years.

Super Oyster crowns introduced in late 1950 and produced for only three years. This style of crown does not screw down and relies on a tapering tube and multiple gaskets to ensure its impermeability. It was not a success.

Upper left: manual Oyster "dimpled" crown, used in the late 1930s and early 1940s on watches such as the Viceroy and Plage. Upper right and below: A pair of "dress" crowns used in the 1940s and 1950s.

Submariner crowns. Top right: "Twinlock" crown, waterproof to only 300 feet and used on the early 5512 models, circa 1954-1958. Top left: prototype "Triplock" crowns, waterproof to 600 feet bearing the legend "Brevet" (patented), circa 1958-1959. Bottom: final development of the "Triplock" used on current model submariners.

CHAPTER 20
From Merchant to Market Leader

There is a saying that "the child is father to the man," meaning that to know the future you must first examine the past. We hope that the previous chapters have helped in this examination. Most of the preceding has come from a wide variety of written or oral sources. When we look to the future, however, there obviously are no sources. All we can do is to understand the paths already taken and try and divine the route ahead.

It is difficult to understand Rolex without understanding Hans Wilsdorf, and understanding Wilsdorf is not easy. To make it easier one must realize that in all of the documents Wilsdorf signed when setting up Rolex he gave his occupation as "merchant." If we look up the dictionary it defines a merchant as "One whose occupation is the wholesale purchase and retail sale of goods for profit" and that describes Wilsdorf perfectly. Nowadays we would call Wilsdorf an entrepreneur and a marketing genius. In the jargon of that trade Wilsdorf could identify a niche in a market and exploit it perfectly.

Wilsdorf was not an inventor. Instead he took other people's inventions, improved upon them and marketed them successfully. In the way that it was John Bardeen from Bell laboratories in New Jersey who invented the transistor (and in fact received the 1956 Nobel Prize for so doing), but it was Akiho Morita of Sony who first made money from it and built an industrial empire, Harwood invented the wristwatch with a self-winding movement powered by a centrally pivoted rotor, but it was Wilsdorf who made money from a similar device.

Wilsdorf could see opportunities where others could see only problems. Despite being possessed of a huge ego (how else do we explain the fact that every advertisement for Rolex up until his death had his name at the bottom?), he had no problems in submerging his ideas and ideals into the Rolex philosophy.

It may seem strange to talk about an inanimate object, such as a corporation, as having a philosophy, but any careful study of the company's history will reveal a pattern of decisions that can only be explained as a philosophy or (perhaps more skeptically) as a perfectly executed long range corporate strategic plan.

This philosophy can best be described in two simple phrases: "Whatever the customer needs, give him something better" and "Never change for change's sake but always improve."

Rolex watches could be described by the skeptical as being "over-engineered." We prefer to say that the products are designed to take everything that the most demanding customer, in the worst possible circumstances, could ever throw at it and still function as it is supposed to. Rolex is analogous to the Mercedes-Benz automobile. When you look at any Mercedes passenger car produced since World War II, you can see the family resemblance and also the very gradual change from one style to another. You need only take one quick glance at a car to tell instantly that it is a Mercedes. In exactly the same way any Rolex Oyster from the post-World War II period is also instantly recognizable as one.

In engineering Mercedes' longevity is so taken for granted that in almost every country in Europe (and often in the third world) they are used as taxis and expected to attain 200,000 or 300,000 miles with little more than routine maintenance. Yet the presidents of these same countries will also be chauffeured in a Mercedes. Mercedes has been able to meld prestige, longevity and respect into an image that is unmatched by any other consumer product in the world, except the Rolex Oyster.[74]

The second half of Rolex's philosophy, "Never change for change's sake but always improve" is an equally important part of the company's success. It is possible to take one of the very first Datejust models from 1945 and place it next to a current model and immediately see the family resemblance. However a detailed examination of the two watches would show that there are no common parts other than the odd screw here and there. Taking another German motoring analogy (this time at almost the other end of the spectrum from Mercedes) the Rolex is, in many ways, similar to the Volkswagen "Beetle". A first post-war Beetle looks, to the general onlooker, not dissimilar to the current "Beetle" still produced in Brazil and Mexico. But a close examination will reveal over 78,000 modifications in the current car. A close examination will probably find a similar number in the two Rolex Datejusts.

Rolex is fortunate in having two major advantages not available to their major watch competitors. First because it were founded only ninety years ago, it was not bound by the tradition and history that most of the long established firms were. Rolex did not even have its own factory during its formative years. It is easy to see how the company felt free to explore areas of horology that the other companies either ignored or felt unworthy of them. The other major companies, including both the mainstream firms such as Longines and Omega and the Grand Marques such as Patek Philippe or Vacheron et Constantin entered their finest pocket watches into the annual Geneva chronometer contests and boasted loudly about their victories. Without a factory or a hundred years of precision watchmaking behind them, this route was closed to Rolex. Instead it took the road less travelled and entered a wristwatch movement in the same competition. Of course, it didn't win but its scores were high enough to make people sit up and take notice.

The other major advantage Rolex has had over every other watch company is that in the ninety years of its corporate life the company has been run by only two men. In the fifty five years from 1905 to 1960 Hans Wilsdorf was Rolex and in the thirty five years since 1960 André J. Heiniger has steered the company to even greater successes. It can not be stressed too strongly how important this clear, distinct management line is in maintaining a corporate philosophy or vision. Most of the competition have had, in the same period, between five and eight people at the helm, and with this constant change it is very difficult to maintain a consistent goal. Rolex was helped in its ability to cling to a chosen course by Wilsdorf's far-sighted decision in 1945 to assign all of his shareholding in the company to the "Hans Wilsdorf Foundation." Besides the obvious charitable and philanthropic causes that it supports so nobly, the major benefit of the Foundation to the company is that it automatically prevents any further sale and dilution of the company's stock.

To celebrate the 50th anniversary of the launch of the Rolex Oyster in 1926, the company decided to institute the Rolex Awards for Enterprise. These "were established to provide help and encouragement in breaking new ground in the fields of Applied Sciences, Invention, Exploration and the Environment." The first awards were made in 1978 and again three years later, and have continued to be awarded every three years since. The most recent awards were made to five laureates who received their prizes for such diverse projects as one for monitoring ozone, solar ultraviolet radiation and atmospheric parameters to another for pasteurizing camel milk and making camel's cheese in Mauritania. The five laureates were flown to Geneva where each received a check for SF50,000[75] and a gold Rolex Day-Date chronometer. A further 36 people were awarded an honorable mention and received a steel and gold Rolex Oyster in a ceremony near their home. While Rolex has always been known for its association with sports and exploration (after all, two of their watches are actually named the Explorer), the addition of awards in the fields of Applied Sciences, Invention and the Environment showed how the company was responding to public concerns.

The very existence of the foundation and its ownership of the controlling block of shares also means that even though Wilsdorf is no longer physically at the helm of his company, his goals are still the company's goals.

If having a goal is important, sticking to that goal is even more important. Sticking to that goal at a time when all around you are moving off in exactly the opposite direction takes real courage. This was the position in which Rolex found itself during the late 1960s and early 1970s. The Swiss watch industry initially responded to the Japanese quartz onslaught by adopting the "ostrich" method and denying there was any such problem. Next it switched tack and fought the Japanese head to head on two fronts. First it tried making uglier watches than even the Japanese had dared. Then it began playing a game it was destined to lose: trying to beat the Japanese at producing digital quartz watches.

It is easy to see now that the Swiss industry was on a losing bet. The signs were there for anyone to read. The Japanese were on a learning curve and they were riding it for all it was worth. They had already decimated the British motorcycle industry, left most of the world famous German camera business bleeding on the ropes, and were gazing covetously at the U.S. automobile market. But identifying these things is being wise after the event. At the time the generally accepted Swiss industry wisdom was that the Japanese had to be met and beaten on a ground that the Japanese had chosen. Rolex declined to accept the industry wisdom and in the end Rolex came out the other side stronger with its image unsullied and its profits still on the upward climb. Most of its erstwhile competitors who followed the industry wisdom and fought it out with the Japanese either went belly up or were "re-organized" and subsumed within the arms of one of the two competing giant ebauche houses[76].

In a rare 1978 interview André Heiniger said of Rolex, "We have been content to eat a piece of the cake, rather than try to have the whole cake. We do not want to be the biggest, but certainly one of the finest in our field." He also expressed the company's goals concisely in the following quote; "The Rolex strategy is oriented to marketing, maintaining quality and staying out of fields where we are not prepared to compete effectively." That is a dazzling concept. Every marketing professional, when asked about his company's goals, will trot out those first two points that Heiniger made, but not one in a thousand would even understand the clarity and brilliance of the final point. Success comes from staying out of areas where you are not guaranteed to win.

Rolex has made a conscious decision to remain apart from (some might say, aloof from) the rest of the Swiss watch industry. In the beginning they had no choice in this matter. Amazingly, the company was not regarded as Swiss by its competitors. It was an alien interloper on the hallowed ground of Geneva. Wilsdorf himself was universally disparaged by most of the Geneva establishment. They could probably have accepted the fact that he was German but that he had actually chosen to become (and to remain) British, this was a difficult bone for them to swallow.

Wilsdorf himself lived his entire life as an "outsider." We have noted that he was born in Bavaria, but the opening sentence of his only autobiographical writings[77] states, "I was born on March 22nd, 1881, of Protestant parents..." Bavaria was the most fiercely Catholic of all German provinces and, as a Protestant, Hans would have been aware of his unacceptance by his schoolmates. The later move to a boarding school would have reinforced the feelings and, from the moment he left the school, he never lived in Germany again, spending the rest of his life as an alien.

These early experiences seem to have had a formative effect on the company. During its early years Rolex sold very few of their watches in Switzerland, and those few mostly to jewelers such as Bucherer and Beyer who then sold them mainly to tourists. To this day, Rolex has chosen to remain in isolation from not only the rest of the watch business but also from the rest of the world. Unlike almost all watch company presidents, André Heiniger is rarely seen at the annual Basel fair. He never gives interviews and this attitude is reflected in all of the company's dealings with the outside world. Even the company's advertising agency for fifty years, J. Walter Thompson, which would normally be promoting its client at every opportunity, will not discuss Rolex, other than through the advertisements it places.

The initial reasons that drove Rolex behind its walls may be in the past, but the company keeps the walls now for its own advantage. Churchill once described Russia as a riddle wrapped in a mystery inside an enigma; many feel that the same could well be said of Rolex. The reality that the company is now completely self-funded, all its residual profits[78] are automatically paid to the Wilsdorf Foundation and both these entities are protected by Swiss privacy laws which are probably the strongest in the world[79]. The company prefers to let its products speak for it.

Over the last few years Rolex has invested heavily in vertical integration, attempting to acquire all of the components and steps in the production and distribution chain. Initially Rolex owned almost nothing. They bought their movements from Aegler and their cases and dials from other sources. They then assembled the watch and sold it to wholesalers who then sold it to the retailers. First Rolex acquired their suppliers, beginning with Aegler, then Genex, where the cases are now made, and, subsequently, many of the bracelet companies. But the supply chain was only one half of the equation. Just as importantly Rolex began to take over their distributors. Now they own almost all of them. The advantages to the company are obvious, they are able to retain the profit that would otherwise go to the parts manufacturer or the distributor, and, of course, there are no longer any worries about supply.[80] Rolex has gone so far with their process of vertical integration that they own almost all of the buildings in which their offices are situated, thereby making sure that no landlord will make a profit from them.

The company's major accomplishment in recent years has been its positioning of its mainstream product as a "lifestyle" purchase. Today a Rolex is valued as much as a statement about the wearer and his or her socio-economic position, as it is as a timepiece. People make their own decision that they are now ready for a Rolex, and woe betide any retailer who dares to suggest a competing timepiece no matter whether more or less expensive. It is worthwhile to stand in a jeweler's shop for an hour or so and watch the differing attitudes between people who come in to buy a watch and those who come in to buy a Rolex. An undecided customer may take 5 to 20 minutes looking at a variety of watches whereas a Rolex customer knows exactly what he wants and is usually out of the door with the watch on his wrist before the undecided customer has even begun. In other words Rolex watches sell themselves, to a large extent. The purpose of the massive advertising campaign is merely to confirm and reinforce the buying decisions the customer has already made. Their major sponsorship decisions are perfectly targeted. Each year the Wimbledon tennis championships are played out in front of a Rolex logo and the company is, of course, a major golf sponsor.

It would seem that Rolex has always been mindful of the supply/demand equation at work in the marketplace. Its watches have been kept just out of the reach of everyone, not by any means unobtainable but just far enough that one needs to stretch. Using this policy Rolex has controlled the supply and the price. While the price rises in the last few years may have seemed excessive to many, it is important to remember that the company pays all of its bills, from raw materials to salaries, in Swiss Francs, one of the fastest appreciating currencies in the world.

It is interesting to watch the activities of the other Swiss watch firms over the last ten years as they have struggled to climb back to profitability and to regain the ground they had lost to Asia at the lower end of the market, and to Rolex at the upper end. Many of the companies reintroduced replicas of older models in limited editions. Others introduced new limited edition models, and yet others brought forth new and wonderful case materials. Rolex stayed clear of all this, followed their own goals, kept their independence and are still pursuing the same goals that Hans Wilsdorf established ninety years ago.

---

[74] The two presidents of Rolex, Hans Wilsdorf and Andre Heiniger, would never be seen in a Mercedes, however. Both were long time Rolls Royce owners.

[75] Approximately £28,400 or $43,500 at December, 1995 rates.

[76] Shortly after the two conglomerates merged. There is now only SMH (the house that Swatch built), which owns Longines, Omega, Rado, Tissot and of course ETA.

[77] The *Rolex Jubilee Vade Mecum*, published in 1946 to celebrate the 40 year Jubilee of the company.

[78] That is, above the sums needed to re-invest in the company.

[79] Remember, this is the country that invented the numbered bank account.

[80] Interestingly the only recent Rolex watch which has ever suffered from a supply problem is the Daytona Perpetual, which is fitted with a movement from Zenith, a company outside the Rolex family.

CHAPTER 21
ODDITIES
W.S.H.

On Saturday, December 17, 1994 the BBC in London transmitted a television program with the title of "WSH." Anyone watching the program would soon have realized that its title was the initials for "Weird Stuff Happens."

Sometimes the unexplainable does happen, and that is as true at Rolex as anywhere else. These oddities are the subject of this chapter, in which we look at some of the stranger and sometimes unexplainable products of the Rolex factories.

When we look at Rolex products from the early days up until the 1940s and encounter uncataloged or previously unknown Rolex items the initial attitude is to dismiss these items as a fakes. This sometimes is true, but just as often an item is in fact an aberration, but genuine. As with postage stamps, where the incorrectly printed pieces are the ones most sought after, so it is with Rolex.

The problems first arose in the earliest of days. One of the watches featured in the Oyster chapter can be seen, in close up, to have a case back where the serial number was obviously incorrectly stamped. Because the number was inside the back (where it would not be obvious to the customer) it was considered expedient to score out the original number and stamp the correct one alongside.

As we have shown, with non-Oyster watches it was common to hand stamp the last two digits of the case number on the inside of the case top. The early silver tonneau watch shown here has a case number of 576083 while the number stamped in the top is 082, a difference of 1. Either the workman who was stamping the tops picked up the incorrect number punch, or (more probably) the incorrect top was picked up during final assembly.

These are the easy ones, the ones we can put down to human error and can work out a logical sequence of events by which we can explain them. But others cause us to stare in disbelief and then shake our heads in wonderment.

The white-colored metal cushion Oyster wristwatch shown is also very strange in that it is neither steel, nickel or sterling silver, but 14kt white gold. There were very few 14kt cushion Oysters ever made (for the simple reason that 14kt is a U.S. market requirement and of course Rolex was not selling in the U.S. at the time) and no other recorded white gold cushion Oysters. A close examination of the inside of the watch raises even more questions. The back is marked "7 World Records" an inscription normally seen in watches made prior to 1924. The watch bears only details of the original Swiss patent, which, as we have seen, was patented by Perret & Perregaux, not by Rolex. Finally it is marked with a serial number of 20,005 when the previously recognized authorities insist that the Oyster numbers begin at 27,000.

Next comes a perfect Oyster Royal in 9kt gold bearing a London hallmark for 1934 and a perfect presentation engraving for 1935. Everything about the watch is over 95% in condition rating, however, the back of the case carries no Rolex or Oyster markings other than an RWC Ltd cartouche that looks as if it had been scratched on with a pin. The watch is stamped "ATO," which we initially took to be the case stamp of Leon Hatot the French watchmaker, who traded under the name ATO. However further research has shown the mark to be that of Albert Thomas Oliver, who at the time was running the family casemaking business in London's Clerkenwell. The Oliver family had made watch cases (primarily for pocket watches) for over 100 years and, when the business finally ceased trading in the late 1980s, their workshop was transferred complete to the watch and clock museum at Preston in the north of England. The hallmark is from London and dated 1934, but Rolex was hall marking all of its watches in Glasgow by 1926. The numbers used inside the case for the serial number are of a different typeface than that used by Rolex, yet they are in the correct sequence for the supposed date of the watch. The watch has fixed bars, as did most wristwatches sold in Britain at that time, and one of the fixed bars carries the last three digits of the case number. The top of the case carries no touch mark, either Swiss or British, though these watches were always stamped on the underside of the lug and often on the side of the case opposite the winder. The Oyster case was supposedly never made outside Geneva, but this watch and the two or three others with similar markings from the same period seem to have come from the only batch ever made anywhere else. Much about the cases seems to confirm their authenticity: the numbers are in the correct sequence, the case is hallmarked London, where Oliver was based, the case material is 9kt which would be correct for Britain, and these cases date from a time when Britain (and its Empire) was still the company's major market. So perhaps Rolex was attempting to make the cases in Britain and thereby reduce the import duties they would otherwise have to pay. If they did make some prototype cases in Britain, who better than Oliver to make them? But once again, who knows?

The 10-1/2''' Hunter was the most long-lived movement ever produced by the company and appeared in a wide variety of guises ranging from the simple "Prima" versions in 1923 to the Kew A versions from 1950. The strangest one yet seen is housed in a simple steel "boy's size" case. The enclosed movement is signed "Hans Wilsdorf Geneva" and bears a number of differences from a normal 10-1/2''' Hunter. It has 17 jewels rather than the normal 15. It is jeweled to the center, has a steel mounting plate for the endstone, and yet is marked "unadjusted." It is also the first Rolex to be signed "Antimagnetic," although the balance does not seem to be anything special. Why this watch was produced, why it bore Wilsdorf's name rather than Rolex's[81] and why it seems to be a unique example are just more of the unanswered questions that will continue to haunt us.

Everyone knows that Rolex never produced self-winding watches with anything other than 360 degree rotors. Their name was made by the "Perpetual" and all their early advertising talks about the advantages of the complete rotor. Yet there exists a solitary steel Athlete with a fully signed Oyster case and a "Harwood" style bumper movement, which is signed "Rolex Geneve" on the top plate (that is to say, under the dial). The history of this watch is completely unknown. It turned up at a NAWCC (National Association of Watch and Clock Collectors) show in the U.S. in 1991 where it was purchased by a major Japanese dealer, in whose private collection it remains today. Just as strangely, there also exists an 18kt Tudor (a strange device in itself) which uses an almost identical movement.

In the 1950s Rolex produced watches with some of the finest dials they have ever sold. The cloisonné dials were discussed earlier, as were the single color enamel dials made at the same time. As the dials themselves were horrendously expensive to manufacture (and therefore to sell) Rolex of course used these dials only on 18kt Perpetual watches, which is to say, on their flagship watches. Shown here is a blue enamel dial on a non-Oyster, non-Perpetual, non-chronometer steel watch, the least expensive grade of watch sold by Rolex at the time. One might say that the dial itself is an aberration, maybe just a one-off, but we also have photographs of an identical watch with a two-tone green/cream enamel dial.

Most of the single color enamel dials turned up on Bombé cases and it is one of these cases, albeit with a conventional dial that is the subject of our next examination. The watch is a conventional steel Bombé case, in itself quite unusual but hardly unique until the case is turned over. On the back is engraved the legend: "18kt Rotor" and upon opening the watch, there it is - an 18kt rotor. This one may be easier to explain. In the early 1950s many of the more prestigious Swiss companies were starting to produce their own automatic wristwatches, and many of these companies, able to charge a premium price for their products, made the rotors of their watches from 18kt (or in some cases 22kt or 24kt) gold. The advantages of using gold for a rotor are easily recognized. It is a much heavier material, and a rotor made of it will swing through a much greater distance than steel for a given wrist movement. This then makes the watch that much more efficient, so it will wind more quickly than an identical watch with a lighter, steel rotor. It is possible that, seeing all these other prestige companies crowing about their new gold rotors, Rolex decided to make some of their own for comparison purposes. It is also likely that Rolex realized that the advantages of using a gold rotor were seriously outweighed by the costs involved, so the watches never went into full scale production. The fact that this watch is in a steel case leans us toward this hypothesis, for it is unlikely that Rolex would have spent the money on an 18kt case for an experimental watch. The hypothesis gains even more weight when we note that Rolex patented a solid gold rotor as patent 279,675 on March 17, 1952, just about the correct date for the watch!

The Submariner is an easily recognizable watch with its heavy case, rotating bezel, and black dial with a triangle at 12, bars for the other quarters, and large luminous dots for the other hour markers. So what are we to make of a regular steel non-chronometer Submariner, ref. 5513, with a dial that seems to have wandered over from an Explorer. Close examination reveals that the dial is obviously factory signed "Submariner" and a cursory glance would show that the Explorer and Submariner dials are so different in size as to be completely non-interchangeable. There are no logical reasons for this watch being produced. It is one of only two or three we have ever seen, and as such remains another small Rolex mystery. In fact two sales leaflets for Rolex published in the U.K. in 1963 and 1966 show this Submariner dial quite clearly.

The strangest Rolex wristwatch we have ever seen is the final subject of this chapter. It completely rewrites many of the truths we hold dear about Rolex. It bears the London hallmark for 1915 and is made of sterling silver. It features the Rebberg 13''' movement normally found in the Rolex man's watches of the period, but, unusually for the period, it features a sweep seconds hand. It really gets strange when we note that the watch is designed to be worn on the left wrist. This is the first Rolex left-handed wristwatch to use this feature, previously thought to have appeared in the late 1940s.

The watch has another feature which was thought to have first surfaced in the 1940s: a calendar. The calendar admittedly is of the sweep variety, but very close examination reveals the bizarre fact that the calendar hand revolves counter-clockwise. In other words, this watch has four hands coming through the center post with three revolving clockwise and one in the opposite direction, not only has this design never been previously seen in a Rolex, it has never been encountered in any other make of watch. There seems to be no advantage in having this system. It is in many ways more difficult to make than one in which all hands rotate in the same direction.

Perhaps the watch was made as a graduation piece by a master watchmaker, but if that is the case why was it hallmarked, something that would normally only be done if the watch was to be for sale? Alternatively, with the hallmark date of 1915, the year Wilsdorf and Davis founded the Rolex Watch Co., perhaps it was made to demonstrate the prowess, skills and abilities of the new company? Most likely we will never know.

---

[81] Although the case is signed both Oyster Watch company on the inside and RWC on the outside.

On the 1913 watch shown here, the case number is 576083 while the number on the case top is 082.

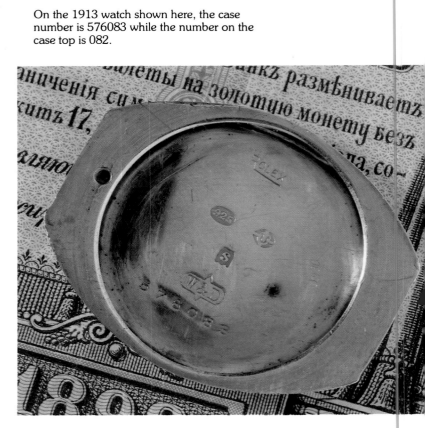

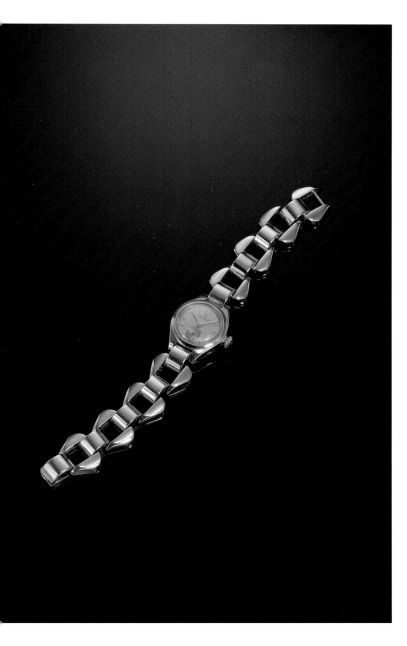

18kt pink gold lady's Oyster Perpetual Chronometer with integral, retro-Deco bracelet. Probably the rarest Lady's Oyster Perpetual in existence. This is the only example known. The existence of this watch confounds the publicity issued by Rolex in 1954 when they proudly announced their first Oyster Perpetual Chronometer for ladies. This watch, which dates from 1945, pre-dates that announcement by nine years.
Model no. 4214
Case no. 56372

On the inside back of a very early cushion Rolex Oyster, the case number was incorrectly stamped and it has been scratched out and restamped alongside as 21640.
Circa 1913

This is one of the earliest known cushion Oysters, number 20005. It is odd in that it is manufactured in 14kt white gold, a material never seen again in a cushion Oyster. It also has only the original Swiss patent number and is marked with the very early "7 World's Records" stamp.
Circa 1926

14kt pink gold Oyster Royal Observatory with the most unusual feature of having both a subsidiary and a sweep seconds. This dubious model is number 3116, case no. 56640
Circa 1938

Rare sterling silver military style wristwatch, 13''' rebberg movement, white porcelain dial with black paint and luminous dial and subsidiary seconds. Original leather strap with attached dial guard and signed Rolex buckle. This is the only model of this watch to surface so far. The case construction is similar to the "Borgel" case in that the movement, dial, and bezel all screw out together. but, unlike the "Borgel," the winding stem is capable of both winding and hand-setting and therefore no "pinset" pusher is required. Not only does this make this watch the earliest known example of a Rolex watch with a moisture-proof case (the case bears a London 1914 hallmark), but it is the only known example of any watch using this style of case construction and winder. The watch becomes even rarer when its condition is taken into account.
No model number
Case no. 634565.
Circa 1914

The previous watch showing the dial guard in place. The guard is sewn into the strap at the top and is therefore not removable from the watch. This style of guard was popular during the First World War, protecting both the fragile glass and dial. It is very rare to find one still on the watch as they do tend to limit visibility.
1914

Rare "Hans Wilsdorf" movement from a "boy's size" Oyster of the late 1930s. The movement is interesting in that it is jeweled to the center, has a stainless steel end plate, is the first known "Rolex" to be signed "Antimagnetic," and has 17 (rather than the usual 15) jewels, yet it is marked "unadjusted" and is contained in a very simple steel case. The movement is based on a 10-1/2''' Hunter ebauche and is so signed under the dial.
Circa 1938

Stainless steel Bombé with 18kt gold rotor. In the 1950s many of the premier watch companies introduced their first automatics, usually with solid gold rotors. Rolex patented a gold rotor, built a few prototypes, then chose not to proceed.
Model no. 5018
Case no. 608472

18kt gold Tudor automatic with Harwood-style hammer automatic mechanism. This is very strange as Rolex always made clear that all of their watches had rotors rather than hammer-style winding weights. Notice the very badly refinished dial.
Model no. 951
Case no. 50768

Stainless steel Oyster Perpetual which has been customized by the master watchmaker Franck Muller at the Ecole d'Horlogerie De Geneve. This involved building and attaching a complete retrograde perpetual calendar and moonphase work and then incorporating it into Rolex's standard design.

Sterling silver, porcelain dialed watch with 1915 London import hallmark. This is bizarre in that it features left hand winding, sweep center-seconds and a peripheral calendar ring which runs counter-clockwise. This means that there are four hands coming through the center hole with three of them rotating clockwise and one counter-clockwise, a feature never previously or subsequently seen on any watch, Rolex or not.
Model no. 5
Case no. 654768
1915

18kt Rolex Oyster Perpetual with moonphase retrograd perpetual calendar. Made by Antonio Preziuso at the Ecole D'Horlogerie De Geneve in 1978-9. Muller and Preziuso were friends and made these two watches simulatneously. This is the only 18kt model and is even more unusual in that it has a star dial. Although not built in the Rolex factory these two pieces typify the attention to detail in a way that would have pleased Hans Wilsdorf and they are still two of the most complicated Rolex watches ever seen.

Chapter 22
PATENTS
Patently superior

The list of patents issued to Rolex from their earliest years until 1990, which follows, is a perfect microcosm of the way the company developed over the years. The very earliest patents are for such basic things as watch cases, dials, and hands. Then, in the early 1920s, we see the first faltering steps along the way to the Oyster. Next come the automatic movement patents, interspersed with ones for the Prince, but by the mid-thirties the focus shifts to gradual improvements of balances, pivots and levers.

It is interesting to note that there seem to be two parallel paths of development at work, cases, winding buttons, and dials from Rolex S.A. in Geneva, and improvements to the movements coming from Manufacture des Montres Rolex, Aegler S.A. in Bienne.

By the 1950s great effort was being put into developing fine adjustment regulators and free sprung balances as the company strove to improve the accuracy of their production watches. While this was going on, the very first attempts at electro-mechanical watches were being patented, extending even to the patenting of a Rolex watch battery!

Looking at the accompanying list of Rolex patents, one is amazed at the developments the company came up with that were never pursued. Sometimes this is for obvious reasons, but other times the patent could have resulted in some of the most desirable Rolex models ever made.

We have chosen to illustrate a few examples from both of these groups. From the first group we show a couple of shaped automatic movements which seem to oppose everything we know as the Rolex Perpetual and from the second group we display all three versions of the World Time watch that Rolex patented between 1948 and 1950[82] as well as a watch with twin contra-rotating bezels.

---

[82] While Rolex produced several World Time pocket watches on the system patented by Louis Cottier, who made the famous Patek, Philippe wristwatches, the three systems outlined in the patents shown all differ from Cottier's.

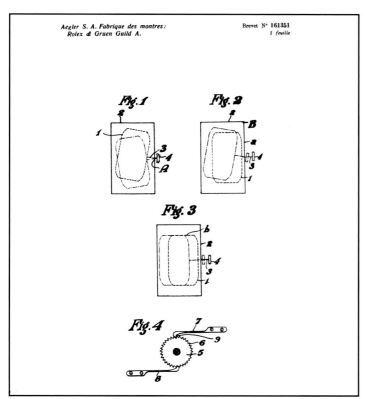

Patent NO. 161351

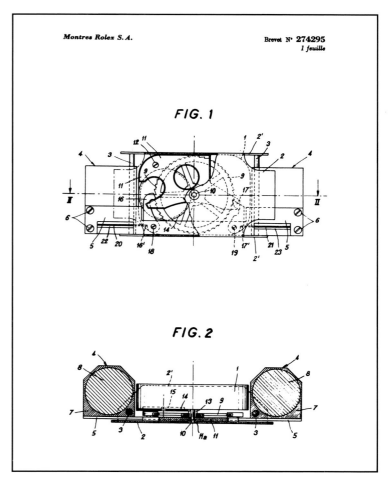

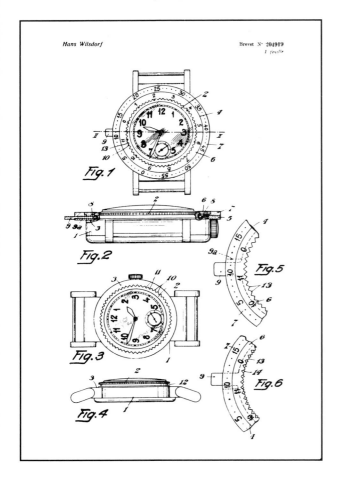
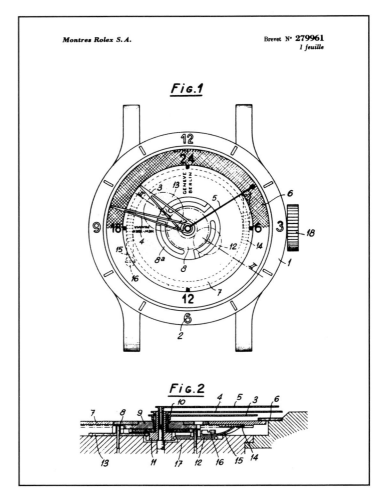
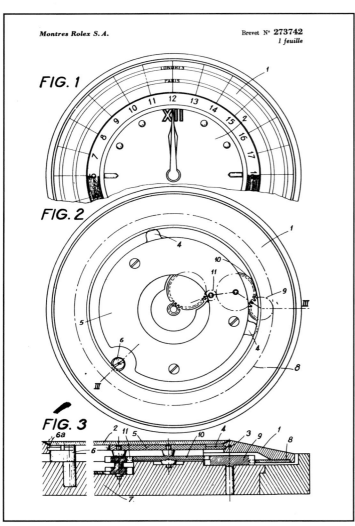
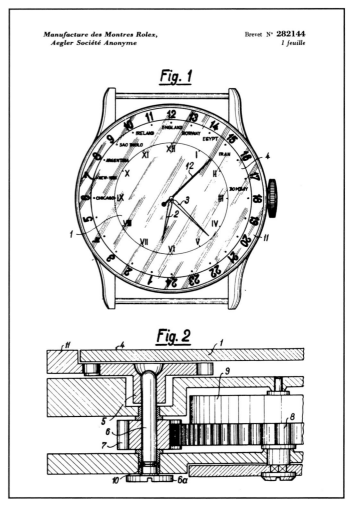

Liste der Muster und Modelle
Liste des dessins et modèles — Lista dei disegni e modelli

N° 6

Zweite Hälfte März 1933
Deuxième quinzaine de mars 1933 — Seconda quindicina di marzo 1933

II. Abteilung — IIe Partie — IIa Parte

Abbildungen von Modellen für Taschenuhren
(die ausschliesslich dekorativen Modelle ausgenommen)
Reproductions de modèles pour montres
(les modèles exclusivement décoratifs exceptés)
Riproduzioni di modelli per orologi
(eccettuati i modelli esclusivamente decorativi)

N° 50933. 23 mars 1933, 18¼ h. — Ouvert. — 1 modèle. — Lunette de boîte de montre-bracelet. — Hans **Wilsdorf**, Genève (Suisse). Mandataire: A. Bugnion, Genève.

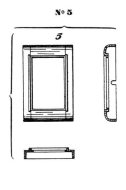

— 12 —

N° 50934. 23 mars 1933, 18¼ h. — Ouvert. — 5 modèles. — Dispositifs de fixation du bracelet aux montres-bracelets. — Hans **Wilsdorf**, Genève (Suisse). Mandataire: A. Bugnion, Genève.

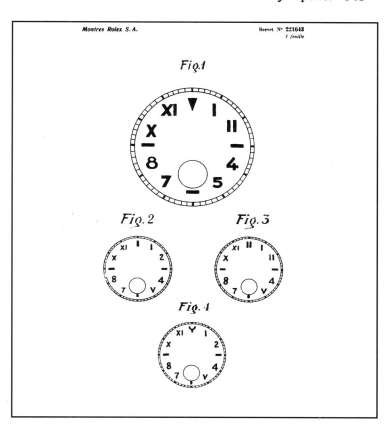

The patent drawings for the Roman/Arabic dial, with three variations which were never produced.

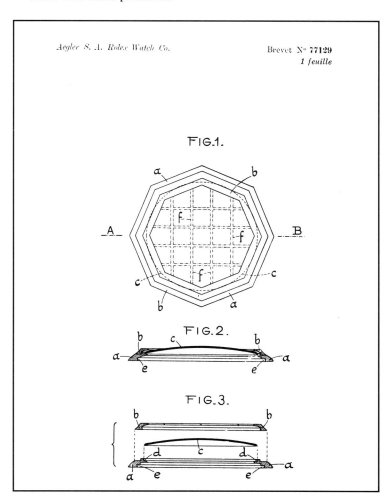

Patent for a protective watch cover for the new hermetic watch, No. 77129. See Chapter 1, page 11-12.

344 Patents: Patently Superior

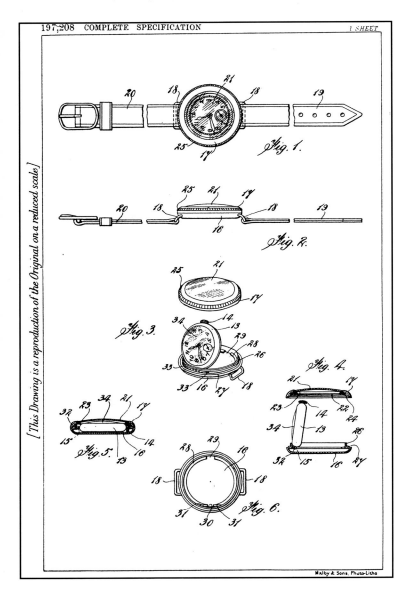

Wilsdorf's patent for the hermetic case, May 10, 1923. *See Chapter 2, page 42.*

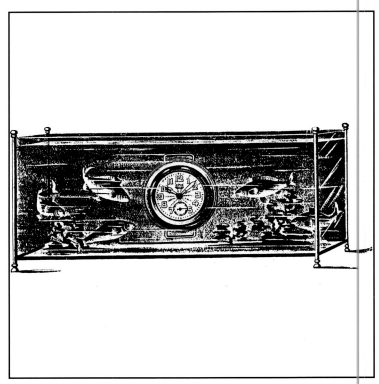

Patent for the aquarium window display for the new Oyster, November, 1922. *See Chapter 2, page 42.*

Patents: Patently Superior 345

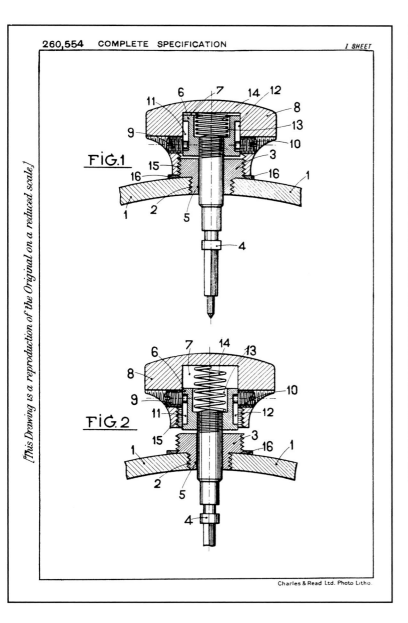

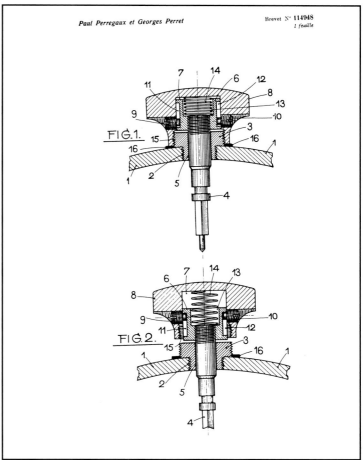

On the right is the original Swiss patent, No. 114948, for a moisture proof winding stem, awarded to Paul Perregaux and Georges Peret, October 30, 1925. They sold their rights to Wilsdorf who obtained a British patent No. 260,554, shown on the left, usually thought of as the original Oyster patent. One can see that they are essentially identical, even to the similarity of the typeface used. For more on this see Chapter 2, page 42.

Patent No. 160492 for an Auto-Rotor movement. See Chapter 3, page 79.

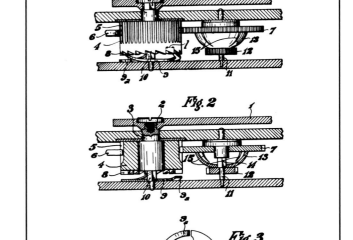

Patents: Patently Superior

Pat. No	Date	Description	French/German	Notes
23,382	02/04/02	Pocket watch Case	Taschenuhrgehause	
47,813	09/06/09	Watch pouch with viewing hole	Zeigereinrichtung an Taschenuhren	
54,712	17/02/11	Pocket watch Case	Taschenuhrgehause	
69,987	27/02/15	Chronometric luminous dial & hands	Zeitmesser-Zifferblatt mit leuchtenden Zeichen	
70,926	27/07/15	Watch	Uhr	1st patent with Rolex name
71,362	23/04/15	Watch Case	Uhrgehause	
77,129	17/11/18	Protective Grill for watches	Lunette de glace pour montres, bussoles, etc	
80,551	30/09/18	Improvements in Illuminating dials, indicators and hands	Perfectionnement aux elements indicateurs lumineux, tels que cadrans, aiguilles etc.	1st patent in Wilsdorf's name
97,101	02/08/21	Winding & setting mechanism	Mecanisme de remontoir et de mise a l'heur	
106,817	04/12/23	Shaped movement with side seconds	Mouvement de montre de forme possedant une aiguille trotteuse	
109,335	03/07/24	Attachment for watches	Attache de montre	Wilsdorf patent
109,521	16/07/24	Adjusting system for watch movements	Dispositif regulateur de mouvement de montre	
109,926	02/08/24	Method of fixing a metal support for balance jewel	Procede pour fixer dans un support metallique une pierre precieuse percee d'un trou destinee a recevoir un pivot de mobile	
111,595	27/11/24	Jeweled bushes for movement wheels	Support empierre pour mobiles de mouvements de montres de compteurs etc	
114,523	19/06/25	Hunter cased watch	Montre savonnette	Wilsdorf patent
115,154	22/07/25	Improvements in and relating to watches	Montre	First patent to mention protecting watch against humidity
120,558	06/08/26	Improvements in and relating to keyless watches	Montre a remontoir, sans pendant	First patent with sprung crown (essential part of Oyster crown)
120,848	18/10/26	Watch & winder	Montre a remontoir	First patent showing screw down crown
120,849	26/08/26	Shaped movement with seconds hand	Mouvement de montre de forme et a seconde	Prince movement
120,851	21/09/26	Shaped watch case	Boite de montre de forme	Oyster Case patent
123,763	27/01/27	Watch case with movable lugs	Boite de montre-bracelet, a anses mobiles	First flexible lug patent
130,191	28/12/27	Waterproof watch and winder	Montre hermetique a remontoir	Second screw down crown patent, but never produced
133,246	07/07/28	Wristwatch	Armbanduhr	Shaped movement & case probably never produced
139,876	06/07/29	Shaped watch movement	Kleinuhrwerk langlicher form	Prince pocket watch movement
140,165	11/07/29	Winding mechanism for waterproof watches	Dispositif de remontage et de mise a l'heure pour montre hermetique	Wilsdorf patent
143,449	03/10/29	Tool kit for opening oyster cases	Outillage pour visser les fonds moletes sur les corps de boites de montre hermetiques	Wilsdorf patent
144,350	06/11/29	Watch movement	Montre	First Rolex automatic patent
146,038	26/12/29	Watch	Piece d'horologerie	Sporting Prince case patent
146,309	10/03/30	Watch movement	Mouvement de montre	
150,978	17/07/30	Shaped watch case	Montre	Styled case
151,574	31/01/30	Winding & setting for waterproof watches	Dispositif de remontage et de mise a l'heure pour montre hermetique	Wilsdorf patent
151,577	17/03/30	Watch case	Piece d'horologerie	Sporting Prince, 2nd case patent
152,147	17/03/30	Watch case	Piece d'horologerie	Sporting Prince, 3rd case patent
153,246	31/10/30	Jump Hour movement	Piece d'horologerie a heures sautantes	
156,808	11/07/31	Protective balance bridge	Mouvement de horologerie	
157,995	14/08/31	Winding & click mechanism	Mechanisme de remontoir pour montres	
160,492	14/01/32	Automatic movement	Piece d'horologerie a remontage automatique	First Auto Rotor patent
160,803	19/02/32	Automatic movement	Piece d'horologerie a remontage automatique	Second Auto Rotor patent
161,351	09/03/32	Automatic movement	Piece d'horologerie a remontage automatique	Wierd rectangular automatic
161,888	25/02/32	Winding mechanism for timepieces	Dispositif de remontage pour piece d'horologerie	
162,509	26/05/32	Shaped movement	Mouvement de montre	
164,259	09/12/32	Watch with digital display	Piece d'horologerie a heures sautantes	Wilsdorf patent
167,229	13/02/33	Pivots for watches	Palier pour pivots d'axes d'horolgerie	
167,848	08/04/33	Balance	Balancier	
168,491	09/05/33	Shaped watch movement	Piece d'horologerie	Further Prince patent
168,628	15/06/33	Shaped watch case	Boite de montre de forme	First new Oyster case
170,143	22/08/33	Winding & setting mechanism	Montre hermetique	Wilsdorf patent
170,500	20/07/33	Winding & setting mechanism	Montre hermetique	Wilsdorf patent
170,803	19/05/33	Automatic movement	Mechanisme de remontoir pour montres	
170,938	19/05/33	Automatic movement	Piece d'horologerie a remontage automatique	Second Auto Rotor patent
171,082	21/08/33	Winding & setting mechanism	Montre hermetique	Wilsdorf patent
171,429	20/09/33	Method of fitting movement in a case	Dispositif de montage d'un mouvement d'horologerie dans sa boite	
174,131	07/11/33	Watch with indirect indication of hours	Piece d'horologerie	Seems never to have been made
176,403	21/07/34	Balance Cock	Dispositif de montage d'une raquette sur le bati d'une piece d'horologerie	
179,154	24/11/35	Lever escapement	Echappement a ancre	Seems to be the HW movement
181,864	29/12/34	Pivot support	Support de contre-pivot	
181,865	21/02/35	Method of fitting movement in a case	Dispositif de emboitage d'un mouvement de montre dans sa boite	Case clips as used in bubbleback
188,077	21/11/35	Balance wheel	Balancier	First attempt at concealed balance screws, like superbalance
188,796	04/03/35	Pivot support	Support de contre-pivot	
189,911	10/06/36	Golf watch	Montre	Wilsdorf patent
193,124	27/08/36	Compensated Balance	Balancier compensateur	Further moves toward the superbalance
195,112	28/08/35	Automatic movement	Piece d'horologerie a remontage automatique	
196,706	25/08/36	Balance wheel	Balancier	Superbalance
204,919 (never made?)	19/05/38	Watch with rotating bezels	Piece d'horologerie	Two contra-rotating bezels to indicate hours & minutes (never made)
205,428	04/02/38	Method of fitting movement in a case	Dispositif de emboitage d'un mouvement de montre dans sa boite	Additional to 181865
206,760	29/07/38	Balance	Balancier pour mouvements d'horologerie	
208,584	17/11/38	Watch case	Boite de montre	Two-piece Oyster royal/bubbleback case
210,108	13/01/39	Regulator & Balance cock	Palier pour le pivot superieur d l'axe de balancier	First patent on fine regulation
210,237	22/03/39	Winding button for waterproof watch	Remontoir etanche	Improved oyster crown
211,693	16/12/38	Winding button for waterproof watch	Piece d'horologerie a remontoir etanche	Improved oyster crown
213,641	12/12/39	Waterproof case	Boite de montre etanche	Only Rolex patent for a rectangular waterproof watch, never made
213,642	12/12/39	Waterproof case	Boite de montre etanche	Variation on above
213,949	12/06/39	Square Waterproof watch	Montre	Two-piece square oyster case, never made
215,706	19/04/40	Vapor absorbing watch case	Montre	Inner case back contains absorbent crystals!!
216,335	13/02/39	Winding button for waterproof watch	Piece d'horologerie a remontoir etanche	Improved Oyster crown
216,723	03/10/40	Bearings for wheel pivots	Coussinet pour axe de mobile de piece d'horologerie et procede pour sa fabrication	
221,643	30/05/41	Watch Dial	Cadran de piece d'horologerie	Arabic/roman dial including 3 unknown variations
225,082	04/12/41	Window display supports for watches	Jeu de supports pour la presentation	Watch stands in the form of the letters R O L E X
228,698	11/08/42	Automatic movement	Mechanisme de remontoir pour montres	Improved bubbleback movement
230,743	11/08/42	Jeweled Lever	Roue d'ancre	
239,050	13/12/43	Balance Cock & Regulator	Montre	

Pat. No	Date	Description	French/German	Notes
241,453	28/06/44	Winding button for waterproof watch	Remontoir etanche	Improved oyster crown
241,953	15/05/44	Watch Bezel	Piece d'horlogerie	Machined bezel
242,198	28/06/44	Winding button for waterproof watch	Remontoir etanche	Improved Oyster crown
243,145	21/08/44	Watch Bezel	Boite de montre	First faceted bezel
247,509	07/11/45	Watch Box	Ecrin	Lady's watch box converts to jewelry box
251,701	10/12/45	Winding button for waterproof watch	Remontoir etanche	Improved oyster crown
251,702	13/12/45	Winding and setting mechanism	Mouvement d'horlogerie	
251,923	20/12/45	Dial and construction method	Cadran de piece d'horlogerie et procede pour sa fabrication	Improved dial
252,577	08/02/46	Attaching movement ring to case	Dispositif de verrouillage d'un cercle d'emboitage a vis d'une boite de montre	
256,275	24/01/47	Watch with date indication	Piece d'horlogerie a quantieme	Datejust
257,185	05/02/47	Exanding bracelet	Bracelet Extensible	First oyster bracelet
258,633	18/08/47	Winding button for waterproof watch	Dispositif de commande d'un mecanisme de mouvement d'horlogerie	Improved oyster crown
260,638	11/06/47	Driving shaft support blocks/bridges	Dispositif de palier pour pivot d'horlogerie	
264,359	28/10/47	Winding button for waterproof watch	Remontoir etanche pour piece d'horlogerie	
270,457	23/03/49	Winding button	Dispositif de commande d'un mechanism de mouvement d horlogerie	Super Oyster crown
271,166	05/03/48	Luminous dial	Piece d'horlogerie	
271,421	14/04/48	Luminous dial	Cadran pour piece d'horlogerie	
271,423	09/03/48	Automatic winding movement	Mouvement d'horlogerie a remontage automatique	
271,424	16/04/48	Automatic winding movement	Piece d'horlogerie a remontage automatique	Very strange automatic mvt
271,426	16/04/48	Automatic winding movement	Piece d'horlogerie a remontage automatique par masse mobile	Very strange automatic mvt variation on above
271,695	26/04/48	Sweep seconds automatic movement	Piece d'horlogerie a remontage par masse oscillante avec secondes au centre	
271,994	29/05/48	Method of mounting rotor on its axis	Dispositif d'assemblage d'un mobile d'horlogerie sur son axe	
271,996	15/03/48	Attaching bezel to a case	Boite de montre	
272,615	10/06/48	Automatic winding movement	Mouvement d'horlogerie a remontoir automatique	
272,877	23/06/48	Automatic winding movement	Mouvement d'horlogerie a remontoir automatique	
272,896	26/07/48	Lady's Flexible bracelet	Bracelet flexible	
273,139	08/07/48	Placing movement in watch case	Disositif d'emboitage d'un mouvement de montre	
273,742	12/08/48	World time wristwatch	Piece d'horlogerie universelle	Never made, different from Patek etc
273,743	26/08/48	3 register chronograph	Piece d'horlogerie avec compteurs d'heurs et de minutes	First rolex patent on a chronograph
274,295	26/07/48	Small automatic movement	Montre de forme de petit calibre a remontage automatique	Wierd rectangular automatic mvt, obviously never made
274,297	01/10/48	Winding staff	Tige de remontoir pour un piece d'horlogerie	
274,903	23/12/48	Automatic movement	Piece d'horlogerie a remontage par masse oscillante	
275,197	06/12/48	Automatic movement	Piece d'horlogerie a remontage automatique	
275,198	06/12/48	Automatic movement	Piece d'horlogerie a remontage automatique	
275,861	17/01/49	Automatic movement	Piece d'horlogerie a remontage automatique	
275,863	23/12/48	Cord band attachment for lady's watch	Montre-bracelet a cordonnet	
276,198	06/12/48	Automatic movement	Piece d'horlogerie a remontage automatique	
277,378	30/04/49	Automatic movement	Montre a remontage automatique par masse oscillante	
278,043	14/05/49	Slimmer automatic movement	Montre a remontage automatique par masse susceptible d'osciller sur un tour complet.	
278,045	04/04/49	Hinged strap fixing for lady's watch	Dispositif de fixation d'un bracelet a une montre	
278,843	08/01/49	Winding crown	Dispositif de commande d'un mecanisme de mouvement d'horlogerie	Super Oyster version 2
279,001	10/08/49	Automatic movement	Montre a remontage automatique par mas mobile	Similar to weight on cal 1030
279,351	14/07/49	Regulator	Raquette de montre	
279,354	09/07/49	Automatic movement	Montre a remontage automatique par masse oscillante	
279,355	16/07/49	Rectangular automatic watch	Piece d'horlogerie de forme allongee a remontage automatique	Bizarre "Prince" style automatic
279,356	09/08/49	Automatic movement	Piece d'horlogerie a remontage automatique par masse mobile	
279,357	09/08/49	Automatic movement	Piece d'horlogerie a remontage automatique par masse mobile	
279,405	10/09/49	Mesh style expanding bracelet	Bracelet	Never made
279,673	10/11/49	Balance wheel	Balancier	
279,675	22/09/48	Rotor for Automatic movement	Masse de Remontage pour montre a remontage automatique	Slimmer weight made of gold, made only as prototype
279,961	29/08/49	World time wristwatch	Montre Universelle	Similar to Tissot "Navigator," never made
280,068	13/08/47	Winding crown	Dispositif de commande d'un mecanisme de mouvement d'horlogerie	Super Oyster version 3
280,559	25/08/49	Attaching bezel to a case	Boite de montre	
280,742	09/08/49	Winding crown	Dispositif de commande d'un mecanisme de mouvement d'horlogerie	Super Oyster version 4
280,743	09/08/49	Winding crown	Dispositif de commande d'un mecanisme de mouvement d'horlogerie	Super Oyster version 4b
280,744	09/08/49	Winding crown	Dispositif de commande d'un mecanisme de mouvement d'horlogerie	Super Oyster version 4c
281,193	28/12/49	Pivot & Jewel	Palier d'horlogerie	
281,493	16/02/50	Calendar Watch	Piece d'horlogier a calendrier	1st attempt at Day-Date with day next to date, just like a Seiko
281,494	28/02/50	Calendar Watch	Piece d'horlogier a calendrier	Variation on 281,493
281,497	06/01/50	Time keeping apparatus	Appareil pour le controle de la marche des montres	Similar to "Vibrograph"
281,799	10/01/50	Pallet set	Roue d'echappement pour piece d'horlogerie	
281,800	26/07/48	Rectangular automatic watch	Masse de remontage pour montre a remontage automatique	Bizarre "Prince" style automatic variation on 279,355, never made!
281,803	23/12/48	Decorated glass for lady's watch	Procede de decoration d'une glace de montre et glace de montre obtenue par ce procede	
281,807	10/01/50	Rectangular lady's Automatic watch	Montre-bracelet	Different square lady's automatic, once again, never made.
282,134	27/01/50	Method of attaching dial to movement	Dispositif de fixation d'un cadran d'horlogerie	
282,141	13/02/50	Method of fitting watch glass	Boite etanche pour montre	
282,144	01/02/50	World time wristwatch	Montre Universelle	Third different version of world-time watch, again never made
282,448	14/01/50	Method of attaching dial to movement	Piece d'horlogerie	
282,777	18/04/50	Lady's watch case with fixing for straps	Boite de montre-bracelet	Became Rolex "Chameleon"
283,489	08/06/50	Jeweled watch case	Montre-bijou	
283,819	19/04/50	Automatic movement	Piece d'horlogerie a remontage automatique par masse mobile	
284,491	02/06/50	Fine adjustment regulator	Raquetterie pour un piece d'horlogerie	
285,527	05/07/50	Movement & construction method	Piece d'horlogerie, procede pour sa fabrication et appareil pour la mise en oevre de ce procede	
286,185	21/08/50	Date ring for Datejust	Disque indicateur des quantiemes pour piece d'horlogerie a calendrier	Strange disc with only 30 dates
286,916	16/06/50	Automatic movement	Piece d'horlogerie a remontage automatique	
287,281	17/08/50	Fine adjustment regulator	Raquetterie pour piece d'horlogerie	
287,287	21/08/50	Date indicator	Piece d'horlogerie	New version of date disc

Pat. No	Date	Description	French/German	Notes
287,619	09/02/51	Waterproof case	Boite etanche pour piece d'horlogerie	
287,936	11/01/50	Bearings for shafts or pivots	Coussinet pour arbre de piece d'horlogerie	
288,975	06/09/49	Automatic movement	Mouvement d'horlogerie a remontoir automatique	Rotor supported by ball bearings
288,976	06/09/49	Automatic movement	Mouvement d'horlogerie a remontoir automatique	As above both are versions of 278,877
288,977	06/09/49	Automatic movement	Mouvement d'horlogerie a remontoir automatique	Rotor not supported by ball bearings
290,045	02/05/51	Winding crown	Remontoir etanche pour piece d'horlogerie	Super Oyster version 5
290,947	05/06/51	Fine adjustment regulator	Raquetterie pour piece d'horlogerie	
292,131	24/06/51	Pallet set	Echappement a ancre pour piece d'horlogerie	
292,132	13/08/51	Pallet set	Echappement a ancre pour piece d'horlogerie	
292,854	10/08/51	Pallet set	Echappement a ancre pour piece d'horlogerie	
293,487	01/12/51	Bearing bridge	Palier pour piece d'horlogerie	
296,065	09/02/52	Fine adjustment regulator	Raquetterie pour piece d'horlogerie	
296,071	18/12/51	Waterproof case	Boite etanche pour montre	
296,723	02/02/52	Free sprung balance	Dispositif a'accouplement a roue libre, en particulier pour piece d'horlogerie	
296,731	07/04/52	Electro-mechanical wristwatch	Montre electrique	First Rolex Electro-mechanical watch patent and first to carry the name of E. Borer
297,082	15/04/52	Electro-mechanical wristwatch	Mouvement d'horlogerie, nottament pour montre-bracelet	2nd Rolex Electro-mechanical watch patent
297,906	03/06/52	Electro-mechanical wristwatch	Montre munie d'une pile electrique	3rd Rolex Electro-mechanical watch patent
298,261	03/06/52	Watch Battery	Pile electrique seche	Electric watches need batteries, so why not make your own?
298,949	06/05/52	Fine adjustment regulator	Raquetterie pour piece d'horlogerie	
298,953	01/05/52	"Cyclops" date lens	Piece d'horloger a calendrier	Solved the only problem with the Datejust
298,954	21/05/52	Watch Bracelet	Montre Bracelet	"Flush-fit" bracelet, first change from bubbleback style straight end bracelet
298,956	13/05/52	Electro-mechanical wristwatch	Montre electrique	4th Rolex Electro-mechanical watch patent
299,776	09/07/52	Winding crown	Remontoir etanche pour piece d'horlogerie	Super Oyster version 6
299,777	12/07/52	Winding crown & stem	Remontoir pour piece d'horlogerie	
300,079	05/07/52	Automatic 1 way winding train	Dispositif d'entrainment a sens unique pour piece d'horlogerie a remontage automatique	Rolex had introduced the 1030 cal with 2 way winding only 2 years previously
300,363	22/07/52	Winding crown	Remontoir etanche pour piece d'horlogerie	Triplock patent
300,381	14/05/52	Coupling system for idle wheels	Dispositif d'accouplement a roue libre	
300,382	22/07/52	Coupling system for idle wheels	Dispositif a'accouplement a roue libre, notamment pour l'horlogerie	
303,005	25/11/52	Watch Bracelet	Montre Bracelet	"Flush-fit" bracelet, variation on 298,954
304,395	08/10/52	Oil dissipating prevention system	Procede pour empecher, dans un mouvement d'horlogier, le lubrifiant de s'epassier	
305,177	05/02/53	Rotating Bezel	Montre	First "Turn o graph" bezel
308,031	18/04/53	Winding crown	Remontoir etanche pour piece d'horlogerie	Triplock patent 2
308,032	18/04/53	Winding crown	Remontoir etanche pour piece d'horlogerie	Triplock patent 3
308,601	18/04/53	Twin bezel watch	Montre	Two counter rotating bezels (never made)
309,964	08/07/53	Clutch for idle wheels	Accouplement a roue libre	
310,284	08/08/53	Decorative lady's watch	Piece d'horologerie	Looks like highly stylised bubbleback
310,556	15/09/53	Push fit bezel	Boite de montre etanche	
310,561	18/08/53	Movable bezel markers	Montre-bracelet	Bezel is fixed but there are 2 movable bezel markers also never made
311,287	08/09/53	Collet for hairspring	Virole pour balancier d'horlogerie	
311,865	12/08/53	Dead beat seconds watch	Piece d'horlogerie a seconds morte	Became "Trubeat" movement
312,285	05/02/53	Rotating bezel	Montre	Further variant on 305,177
313,709	13/09/53	Rotating bezel in gold	Montre	Two-tone version of above
315,933	05/05/54	Fine adjustment regulator	Raquetterie	
317,198	28/09/54	Anti-magnetic case	Boite de montre antimagnetique	Became "Milgauss"
317,532	26/10/54	Pallet & production method	Ancre d'echappement a ancre et procede pour s fabrication	
318,872	21/09/54	Level using ball bearings	Niveau a bille	
320,417	16/05/55	Watch case with hidden lugs	Montre-bracelet	
320,664	05/04/55	Lady's watch case with fixing for straps	Boite de montre-bracelet	Became Rolex "Chameleon"
320,800	04/06/55	Waterproof testing machine	Appareil pour eprouver l'etanche des boites	
320,817	16/05/55	Rotating bezel insert & assembly	Montre et procede pour sa fabrication	
321,190	16/05/55	Watch case with hidden lugs	Montre-bracelet	Variation on 320,417 above, was made
321,191	16/05/55	Watch case with hidden lugs	Montre-bracelet	Variation on 320,417 above
321,955	16/05/55	Instantaneous calendar change	Montre-calendrier	
322,341	19/08/55	Calendar disc	Piece d'horlogerie a calendrier	Datejust variant
322,541	28/01/55	Rotating bezel	Montre	Plastic bezel insert (like 1st GMT)
323,978	29/01/55	Rotating bezel	Montre	Plastic bezel insert (like 1st GMT) variant on above
323,982	23/07/55	Day-Date calendar watch	Piece d'horlogier a calendrierDay-Date variation	
324,229	23/06/55	Wrist depth gauge	Monometre Bourdon	Diver's depth gauge (to go with Submariner?)
324,267	03/12/55	Potence (or bridge) for movement	Palier pour piece d'horlogerie	
326,628	23/06/55	Watch glass & manufacture method	Procede de fabrication d'un verre pour montre et pendulettes	
326,629	07/09/55	Potence (or bridge) for movement	Palier pour piece d'horlogerie	
327,839	24/05/56	Watch bracelet	Bracelet	Rubber strap!
330,190	09/05/56	Automatic movement	Piece d'horlogerie a remontoir automatique	
330,197	23/04/56	Shaped lady's watch	Piece d'horlogerie	Unusual shaped movement
330,199	08/05/56	Shockproof balance	Palier amortisseur de chocs pour mobile d'horlogerie	
330,903	16/05/56	Potence (or bridge) for movement	Palier pour piece d'horlogerie	
330,904	23/05/56	Shockproof balance	Palier amortisseur de chocs pour mobile d'horlogerie	
332,203	20/08/57	Regulating screws for balance wheel	Vis reglante pour la balancier d'horlogier	
332,546	09/02/57	Regulating screws for balance wheel	Vis reglante pour la balancier d'horlogier	
332,559	01/03/57	Shockproof balance	Palier amortisseur de chocs	
332,560	17/04/57	Balance cock & regulator	Piece d'horlogerie	
332,594	20/08/57	Tool for turning screwed movement parts	Outil, notamment cle pour l'entrainement de pieces visees	
332,885	15/03/57	Balance wheel	Balancier d'horlogerie et procede pour sa fabrication	
332,887	30/07/57	Fine adjustment regulator	Organe regulateur pour piece d'horlogerie et procede pour sa fabrication	
332,890	23/04/57	Automatic movement	Piece d'horlogerie a remontage automatique par masse rotative	
332,894	27/02/57	Shockproof balance	Palier amortisseur de chocs pour mobile d'horlogerie et procede pour sa fabrication	
332,900	29/04/58	Calendar movement	Piece d'horlogerie a calendrier	
332,930	17/04/58	Flexible metal bracelet	Bracelet avec dispositif de fermeture	Strange dress metal bracelet
334,376	14/11/57	Fine adjustment regulator	Raquetterie pour piece d'horlogerie	
334,377	14/11/57	Fine adjustment regulator	Raquetterie pour piece d'horlogerie	
334,702	03/12/57		Piece d'horlogerie	
336,019	08/01/57	Rotating bezel	Piece d'horlogerie	First patent to list M. Huguenin as inventor
336,024	29/01/57	Shockproof balance	Palier amortisseur de chocs pour mobile d'horlogerie	Patented by E. Borer
336,759	08/01/58	Automatic movement	Piece d'horlogerie a remontage automatique par masse rotative	Patented by E. Borer
337,784	03/04/58	Mainspring Barrel	Barillet pour piece d'horlogerie	Patented by E. Borer
338,609	04/06/55	Wrist compass	Boussole	Seems to match 324,229

Pat. No	Date	Description	French/German	Notes
339,574	16/06/58	Balance wheel	Balancier pour piece d'horlogerie	Patented by E. Borer
342,898	08/08/58	Balance wheel	Balancier pour piece d'horlogerie	Patented by E. Borer
343,889	29/08/56	Manufacturing method for "ribbon" form mainspring	Procede de fabrication d'un ressort moteur en forme de ruban et ressort moteur obtenu par ce procede	Patented by E. Borer
343,897	06/04/59	Movement	Piece d'horlogerie	Patented by E. Borer
343,898	15/04/59	Escape wheel & pallets	Echappement de piece d'horlogerie	Patented by E. Borer
343,904	03/06/59	Balance wheel	Balancier pour piece d'horlogerie	Patented by E. Borer
343,907	12/02/59	Autocompensating balance wheel	Balancier a serge monometallique pour piece d'horlogerie a spiral, dit "autocompensateur"	Patented by E. Borer
343,909	12/05/59	Balance wheel	Balancier pour piece d'horlogerie a bras elastique	Patented by E. Borer
343,931	06/07/59	Potence (or bridge) for movement	Coussinet de palier, notamment pour arbre de mobile d'horlogerie	Patented by E. Borer
343,943	13/04/59	Watch case	Boite de montre	Patented by Hans Wilsdorf
343,947	29/06/59	Lady's watch case	Montre-bracelet	Patented by Hans Wilsdorf
345,600	20/03/59	Balance wheel	Balancier pour piece d'horlogerie	Patented by E. Borer
348,929	07/09/59	Lady's watch case	Montre-bracelet et procede pour sa fabrication	Patented by E. Borer
365,671	23/12/60	Pallet set	Echappement a ancre pour piece d'horlogie	Patented by M. Huguenin
367,125	11/04/61	Electro-mechanical wristwatch	Montre electrique	Patented by E. Borer & JP Osterwalder
371,993	06/09/61	Rotor	Masse de remontage oscillante pour piece d'horlogie	Patented by M. Huguenin
419,977	24/08/64	Electro-mechanical wristwatch	Piece d'horlogerie electrique a organe regulateur moteur	Patented by M. Huguenin
423,638	11/06/64	Screwed chronograph pushers	Dispositif etanche de commande, a poussoir, pour piece d'horlogerie	First used on model 6285. Patented by Jean Perret
442,155	28/06/65	New calendar mechanism	Piece d'horlogerie quantiemes	Patented by M. Huguenin
451,825	26/05/66	Fine adjustment regulator	Raquetterie pour piece d'horlogerie	Patented by M. Huguenin & A. Zibach
460,641	24/06/66	Clutch for idle wheels	Dispositif d'accouplement a roue libre pour piece d'horlogerie	Patented by M. Huguenin & A. Zibach
465,504	09/12/66	Indirect minute drive train	Piece d'horlogerie a entrainement indirect de l'organe indicateur des minutes	Patented by M. Huguenin & A. Zibach
487,440	17/02/67	Coupling system for limited or idle wheels	Dispositif d'accouplement limiteue de couple	Patented by M. Huguenin
492,246	06/11/67	Gas escape valve	Montre etanche	Used for Sea Dweller, patented by A. Zilbach
498,436	18/02/69		Montre-bracelet	Patented by J. Perret
502,623	25/06/68	Fine adjustment regulator, balance and hair spring	Raquetterie pour piece d'horlogerie balancier et spiral	Patented by M. Huguenin
513,446	23/06/69		Mouvement de montre	Patented by F. Gasser and P. Girard
519,747	13/06/69	Quartz watch	Piece d'horlogerie electronique	Patented by F. Gasser
530,667	20/10/70	Quartz digital timepiece	Piece d'horlogerie a affichage digital	Patented by P. Girard
532,811	11/05/70	Liquid crystal display timepiece	Piece d'horlogerie comprenant un dispositif d'affichage avec moins une cellule a cristal liquide	Patented by P. Girard
533,329	27/11/70	Energy cell & fitting method	Dispositif de fixation d'un generateur chimique d'electricite dans un mouvement Centre Electronique Horloger SA d'horlogerie	Patented by D. Paratte.
533,785	15/11/71	Method of converting an oscillating movement into a rotational one	Dispositif pour transformer un mouvement oscillant en un mouvement rotatif	Patented by Claude Challandes for Rolex & Le Coultre
546,977	11/05/70	Quartz digital display watch	Dispositif d'affichage a cristal liquide pour piece d'horlogerie	Patented by P. Girard
567,299	21/10/69	Quartz digital display watch	Piece d'horlogerie d'affichage digital	Patented by P. Girard
569,320	06/03/70	Multi-display quartz digital watch	Piece d'horlogerie d'affichage digital electronquement elabore	Patented by P. Girard
573,621	13/11/72	Waterproof case	Boite de montre etanche	Patented by M. Schmitt
574,639	21/10/69	Quartz digital display watch	Montre electronique dont l'affichage est realize par au moins une cellule a cristal liquide	Patented by P. Girard
574,640	27/09/73	Case opening machine	Appareil pour visser et devisser les fonds de boites de montres	Invented by R. Addor
577,703	21/10/69	Digital quartz watch	Utilisation d'elements electro-optiques	Invented by P. Girard
583,438	21/09/73	Watch case with sapphire glass	Boite de montre	Invented by R. Addor
598,648	25/03/76	Machine for lubricating escapements	Appareil a lubrifier les roues d'echappement	Invented by T. Huber
600,859	11/05/76	Method for analysing the circulation of blood in living tissue	Procede et dispositif d'analyse de la microcirculation sanguine dans un tissu vivant	Invented by R. Grandjean. We checked this one thoroughly. It really is from Rolex
604,375	21/11/75	Method of fine adjustment of the thermal properties of quartz crystal	Procede d'adjustment fin des proprietes thermiques et/ou electriques d'un resonator a quartz d'horlogerie	Invented by J. Herrman and J-C Martin for Rolex & Manufacture Complications SA
609,524	10/03/76	Machine for removing bezels & glasses	Appareil pour separer l'un de l'autre deux elements d'une boite de montre	Invented by J-L Vuille
623,981	09/06/78	Electronic dive time calculator	Compteur electronique de temps pour la plongee sous-marine	Invented by D. Strubin
642,808	05/01/81	Watch battery condition display	Procede pour determiner l'etat de decharge d'une pile electrique et dispoitif pour la mise en oeuvre de ce procede	Invented by E. Zaugg
642,811	06/04/81	Seconds display for electronic watch	Piece d'horlogerie electronique a indicateur des secondes	Invented by E. Zaugg
661,832				
661,834	19/08/85	Waterproof winding button & stem	Boite de montre etanche	Invented by R. Duffey
662,237	22/11/85	Yachting chronograph	Chronometre de depart de course, notamment de regate	Invented by R. Besson
662,697	26/03/86	Waterproof winding button & stem	Boite de montre etanche	Invented by R. Duffey, A. Zwahlen & A. Bonney
669,082	25/05/87	Method of correcting date display	Dispositif de commande ou de correction de l'affichage du jour ou de quantieme pour une montre, notamment une montre-bracelet	Invented by R. Besson
671,135	11/12/87	Method of controling output of an electric motor in a watch	Procede de commande d'au moins un moteur electronique et montre analogique comportant au moins un moteur electrique commande selon ce procede	Invented by R. Besson
672,222	11/11/87	Method of altering perpetual calendar display without pushers	Procede de programmation du calendrier perpetual d'une montre et montre pour la mise en oeuvre de ce procede	Invented by R. Besson & J. Ortelli
672,317	11/11/87	Plastic material usable as dial illumination	Composition thermoformable pour le moulage d'objets sensible a la chaleur	Invented by P. Matthey & A. Rossel
673,555	19/04/88	Perpetual calendar with electro-mechanical display	Montre du type a calendrier perpetual et a affichage analogique du temps	Invented by R. Besson
673,924	19/04/88	Electronic watch with analog display	Montre electronique a affichage analogique du temps	Invented by C-E Leuenberger & J-J Burri
677,713	03/01/90	Watch case with rotating bezel	Boite de montre munie d'une lunette tourante	Invented by R. Addor & A. Jaussi
678,908	04/04/90	Quick setting of both Day & Date	Dispositif de commande et de correction de l'affichage du jour et du quantieme pour une montre, et montre-bracelet equipee d'un tel dispositif	Invented by R. Besson
678,909	06/03/90	Watch case with decorative bezel	Boite de montre	Invented by R. Addor & A. Jaussi
678,910	06/03/90	Chronograph resetting mechanism	Dispositif de reglage pour bascule a double marteau de ramise a zero des mobiles indicateurs de chronographe	Invented by R. Addor & M. Schmidt
680,409	03/11/89	Electro-mechanical display for perpetual calendar	Procede d'initialisation automatique au moins de l'affichage du quantieme	Invented by R. Besson & C-E Leuenberger
680,696	03/11/89	Method of determining zero on a quartz analog watch	Procede pour determiner la position du zero d'une aiguille d'une montre quartz a affichage analogique et dispositif pour la mise en oeuvre de ce procede.	Invented by C-E Leuenberger & R. Besson
681,673	14/12/90	Combined analog/digital display of Day/date	Montre-bracelet a affiche dissimule	Invented by C-E Leuenberger & G. Schweighauser

Chapter 23
COLLECTING ROLEX WATCHES

col•lect (ke-lèkt) verb
col•lect•ed, col•lect•ing, col•lects verb, transitive 1. To bring together in a group or mass; gather. 2. To accumulate as a hobby or for study[83]

As the journey of a thousand miles begins with a single step, so any collection of Rolex watches must begin with a single watch. Often the decision of which should be your first Rolex is the hardest and it is a decision we will try to help you with. First, decide whether the watch will be worn every day, or just occasionally. If it is to be an everyday piece then a Perpetual Oyster makes the most sense. If it is not for daily use, then an early Rolex may be chosen.

Next, choose where to buy your watch. If you enjoy the hunt and are willing to spend the time then by all means start to haunt flea markets, pawn brokers, junk shops, jumble sales, yard sales and garage sales. However, you should be aware that there are people out there who are going to exactly the same places that you are and who know what they are looking for. They are doing it for a living. These people also have built long term relationships with the suppliers. This is a course we do not recommend for a beginner.

Instead, rely on the knowledge that someone else has acquired. Possible sources include a jeweler who sells older watches. He may have taken in trade-in against new watches, a department store that may have leased a counter to a vintage watch specialist, major auction houses, antique malls, an antique store that stocks a few watches as well as general merchandise, a specialist who takes a stall at one of the movable antiques markets, or one of the specialty shops that deal exclusively in vintage timepieces. Our recommendation would always be to go to the latter, however this is not always a viable option unless you live in or close to a major city. The auction houses have the widest selection of watches amongst all the above listed sources, but they are not ideal places for the beginner. As you collect and learn, they will become one of your major sources. Jewelers and department stores will give you the best warranties and will accept your credit card, but they are most likely to have the highest prices.

Initially it is always best to stick with the specialists. When looking for a dealer there is only one important question, can you trust them? Examine their stock. Do the watches look in good condition, do a large number have restored dials, are they on new straps or on the old, ratty ones, and most importantly do they have watches you want?

Then of course comes the hard part, which of the models does one choose?

Again it is important to break the decision process down into individual steps. First and most important is cost. With the first watch in a collection you should plan to spend no more than the current cheapest production Rolex, the Air-King Oyster Perpetual, costs at retailer's list (at the time of writing, late 1995, this was around £1,200.00 or $1,900.00). This figure should allow you to purchase a collectable, serviceable watch. Second, look at yourself. Someone who is of a slim build will not want to purchase a Submariner, nor will your average body builder choose a bubbleback. Third, discard the more common watches, Datejusts and Dates, not only are they almost identical to the watches on sale today but being so familiar they do not have a visible collector identity. Fourth, ignore Oysterquartz models, Rolex is famous for its mechanical movements and no quartz Rolex has yet achieved aftermarket desirability.

Having discarded all of the above, we would suggest that for an initial Oyster Perpetual the ideal model would be an Explorer 1, in particular the 1016 model. The Explorer 1 is one of the most universally recognized Rolex models and is characterized by its high visibility dial. The 1016 Explorer 1 comes in two versions, the earlier one with a gloss dial and the later one with a matt dial and a hack seconds feature, either is desirable and both are eminently usable everyday watches.

If you are looking for an earlier watch to wear only on odd occasions, then you have a much wider choice; we would advise against the purchase of an early Rebberg Rolex as a first watch although the price will certainly be right. We would also suggest that you also ignore the subsidiary brands, Rolco, Unicorn, or Marconi. The most desirable watch in this group, the Prince, is outside the price guides we have suggested above.

These factors leave a couple of watches standing head and shoulders above the rest: the original Oyster, in both octagon and cushion case styles; and the rectangular dress models, particularly those with the HW movement. When looking at early Oysters there are several things to look for. The watches with milled bezels are much more desirable than the later ones with plain smooth bezels.

Ranking dials in descending order of desirability, the most desirable are 1) enamel dials with roman numerals, 2) any enamel dial, 3) black painted metal dials, 4) metal dials with applied numerals, 5) metal dials with luminous numerals, and 6) the remainder. With these watches sweep seconds are rarer and therefore much more desirable. The case desirability order is octagon then cushion (while this is a matter of personal taste, the market has decided that this is the current order). Any case made of stainless steel is better than one made of silver, which in turn is better than any of the nickel, or nickel based cases. You will find some marked "Snowite;" these are basically chromed nickel. The reason that steel is more desirable than silver is because steel cases came late in the run of the Oyster and are therefore much less likely to be marked and scratched. When the watches were available in both silver and steel, the steel always costs more.

If you are looking at a rectangular watch, try and get one with a tonneau shaped movement, known as the HW calibre. Because these rectangular watches are still not as strongly collected, it will be possible to get a gold watch within the price. It will almost certainly be 9kt gold. If it is available in silver there should be very little price difference. Due to the lack of moisture and dust protection in these cases there is a strong chance of finding that the dial will been reprinted. Try to ignore these watches as any watch with a refinished dial will never be as desirable as an original one. Look carefully at the case, particularly inside the top part of the case. Examine the areas around

350

the corners looking for solder, and look at the point where the lugs join the case looking, again, for solder. Make sure that the number of the case stamped on the back is repeated inside the top (usually only the last 2 or 3 digits). When purchasing any watch, Rolex or not, wind the watch for a couple of turns moving the button back and forth. If the watch was not running this should be sufficient to get the movement going. It is easily observed by examining the seconds hand. Do not wind it all the way to the stop. Pull out the winder and move the hands clockwise (very important) through two or three complete revolutions of the hour hand. When you do this, make sure that the hands do not catch the seconds hand, touch each other, or scratch the dial.

Ask the seller to open the watch and, holding the dial and movement by the edge of the dial, watch the balance wheel rotate. It should move so swiftly that you cannot count the swings. Next rotate the movement through 90° and make sure that the balance swings just as briskly in this position. Do this another four times thereby observing the watch in all six possible positions (dial up, dial down, dial forward, dial back, winder up and winder down). In all these positions the balance should move exactly the same. Make sure that the movement is correctly signed. Examine the pictures of movements in the book and make sure that the movement corresponds to one of them. Then ask the seller to put the watch back together and, as he closes it, make sure that it does so with a noticeable click. This is your re-assurance that the case closes correctly.

If the watch is one you want and can afford, then buy it. Welcome to the world wide fraternity of vintage Rolex watch collectors. Of course you will not stop with just one Rolex. Sooner or later you will want to expand the collection and the whole decision process will have to be undergone once again. However this time you will be armed with the knowledge gained on your first foray into the market and will, hopefully, be a little more ready.

The decision as to your second and subsequent watches should not be as difficult as the decision over your first watch. We would like to offer our suggestions as to the way a collection should proceed, but please realize, these are just suggestions. We strongly suggest that the second watch in any collection should be an original Oyster, either the cushion shape or the octagon. This is a very important watch for a collector as it was the watch that made Rolex. It differentiated them from all the other companies and it inaugurated a name that is still printed on the dial of almost all Rolex watches today. Just as importantly it is a beautiful watch, slim enough to slip under a shirt cuff and sporty enough to be worn with Levi's. For your first early Oyster purchase we suggest a sterling silver model, mainly because it is cheaper than gold, but, just as importantly, also because it looks less ostentatious. The most desirable early Oysters have enamel dials and milled bezels. Expect to pay more for these features. A sweep second on these models is very rare and you will pay for this feature if you find it. If you do find one of these watches remember that while it was initially waterproof, it probably isn't now, and it was never shockproof.

The next purchase in a collection should be the first Perpetual, usually known as the "bubbleback." These watches were produced in quite large numbers from the early thirties until the mid-fifties and so should not be so difficult to locate. However, due to the complexity of the movement and the fact that the movement is completely covered by the rotor assembly, it is much more difficult to detect any mechanical problems with one of these watches than it is with the more simple manual movements.

Hopefully by now you will have determined which supplier you can trust. It is from this supplier that you should purchase your first bubbleback. Again, look for an original dial. Try to find a steel watch with gold bezel and winder. The reason for suggesting a mixed metal watch is that this configuration is now a Rolex feature and it was with the bubbleback that Rolex first introduced the mixture.

Mixed metal watches are not very much more expensive than plain steel ones, but examine the case closely. There is a long established tradition by unscrupulous dealers of adding gold bezels and winders to all steel watches. If the bezel is easily removed, it is likely that it has been recently added. Likewise, if the condition of the bezel is better than the steel portions of the case then it is also likely to have been recently added. Remember that gold is a much softer material than steel and as such will dent, scratch, and wear much more quickly. This softness means, on the other hand, that it is quite easy to polish marks out of gold but this, of course, results in the bezel being worn very thin by polishing.

You must also look carefully at the rear of the watch as the earlier bubblebacks were made of a steel that was not truly stainless and some are susceptible to case corrosion. This corrosion usually shows itself as small pinpricks on the rear of the watch and around the area where the back screws into the case. In some cases, the corrosion is bad enough to make the watch impossible to waterproof, so these watches should be ignored.

Key points when selecting a bubbleback are:
1) A small (or subsidiary) seconds dial is more desirable than a sweep seconds dial;
2) Any watch with the early flat back and movement signed "Auto-Rotor" is much more desirable than a later one, particularly if the movement is held in a movement ring;
3) Any black or two-tone dial is very desirable;
4) Any dial signed "Rolex Oyster" on the top line with the word "Perpetual" on a line below rather than the more common "Rolex" then "Oyster Perpetual" is to be sought.

There comes a point in any collection when you have to make a decision over which way the collection is going. Any collector who now owns two or three vintage Rolex watches must decide whether to concentrate on a single model, such as the Prince or the bubbleback and collect across the wide variety of watches made in the particular model, or to collect one of all the different models made by Rolex. These two different collecting strategies are known as horizontal and vertical collecting, respectively. Vertical collecting is much easier than the horizontal kind, but, as always, the last one or two items to complete a collection (of either variety) are always the hardest to find.

It is at this point, when you are deciding which way your collection is going, that you will have to begin thinking about trading one of your collection for either a better example or for a model you would rather own. When looking to exchange a vintage Rolex for another, it helps to put yourself in the shoes of the person with whom you are dealing. You know the advantage you will receive from the exchange (you will finish up with a watch you want), so ask yourself what the other party has to gain from the deal. If you are the one who wants to initiate the exchange then you are the one who must offer the enticement to the other party. Do not attempt to intimidate the other party by pointing out any possible defects in his piece while talking up yours. The sensible retort to this would be to ask you why you are so interested in replacing your watch.

Simply note the pros and cons of the two watches. Decide for yourself what you are prepared to pay and (without stating your budget) ask what the other party wants in cash to complete the exchange. If the distance between what you want to pay and what they want to receive is too wide, do not be afraid to walk away. If the distance is not too wide then you should counter with an offer that places your next offer halfway between his initial request and your first offer. Their response will tell you immediately whether there is any chance of making the deal or not. If not, walk away. Do not get emotionally involved in the negotiating process. We are not talking about warmth or food and shelter. This is not something vital to your existence. Just walk away. Even if you regret it later, remember the old saying "If you have nothing to regret, you have lived a boring life."

With this information under your belt, it is now time to embark on to one of the many varied Rolex collecting routes. We are setting out below a hypothetical vertical Rolex collection and three possible hypothetical horizontal collections. We have taken each collection almost to its possible extremes and are assuming that the potential collector has unlimited money and time. We have also indicated possible times when pieces should be traded or upgraded. Please remember that these are personal suggestions, they are not engraved on tablets of stone.

Vertical Rolex Collection

1. Early Oyster (silver)
2. Bubbleback (mixed metal)
3. Early Rolex Rebberg (silver, hinged back)
4. First model Datejust (4467)
5. Prince (silver flared 971 or 1491)
6. Steel Rectangular (H W movement)
7. Full or Demi Hunter Rebberg (silver)
   Trade in 3
8. Steel 3-piece case, first model bubbleback (1858)
   Trade in 2
9. Two-tone flared Prince (971 or 1491)
   Trade in 5
10. Steel non-Oyster chronograph (2508)
11. Steel Athlete (4127)
12. Early Explorer (6610 or 5504)
13. Two-tone hooded bubbleback (3065)
14. Early GMT-Master (6542)
15. Gold 3-piece Viceroy/Imperial (3116/3359)
16. White dial Explorer/Explorer Date (5504/5700)
17. Prince with "Jumping" hours (HS)
18. Milgauss (1019)
19. Submariner with British military markings (5513)
20. Steel Oyster chronograph (4500)
    Trade in 10
21. "Comex" Sea-Dweller (1665)
22. "Quarter Century" 14kt Prince (3937/2541)
23. Gold bubbleback with mixed Arabic/Roman dial
24. Steel Non-Oyster Moonphase (8171)
25. Two-tone ribbed hooded bubbleback (3595)
26. "Comex" Submariner with gas escape valve
27. Gold Perpetual with enamel dial (6085)
28. Early Milgauss (6541)
29. Steel Daytona "Paul Newman" (6239/6241)
30. Tru-beat/Metropolitan (6566/1020)
31. Gold or silver "Hermetic" (356)
32. Steel Hooded bubbleback (3599)
33. Oyster Moonphase (6062) Trade in 24
34. Gold Scientific with enamel dial (2941)
35. Gold Perpetual with cloisonné dial (6085)
    Trade in 27
36. Kew A Speedking
37. Platinum Prince
38. Hunter cased Prince (1599)
39. Centregraphe (3346)
40. Unique piece: 1915 calendar/Mixed dial Zerographe/Split second chronograph.

Horizontal collection based on the Rolex "Prince"

1. Silver classic (1541)
2. Silver flared (1491)
3. 9/14kt Flared (1491)
4. Steel Railway (1527/1768)
5. Gold Two-tone flared (1491/971)
   Trade in 2
6. "Quarter Century" 14kt (3937/2541)
7. Gold Filled Canadian market model
8. Any of the above with box & certificate
   Trade in original
9. "Striped" two-tone flared (1941/971)
10. Any of the above in "New, Old Stock" condition
    Trade in original
11. Asymmetric (3360)
12. Asymmetric sweep seconds (3361)
13. Platinum Prince
14. Hunter cased Prince (1599)
15. Platinum cased "Half-Century"

Horizontal collection based on the Rolex "bubbleback"

1. Steel sweep 2-piece case (5050/5011)
2. Two-tone sweep 2-piece case (most models)
3. Two-tone sweep 2-piece case with engine turned bezel (most models)
4. Gold 3-piece case with small seconds (3132)
5. Early flat back 3-piece case with small seconds (1858)
6. Two-tone hooded sweep seconds (3065)
7. Steel Athlete perpetual (3548)
8. Gold Imperial perpetual (3176)
9. "Army" perpetual
10. Two-tone sweep 2-piece case with mixed Arabic/Roman dial (most models)
    Trade in 3
11. Ribbed hooded two-tone (3595)
    Trade in 6
12. Boys bubbleback with wide bezel
13. Scientific dial
14. Steel fluted hooded (3801)
15. Steel hooded (2490)
16. Hooded Athlete (3536)
17. Athlete Bumper automatic (3591)
18. Cloisonné or enamel dial

Horizontal collection based on the "Explorer" Models

1. 1016 non hack seconds
2. 1016 hack seconds
3. 6610 from 1960-63
4. 16650 white dial with white circles around tritium
5. 5500 precision from 1960-68
6. 8044/5 black dial from 1954
7. 8044/5 white dial from 1954
8. 1655 Explorer II with orange 24 hr. hand
9. 16650 white dial with red enamelled bezel
10. 6150 Regular Dial from 1956/59
11. Model 6350 Checker dial from 1954/5
12. Model 6150 "Dress" style dial 1956/9
13. 5504 Black dial precision from 1958
14. 1002 White dial precision from 1957
15. 5504 White dial precision from 1957
16. British Military
17. 5700 White regular dial Explorer Date from 1959
18. 5700 White checker dial Explorer Date from 1958
19. 5513 or 6200 Submariner precision with "Explorer" dial from 1960s

---

[83] *The American Heritage Dictionary of the English Language,* Third Edition is licensed from Houghton Mifflin Company. Copyright © 1992 by Houghton Mifflin Company. All rights reserved.

Collecting Rolex Accessories

There has always been strong interest in collecting Rolex related items. When the Rolex "fever" began, the items most sought after were original boxes for the "Bubbleback" and Prince models, the most desired watches at the time. As interest in Rolex spread, people began to collect other items of Rolex material, including original catalogs, point-of-sale materials, and other displays.

Rolex has always been a marketing driven company and point-of-sale materials, such as window and store displays, were always beautifully designed and made. Almost all of the classic displays were made by the famed Schmidt studio of Lausanne, whose relationship with Rolex extended for almost 50 years.

Rolex always considered it vitally important that the window and in-store displays should stand out and differentiate Rolex from competing brands. To this end they would spend more on these items than other companies and would change them more often. It was here that the problems began. Putting it simply, most of the these display materials were supplied to the original retailer by Rolex on the basis that they would be used for their intended purpose and then returned to Rolex. Of course, most retailers would use the displays until they were replaced by newer versions. They did not bother to return the old displays to Rolex, usually just because of the problems involved in packing and returning the items. The materials were the property of Rolex, and the jeweler had no right to sell or dispose of them, but life being what it is, many retailers would sell the surplus items. They would often end up with non-Rolex retailers who would utilize the displays and watch stands to display pre-owned Rolex watches. This, of course, caused confusion with the public. They were unsure as to who was an authorized agent and who was not.

So that we will not be guilty of causing (or compounding) any such fear, uncertainty, or doubt in the readers of this book, we have made a very conscious decision not to show any display materials less than 25 years old. It can be safely assumed that no materials more than 25 years old could be confused with current displays. This, of course, does not apply to such items as boxes, etc., which immediately become the property of the watch owner.

Rolex has always made a point of emphasizing their unique selling points in their display materials. The advent of the first Rolex hermetic watches was promoted with the company's first dedicated window display. This was registered in Switzerland, as 52986 on November 9, 1922, and comprised a classic rectangular aquarium in which swam a number of tropical fish. Suspended in the water was one of the new Rolex watches proudly bearing the name "Aqua." Four years later, when the Oyster came to market, it was logical to use this same display for the new watch. Over the next 30 years it developed from its original rectangular shape into a round fishbowl, always with a device for suspending the watch.

After World War II Rolex introduced the final, perfected version of the fishbowl. It featured a watch hanger bearing the legend "Rolex Oyster Perpetual, Exact Time," because now the Perpetual was being heavily promoted as the latest unique Rolex feature. Rolex introduced a window display which emphasized the perpetual movement and completely ignored the Oyster side. This new display comprised an "exploded" PA (or lady's bubbleback) encased in a perspex (or Lucite) cube, standing on a pedestal bearing the legend "The 157 parts of the finest self winding watch in the world." Once again Rolex was in the lead. The technology involved was in its early days and this display was one of the most complex yet attempted. Very few were made and even fewer have survived. An initial viewing of the display quickly sorts the collectors from the watchmakers. The first group comments on its beauty and rarity, while the second group is initially silent before examining the display closely and pointing out one of the items before declaring "I need one of those.....".

In the mid 1950s when Rolex began to sign their watches "Superlative Chronometer Officially Certified" the emphasis of the company's promotion began to switch to the accuracy of the watches. A new line of displays was introduced. These consisted of a beautifully made clock contained in a wooden box which was a copy of a marine chronometer box. The clocks themselves were brass painted in green and black enamel to resemble marble. They bore the legend "time to the second" on the front and the company's name and crown logo behind the dial. The clock was powered by a mechanical movement driven by an electrically powered "remontoir," which rewound the mainspring every four or five minutes. This technique, which meant that the mainspring was always at a constant tension (and so produced constant force) replicated the system in the Perpetual. The shape of these clocks, like a horse's hoof, gave them their name in French "sabot," or hoof. These clocks remained in use for over 30 years and many Rolex agents still display them. During this period the company introduced a line of display clocks, which were essentially enlarged mantel clocks. They were simply styled in a classic 1950s way and were unlike any other Rolex clocks in that they were sold to other companies. The brown one shown bears the "Maharajah" logo of Air India.

It was in the mid-1980s that Rolex introduced a replacement for the sabot that maintained the same general shape but now replicated a gold Submariner model. This new clock was powered by a quartz movement. It is still in daily use, and because it is less than 25 years old we have chosen not to show pictures of it.

Stainless steel "Datejust" awarded to members of the winning Polo team at Guards Polo Club in Windsor near London (the club that both Prince Philip and Prince Charles belong).
Model no. 1522
Case no. 7187653
1982

Stainless steel paper knife and scabbard. The knife is a one-piece item with a flat handle stamped with the crown and engraved with the company name along both edges. The scabbard is Rolex green and is embossed to resemble crocodile skin. It is also embossed with the crown near the tip.
Circa 1980

"Sabot" (or hoof) clock from the 1950s, made from a brass casting which is then painted with green enamel in an imitation of marble. Metal dial with applied roman numerals and subsidiary seconds. The rear elevation of the clock has applied gilt Rolex logo and name, while the front bears the legend "Time to the second". Mechanical movement wound every four minutes by a battery operated remontoir. The clock is contained in a heavy wooden carrying case which also bears the Rolex logo, inside and out. Movement serial number 1407.
Circa 1950

Collecting Rolex Watches 355

Another sabot clock, this time made for a Spanish-speaking market. It now carries the legend "Le hora al segundo," which has the exact meaning of the previous version.
Circa 1950

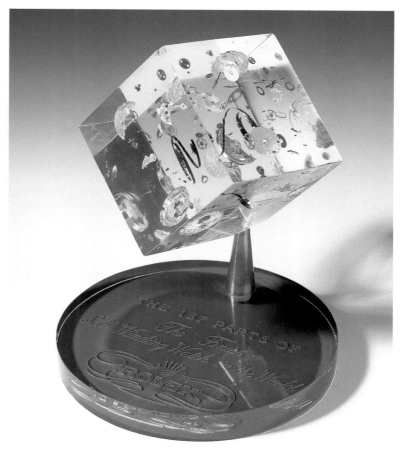

Store (or window) display with an "exploded" AR movement encased in a Perspex or Lucite cube and standing on a base bearing the legend "The 157 parts of the finest self-winding watch in the world"
Circa 1948

Rare final version of the "fishbowl" window display, this one dating from the late 1940s. Rather than the simple bowl of the earlier period this smaller bowl is carried on a brass holder which bears lacquered plates representing a water lily, its leaves, and bloom and also a hovering dragonfly. The watch is suspended from an enameled holder surmounted by the Rolex crown and bearing the legend "Rolex Oyster Perpetual" and "Exact time." This unit is probably the finest window display the company ever produced.
Circa 1950

Sterling silver and ebony(?) plaque awarded at the Rolex Cup at the Guards Polo Club in July, 1988. The plaque is made by and contained in a box from Garrards, one of London's finest jewelers.
1988

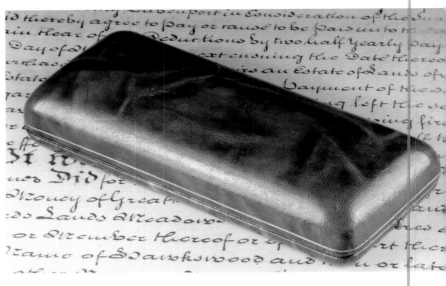

Rectangular "coffin" style bubbleback box from the late 1930s, one of the first to use the later style Rolex crown logo.
Circa 1938

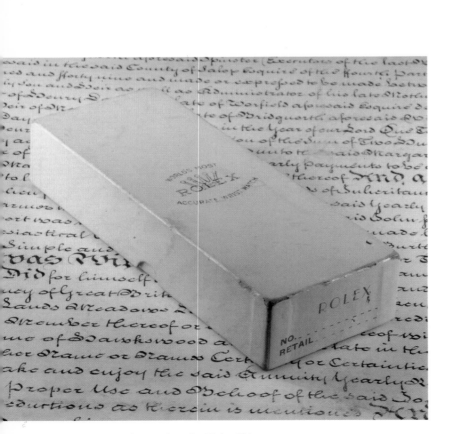

White cardboard outer box for Rolex Prince, with one of the earlier examples of the Rolex Crown, the Rolex trademark. Ca. 1931.
Circa 1935

Brown leatherette chest-style box with gilt embossing for first model "bubbleback" still using old model crown. Mid-1930s.
Circa 1937

Green oyster-shaped box for lady's Oyster "bubbleback" early 1950s.
Circa 1950

Rectangular green box for boy's "bubbleback," mid-1940s.
Circa 1945

Green box made for Rolex "Ovetonne." This Rolex trademark was in use until the mid-1960s.
Circa 1946.

Unusual "treasure chest" motif on box made for a lady's watch called the "Golden Dream." Ca. 1960s.
Circa 1962

Collecting Rolex Watches 359

Box adorned with unusual open metalwork, made for 18kt Datejust in the early 1970s. Circa 1975

Early 1980s box for the mid-size Date model.

Wood-lined GMT-Master box commemorating the 50th year of the Rolex Oyster. 1976

Gastronomical guide to fine watches and fine dining in Italy. Rolex certainly knew their markets well. Good food and fine watches. Circa 1958

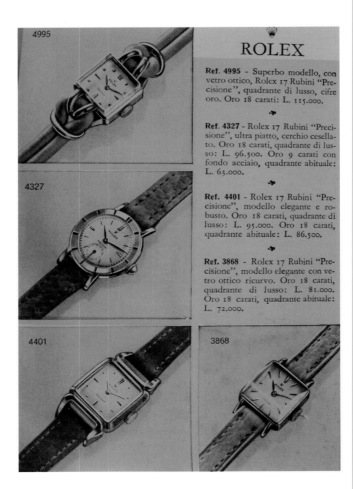

ROLEX

Ref. 4325 - Rolex Cronometro, ultra-piatto, 18 Rubini, cerchio finemente cesellato. Oro 18 carati: L. 129.500. Oro 9 carati, fondo acciaio, 17 Rubini: L. 59.400.

Ref. 3737 - Rolex Cronometro, 18 Rubini, con cerchio finemente cesellato. Oro 18 carati: L. 147.600.

Ref. 3667 - Rolex 17 Rubini "Precisione", ultra-piatto. Oro 18 carati, con quadrante di lusso: L. 99.000. Oro 18 carati, con quadrante abituale: L. 82.800.

Ref. 8094 - Rolex Cronometro, ultra-piatto, 18 Rubini. Oro 18 carati: L. 141.300.

ROLEX

Ref. 4062 - Rolex Cronografo di alta "Precisione", 17 Rubini, *Antimagnetico*, Telemetro, Tachimetro. Oro 18 carati: L. 126.000. Acciaio: L. 60.300.

Ref. 5036 - Rolex "Oyster" Cronografo Calendario Totalizzatore, 17 Rubini, *Antimagnetico*, indicante il giorno, la data e il mese. Impermeabile. Oro 18 carati: L. 224.000. Acciaio: L. 93.600.

Ref. 4313 - Rolex Cronografo Totalizzatore, 17 Rubini "Precisione", *Antimagnetico*, Telemetro, Tachimetro. Controllo sino a 12 ore. Oro 18 carati: L. 157.500. Oro 9 carati, fondo acciaio: L. 87.300.

Ref. 4768 - Rolex Cronografo, Totalizzatore, Calendario, 17 Rubini "Precisione", *Antimagnetico*, indicante il giorno e il mese. Tachimetro. Oro 18 car.: L. 166.500. Acciaio: L. 86.400.

ROLEX
"Il Principe degli orologi"

Ref. 4858 - Rolex 17 Rubini "Precisione", superbo modello con bracciale serpente snodabile, incastonato con 10 brillanti prima scelta. Oro 18 carati, quadrante di lusso: L. 331.000.

Ref. 4457 - Rolex come sopra descritto ma senza brillanti, attacchi leggermente differenti. Oro 18 carati, quadrante di lusso: L. 231.300.

GENOVA
E LA LIGURIA
San Remo, Portofino, Santa Margherita, Rapallo, ecc.

★

Nella cucina genovese l'olio d'oliva della Liguria, giustamente famoso in Italia e fuori, la fa da padrone. Con l'impiego sussidiario di vegetali aromatici (origano, basilico, ecc.) che danno sapori e profumi spiccati, si determina la maniera di cucinare che in tutte le città marittime è nota come "alla genovese".

DOMANDATE:

IL PESTO, cioè il condimento di magro più caratteristico di questa cucina: foglie di basilico, formaggio pecorino sardo grattugiato, prezzemolo, olio d'oliva, aglio, pinoli e noci.

FRA LE PASTE: le *trenette o' pesto*, che sono le celebri linguette di pasta asciutta, nonché le *tagioi* (tagliatelle) e le *ficaggie* (fettuccine). Capolavoro della cucina genovese: i *ravioli*; caratteristici anche gli *gnocchi alla genovese*, e gli *gnocchi verdi*, in cui entrano gli spinaci.

LE PIETANZE AL FUNGHETTO, melanzane o zucchette tagliate in modo da sembrare funghi, cotte in padella con i consueti ingredienti.

LA TORTA PASQUALINA, di pasta sfoglia, ripiena di un delicato e ricco "paté" di carciofi, o di bietole, nonché ricotta, uva e farina: piatto primaverile di alto rango, che non manca mai alle tavolate pasquali della Dominante.

LE LUMACHE ALLA GENOVESE, lavate nell'acqua marina e nell'aceto, bollite nell'acqua con aceto e aromi, cucinate in casseruola.

FRA LE CARNI: la *vitella all'uccelletto*, scaloppine di vitello "ad arcipelago", cioè simili a piccole isole frastagliate cotte in teglia al burro e all'olio; la *cima*, una specie di galantina costituita da una punta di vitello disossata e riempita con un battuto.

FRA LE INSALATE, il *cappon marro*, piatto solenne, stimato "la regina delle insalate". Si tratta di una specie di grande sformato.

FRA I DOLCI: i marzapani di Pasqua, fatti di pasta di mandorle e rivestiti di zucchero glassato; le insuperabili *frutta candite*, i *biscotti del Lagaccio*, del tipo Zwiebach.

362 Collecting Rolex Watches

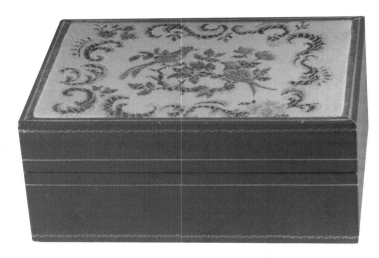

Decorative faux needlepoint lady's "President" box 1980s.

The "pocket watch" shown is, in fact, a perfume bottle containing a scent aptly named "Perpetually Yours." The pendant and crown form the bottle cap. This is the original fitted box. It is not known whether this perfume, which dates from the early 1960s was a promotional gift or was actually sold.
Circa 1966

A 1920s "guarantee" form from Rolex that shows that even at these early dates Rolex was testing its watches for a period of 21 days.
1920's

Collecting Rolex Watches 363

As previous item but bearing the logo of Air India.

Wooden and metal display clock with silver dial and hands, from the 1950s.

Completely fake Oyster Perpetual Moonphase; this watch is a simple steel oyster perpetual to which has been added a dial and date discs; however none of these additional functions are operational. The case has had a gold bezel (in a style Rolex never made) added and the back has been cut into two parts to give the additional depth needed. This watch is unlikely to fool any but the most naive.

CHAPTER 24
FAKES
Imitation Is the Nastiest Form of Flattery

One of the major problems that has arisen since Rolex watches became collector's items is that certain unscrupulous characters have spotted a way to turn a fast buck by faking the more collectable watches. In an attempt to help the collector we are setting out below some of the fakes we have seen. But be very careful. This is a moving target. The fakers, and those trying to discover them, are always raising their games.

The earliest model made by Rolex that is likely to be subject to fakery is the cushion or octagon Oyster. It is common to find these watches with gold bezels on steel or silver watches. This configuration is wrong. We have access to all the catalogs from Rolex over the complete period that the original Oyster was produced, and in none of them does Rolex show two-tone oysters. The two-tone configuration was only introduced with the arrival of the "bubbleback." Another version of this two-tone enhancement involves adding gold hour markers to a silver or steel bezel. Again these watches should be avoided.

It used to be that the only two certainties were death and taxes, but they were followed closely by the assurance that no-one would fake a bubbleback. Sure, people would "upgrade" them with all kinds of embellishments, gold bezels, engine turned bezels, and hoods were the most prevalent. Real hooded bubblebacks have two features distinguishing them from regular models. First, on a normal bubbleback the lugs do not curve down toward the wrist anywhere near as much as on a real hooded bubbleback, and, second, all real hooded bubblebacks have both model and serial numbers on the outside back of the watch, not between the lugs.

The reason we knew no-one would fake a complete bubbleback was simple, it was not economically viable. It is actually harder (and therefore more expensive) to make a bubbleback case than it is to make a current President case. And we all know which one sells for more. Q.E.D.

Everything changes. In 1988 you needed $1.60 to purchase 1 Soviet ruble. At the start of 1996 you got 4,600 Russian rubles for $1.00 officially (the black market rate was even better). Pretty much the same thing was happening all over the old Soviet bloc. Suddenly all ideas as to what was too labor intensive to make fell by the wayside, and completely faked bubbleback cases started turning up.

The distinguishing feature of the three we have seen is that they are near perfect; the quality of case workmanship is fabulous. In fact, they are too good. In all the examples we have seen, the dials and movements do not match the cases in terms of wear. Please remember the name of these watches. They are "Oyster Perpetuals" and the sole purpose of the Oyster case is to protect the case and dial from wear. So it is not logical to have a watch in which the case is in perfect condition and the movement and dial are not. If the case has done its job correctly (and they usually do), the logical thing would be for the case to have excessive wear and the movement and dial to be in fine condition. The examples we examined were all hooded 18kt pink gold models, and all carried the reference #3065.

In addition to the case/movement discrepancy, these bubblebacks had one major mistake in their case construction. The underside of the hoods were frosted, as it came from the casting mould. We have had the good luck to own a couple of new old stock hooded bubbles in our time and with one of them one of the authors still retains visiting privileges. On the real one, the underside of the hood was machined, not frosted.

On all of these fakes there were problems with the dials. One of them had a refinished dial (again illogical on a N.O.S. watch), another had a dial which came from a much later watch, and the third had a dial which was too tight for the case, the outer minute track being almost hidden by the edge of the case.

The next problem is that we have had reports of non-hooded bubblebacks coming from the same source, so beware.

The first Rolex to be faked was undoubtedly a Prince and it retains its position as the most faked (and "enhanced", which is the same thing) Rolex model.

The "enhanced" models are the easiest to identify. If the case material is inconsistent throughout, it is not real. Rolex did not make any Prince models with a silver body and gold flares. Any watch fitting this description is usually a straight sided "classic" model no. 1862 with flares added later. The flared Prince always has the model number either 971, 1490, 1491, 2540, or 2541 on the back. *Be very careful of any Prince on which the numbers have been polished down and are no longer visible.*

All pre-World War II original Prince watches had solid strap bars, which were always made from the same material as the watch. On a gold or silver watch, the bars would also be gold or silver. On a British watch (and most Princes were sold in Britain) they would be hallmarked with the precious metal content. The above information also applies to the very rare steel model with gold flares. On this model, the bar and the flare opposite the winding crown both carry hallmark stamps, either English or Swiss, the former being either the number .375 or .75 (signifying 9 or 18kt gold respectively), and the latter being a cartouche showing a flower, squirrel or lady's head (signifying 9, 14 or 18kt gold respectively).

While the addition of flares to a "classic" is comparatively simple to detect, the conversion of a steel classic to a steel and gold one is quite difficult to detect, as the only gold on this model is a strip down each side of the dial and gold loop lugs. The up side to this problem is that these watches are very rarely converted because the value added by the conversion is usually about the same as the cost of the work involved, so there is no profit in the conversion.

There is another kind of "enhancement" that is more dangerous and difficult to spot, where an original back is used with a completely new upper. Again one may find a back numbered as a "classic" or as a "railway" but with a flared top. With this type of watch, matching case condition is everything. Look inside the case back and top. Check for finishes that match and to see that the last two or three numbers of the case number are repeated inside the top with a typeface that matches. Check the straight inside edges of the flares, where they become the lugs. This is a wearing area, so if they are sharp then you can expect the rest of the case (including the back) to be in pristine condition, probably still showing machining marks. This type of conversion is often done when converting a non-jump hour watch into a jump hour. This is done by taking a Gruen jump hour movement and adding the specialized pieces to a normal Rolex Prince movement, as the two watches were both manufactured on the same production line, this is a simple task. However the higher movement will not fit in a normal Prince case and so a new top has to be used. To check for this type of conversion, check that the model number is correct for the type of case and that the letters HS follow it (HS stands for Heures Sautantes or jumping hours). A jump hour Prince will always have these letters.

The Prince that some deny ever existed is our next subject, the "striped" Prince. Rolex actually did make some of these, as seen in the 1927 newspaper advertisement shown on page 114 featuring the watch. The aftermarket has made very many more than Rolex ever did, making it the most common enhancement to solid gold models. To recognize a correct one it is important to understand how these cases were made. A normal case had a number of channels machined in the top and a thin strip of white gold was laid into each of these spaces and then soldered. The gold inlaid strips only went down to the second facet of the case top, *they did not extend all the way to the join with the case bottom.* You would also never see the white gold from inside the case. This model often is faked using a correct back and a completely new top, so see the information in the paragraph above.

Before moving to the subject of completely fake Princes, it is useful to clear up some common misconceptions. First, there were some genuine Rolex Prince models made with gold filled cases. These models almost without exception come from Canada. They are mostly straight sided, the backs are marked inside with the phrase "Rolex Quality Guaranteed." The markings inside the case are simply engraved, so to many people they seem fake; they are not. Sometimes the words "Cheshire Quality" appear as well but no mention of World's Records, etc. They often (but not always) have the 15 jewel movement or the later 17 jewel "oak leaf" engraved movement and very simple dials. The quality of these cases is much lower than the gold filled Gruen cases of the same era.

With Rolex watches of this era there is always an exception. At the same time these simple gold filled watches were being made, Rolex also made a two-tone flared gold-filled Swiss case with engraving along the length of both sides of the dial opening. This watch is of the normal Swiss high grade and, in fact, is rarer than solid gold Princes of the same period.

With fully faked Prince models there is still some good news: as yet no-one has faked steel, silver or G/F watches, or any of the late models, the asymmetric (models numbered 3361 and 3362), or the curled lug model 3937, so if you are confronted by any of these you can be fairly confident they are authentic.

To our knowledge, the fully faked Prince watches have come from three sources: London, Manchester, England, and Germany. The Manchester cases are the easiest to dismiss, they are cast and as such weigh 20% to 30% more than a normal case, which in the 18kt "flared" model should be 31 grams. The general standard of finish is such that it would be difficult to imagine any competent person being taken in by them and frankly we have not seen one for more than three years leading up to this writing. The London made cases are better, but are betrayed by the inner finish of the upper case. This is a rough "frosty" finish rather than a clean machined finish. They are also betrayed by their case stamping and hallmarks because they are not original. They are in fact the result of casting and therefore have a different finish.

When a case is stamped at the factory, it happens after the construction of the case is complete. When the die is hammered into the virgin surface, two things tend to happen. As the die leaves its impression in the surface, surplus metal is forced out around the edges and the bottom of the impression is perfectly smooth. Both of these effects can be seen with a diamond loupe.

Neither of these happen with casting, no matter how fine the casting material is. There will be no sign of the metal being pushed up and the bottom of the stamping will be rough.

The most recent cases to have surfaced are from Germany and are by far the best cases yet to have been made. They are constructed in the same way as the originals, being fabricated from sheet metal with the cast flares then soldered to the finished case. They weigh exactly the right amount. While they have excellent stamping marking, it is also their failing. They lack the casemaker's mark inside the cases and there are no national hallmarks, neither Swiss nor British. On the outside of the case back the model and case numbers are stamped in the correct positions but are in the wrong typeface.

The overall quality and finish of these cases is so high that they appear brand new, factory fresh and perfect. Look closely at any Prince in this condition and examine the dial and movement. If the watch is genuine then the dial and movement should also be perfect. If the condition of both matches then you may have a new old stock piece which does turn up. We have had one per year in the last 6 or 7 years. But it is much more likely that you will have a newly manufactured fake. As in most things to do with old watches it is very much a case of *caveat emptor* (let the buyer beware) and in the end the more you see, the more you know. To advise you in what to look out for, we are examining one of the fake German cases below.

Report on platinum wrist-watch case numbered 14030
The case of this watch is evidently incorrect, having been recently made. The evidence is manifest, there are over 20 errors in the construction and finishing of the case to say nothing of the most obvious flaw, the disparity in condition between the state of the movement and dial in comparison to that of the case.

The French architect Le Corbusier famously said that "God is in the details," and in examining this case it is necessary to look at what sometimes seem the tiniest details.

We shall examine the top and the back individually, starting with the back.

1) The typeface used in the numbers on the back are of a different typeface than that used by Rolex, which was always a "serif" typeface. That used on the fake watch is a "sans-serif" face. The number is written as 14030 whereas it should be written as **14030**. Note the lack of a foot to the one and the round top to the three.

2) When Rolex stamped the numbers for the case and the case style, they were done in a frame that held both numbers. This resulted in the numbers being parallel as these so obviously are not.

3) See item 2, it is obvious that they have been done at different times, as evidenced by the different impression depths of the two numbers.

4) The width of the stamping is completely wrong. This is because the stamp used to make the impression has sloping sides, which mean that the heavier the impression the wider the top of the impression will be.

5) The stamping inside the back is obviously incorrect, the RWC L^TD mark does not match that normally used by Rolex. The underlining on the TD should begin under the downstroke of the T and end in the center of the D, whereas on the "Prince" it begins after the downstroke and continues almost to the end of the D.

6) The typeface used in the "RWC" stamp is wrong, Rolex uses a condensed typeface in which the letters are compressed horizontally, this causes the right downstroke of the R to be almost vertical, whereas the "Prince" uses a normal typestyle with the stroke at approximately 60°.

7) The measurements of the oval cartouche in which the RWC Ltd appears are wrong. While it should measure 5.50 x 2.24mm in fact it measures 5.75 x 2.52mm.

8) All of the stamps used for the back of the case (apart from the rear numbers) have been made by engraving a plate of steel. An impression is then taken from this and then used to cast a stamp. The inherent problem with this method is that it is impossible to produce a perfect 90° turn with an engraving tool, resulting in the top left corner of the letter R and all the corners in E being rounded. Rolex would produce a photographic plate that would be used to acid-etch their stamp and which, of course, produce a perfect angle.

9) This phenomenon is also evident in all the corners of the rectangular "Platin" cartouche.

10) The underlining above Geneva Suisse also displays this problem perfectly.

11) The hand engraving involved in the process can also be seen in the line under the TD of LTD under sufficient magnification it is quite evidently uneven.

12) The back shows only three watchmaker's repair marks, covering a notional 46 years (1948..1994), yet somehow they are all in the same hand and evidently recently inscribed.

13) It is impossible for a case to have been used as heavily as this watch nominally was, looking at the dial and movement. There are no signs of dirt in the engravings, an impossibility no matter how often or recently the case had been cleaned.

This concludes the summary of problems with the case back, we will now examine the top of the case.

14) The inside lower edge of the flares retains a very sharp edge, impossible if the watch actually is as old as it pretends to be. This is an edge that would wear against the wrist, even on a material as hard as platinum.

15) The case has been made so recently that the holes for the spring bars still show some of the tiny particles of platinum from the drill.

16) The holes for the spring bars show no signs of dirt (or even the remains of dirt), once again, impossible in a watch case of this purported age.

17) In an attempt to give some semblance of age to the case, the area around the winding crown has been given a circular milled finish, similar to that which would occur by the friction of a winding crown. However, platinum is harder than steel and would not mark in such a way.

18) These milling marks extend all the way to the winding stem hole. This, of course, would not happen with a winding crown, which would mark in one area only.

19) The hole for the winding stem is so new that there are still visible burrs on the inside of the case.

20) Inside the case, around the glass aperture a true Prince would have a raised "frame" into which the glass would sit. This feature is missing on this case.

21) The dial has a mark all around the perimeter, caused by the frame mentioned above in point 20. As stated above, there is no frame to cause this mark.

22) In a correct case the vertical long sides of the inside case close to the four corners are marked with a soldered X, in this case they are missing.

Now let us look at fake or replaced dials

The Rolex Daytona made from 1962 to 1988 came with two dial configurations, a basic one and an exotic one; the exotic (or "Paul Newman") dial is by far the rarer and commands an 80% premium over a watch with the basic dial. This means, of course, that people are now remanufacturing the dial in order to profit from the situation. There are some things you can look out for when confronted by one of these dials.

First, examine the luminous markers carefully. On a correct dial there are markers at each of the hour points except at "12". Next examine the three subsidiary dials paying close attention to the concentric circles. A correct watch will have them so close together that it takes a detailed examination to see the spacing between the circles, while most of the fake ones have them much more widely spaced.

The problem with these fake dials is that there is now so much money to be made from them that the quality keeps getting better. While the information contained here should help you to identify a large proportion of the fake dials, it will not weed out all of them. To obtain the final confirmation it is necessary to remove the dial. Of course not all watch sellers are in a position (or will want) to remove their dial, in these circumstances purchase of such a watch should be made provisional upon having the watch disassembled and examined by a competent watchmaker. When the dial is off you should look for two things on the rear of the dial, first look for the name of the dial manufacturer, Singer, and their logo, an intertwined S and L, these markings should be stamped into the rear of the dial, not engraved. It is also important to look at the areas where the hour markers are attached to the dial. If the watch has been converted from a regular Daytona dial to a Paul Newman, the faker will have had to remove the markers to do so. When they are reattached there will be evidence of this on the rear of the dial, usually in the form of poor soldering.

Fakers sometimes do make the simplest of mistakes. You may see a dial marked "Oyster Cosmograph" on a Daytona with non screw-down pushers. If so, the watch has had a dial swap, as the designation "Oyster" was only on watches with screw-down pushers. Obviously the reverse is true also, if you see a watch with screw-down pushers but it is not signed "Oyster" then once again there is a problem.

The other problem we often see is one of repainted or "refinished" dials. The quality of these dials varies greatly. Many are easily spotted by the most simple of mistakes, such as the lack of "Swiss" or "Swiss Made" down below the "6". However, prior to the mid-1920s, it was not necessary to print "Swiss" on a dial, although most did. Many others are so good that detailed examination is vital.

When dealing with chronographs and other watches with subsidiary dials, always examine the edges of these sunken dials with great care. When the dials were originally made, the subsidiary dials were made by being drilled. This had two results. First, the flat bottom surface would be marked with tight concentric circles. Second, there are two parallel surfaces (the dial and the bottom of the sunken dial) joined by a short (less than 1mm) vertical surface, leaving two 90° angles at the top and bottom. When a dial is refinished, it is first polished on a buffing wheel to produce a flat surface prior to printing. This removes the sharp edges at the subsidiary seconds dial and rounds them. It also often removes the concentric circles.

There has been a problem with the early Rolex chronographs for several years now and once again attention is vital. This is logical really, as they were never made in huge quantities and therefore are both valuable and rare enough for most people not to have seen a large enough sample of the real thing.

The faker's greatest advantage with these watches is that the movements were not made by Rolex, but by Valjoux, who made movements for almost everybody from Baume & Mercier to Vacheron et Constantin. This means that the faker merely has to engrave or change a bridge and suddenly a Girard Perregaux is a Rolex.

We shall concentrate on just two of the watches we have seen recently. They provide excellent examples of the sort of thing that is making the rounds at the moment.

The first item is a non-Oyster chronograph bearing the model number 2508. The first thing that caught the author's eye was the overall condition: it was in just too good a condition for a 1940s watch. Looking at the outside of the case, the "2508" was in the wrong typeface. It was in an Italic typeface, which, of course, was not used. Rolex used the same typeface for all their number stamps on watches from the start until the late 1940s. Inside the case the marks looked fine, until looked at under a 10 power diamond loupe. It was then obvious that, instead of being stamped, they had been engraved, but very professionally, using a pantograph. The customary "burring" which is always seen on engraving was just visible.

The movement was in excellent condition, with classic "Geneva striped" finish, the top bridge with its "Rolex Geneve" and "17 jewels." The engraving was very, very good and may even have been from an original watch. On the main plate of the movement was the inscription "Swiss Made" and this was in a different typeface from that on the top bridge. This fact caused suspicions to rise further and they were confirmed totally when we operated the pusher and one of the levers moved, revealing the inscription "15 jewels" below it.

The next watch to come our way was a very small 18kt Oyster chrono bearing the reference 3481. The dial had been refinished and so it was difficult to make any decisions about that, but the case and movement were enough to seal its fate.

On first view it seemed authentic but after serious examination it was obvious that the piece had never seen the inside of a Rolex factory or dealer.

First, the top bridge had been ground down to remove the original name and then re-engraved with the legend "Rolex Geneve." In doing this three things had happened: 1) the uniform parallel machined finish evident on the rest of the movement had been replaced with a finish that was completely random and pointed off in many directions; 2) the bridge had been machined so much that the two jewels in the bridge no longer were seated properly and their position in the bridge relative to all other jewels in the movement was much higher; and 3) the quality of the work done was so poor that particles of metal were still to be seen floating around in the movement.

The engraving of "Rolex Geneve" on the bridge did not match the engraving of either "Swiss Made 17 Jewels" by the running seconds wheel, or of "A R / F S" on the balance cock. The correct engravings should have been done with a pantograph and would reveal the brass beneath the rhodium finish and the typeface would be a cursive sans-serif. The Rolex engraving on this watch was not done with a pantograph (for example the legs of the "N" in "GENEVE" were not parallel), they did not show brass beneath and were in a much more condensed non-cursive typeface.

But the clincher was that the watch was shockproof and used the classic Incabloc system, Rolex was one of the owners of Incabloc's main competitor Kif, and always used the Kif-flector method of shock absorption.

While the case had a model number inside the back, it did not have a serial number, either between the lugs or in the back. Under a 10 power loupe, it was obvious that there was porosity in the edges of the threaded areas of both the case back and the main body of the case. This and the heaviness of the case pointed to it having been cast. This was confirmed upon examining the Swiss 18kt hallmark. It should have had a perfectly flat surface, but was, in fact, granular, displaying the surface of the casting medium.

While some fakers are expert, some make such basic mistakes that one wonders if they have ever seen a real Rolex. We all know that the "street corner" Day-Dates with quartz movements in "two-tone" cases are fakes of a model never made by Rolex. What are we to make of a fake Rolex pocket watch in a base metal case with some of the worst stamping ever seen? Once again it is attempting to pass itself off as a watch that Rolex never made!

Do not make the assumption that the fakers are only making Rolex watches. In late 1994 a huge operation was uncovered which was making fake Rolex bracelets. The story behind this operation is worth studying and so we shall examine it in detail.

In late November, one of London's main dealers in vintage Rolex watches received a telephone call from a dealer with whom he had done some small business over the years. The substance of the telephone call was that this second dealer was in contact with someone who had uncovered a lot of 1930s beaded (or grain of rice) bracelets. Over a couple of weeks subsequent telephone calls he was told that the bracelets were from the Caracas, Venezuela retailer Serpico y Laina, which company had recently been purchased by the second dealer's contact. It was made clear that there were only a small number of the bracelets, less than 150, and that they were in a variety of metals: steel, steel and gold and 18kt pink and 18kt yellow. The bracelets could only be purchased as a lot, not individually.

The London dealer liked the sound of the offer. Buying them all would allow him to control the rate at which the bracelets would be released on to the market, thus avoiding having the market swamped, which might happen if the bracelets were to be sold to a number of dealers. But, as always when confronted with a gift horse, he looked it closely in the dentures and made one small (but vital) request: he would like to obtain a sample of each of the bracelets so that he might verify their originality. He would, of course, pay for them.

This proved to be a problem, and, of course, the London dealer began to suspect a rat. It took over two weeks for the items to turn up and when they did the circumstances of the handover were bizarre (to say the least). The second dealer refused to come to the store, saying he wanted no-one to know about the bracelets. Could they both meet on a street corner and do the transaction in a car on Saturday evening? Despite these conditions the meeting occurred and the bracelets were indeed present, wrapped in oiled paper which seemed old enough to be genuine.

On initial examination they looked good, but subsequent examination showed up a couple of flaws. First, the bracelets did not bear the stamp "GF," for Gay Frères, the original manufacturer of these bracelets. Second, and most tellingly, they were stamped "Swisse Made," when the two possible stampings would be "Swiss Made" or "Fabrique en Suisse".

Our dealer was allowed to take only a steel bracelet with him and was told that the transaction to purchase the samples had to be made by Monday afternoon, as his contact was going out of the country, and that the balance of the deal must be completed within two weeks. When he had the steel bracelet back in his possession he examined it much more closely, using the facilities available to him on a Saturday evening at home. He was left feeling that there was probably a 50/50 chance the items were correct, and was unsure what to do. He discussed the matter with a colleague and the decision was made to go ahead with the purchase of the samples package (1 each of steel, steel/gold, pink gold, yellow gold and yellow/white gold) conditional upon certain guarantees being given. The costs involved were about £8,000 (or $12,500).

On Monday the transaction was completed. The other bracelets were handed over and then taken back to the store, where a much closer examination was possible. Here a number of flaws began to show up. On initial examination it was obvious that the standard of fit and finish was much inferior to that which would be expected from such a well known firm as Gay Frères. Then it was noted that the stampings on the gold bracelets were wrong. At this point bells began to ring and a decision was made to turn the items over to the author for close examination.

The author received the offending items on Wednesday and had a chance to have a good look at the bracelets before he flew off to Milan, Italy for a watch show the following day. He was unhappy with a number of points and decided to examine them in greater detail upon his return.

Interestingly, while in Milan, he was approached by a couple of Italian dealers who offered him more of the bracelets, up to 30 of them (remember that the London dealer claimed he had been promised all of the bracelets). Subsequent telephone calls to other dealers in the U.S. confirmed that major dealers had purchased significant quantities of the bracelets and that they had also begun to turn up in Japan. It became apparent that there were at least 500 of them, and as his sources are by no means comprehensive, he guessed that there may be as many as 2,000 of them in circulation.

With this knowledge in mind, he began his close examination of the bracelets and problems began to jump out at him. He had in the interim obtained access to an original Serpico y Laina bracelet and was therefore in the privileged position of being able to make direct comparisons:

1. They were marked S&L for Serpico y Laina whereas the originals are marked SyL.
2. The links on the original bracelet were closed at the rear whereas these were bent over and still open
3. The Rolex cartouche (the RWC Ltd. stamp) was obviously made from an engraving as it had none of the clarity of detail one would expect from a stamping.
4. The gold marking was 18kt 750, whereas the correct marking is simply 750.
5. The closing faces on the top part of the buckle still showed file marks.
6. The stamping of "Swisse Made" was very bad with the initial "S" almost invisible, again no way would this have passed inspection.
7. The misspelling of "Swiss" could not have been done by anyone in Switzerland, as this was a stamp that was being used every day in the factory.
8. The "9" stamped on the gold/steel bracelets was much larger that normally seen on bracelets from GF.
9. The RWC Ltd. and Swisse Made stamps were in exact alignment on all the bracelets, meaning that they were together on one stamp. On the correct bracelet they are separate stamps and thus their relative positions vary.

It was by now obvious that the samples, and by deduction the whole shipment, were faked. As you may recall, earlier we spoke of certain conditions that were laid down prior to the purchase of the samples, one of those conditions was that the original supplier (who remember, was supposedly in the process of buying Serpico y Laina) should supply a letter of authenticity on Serpico y Laina stationery. Obviously this condition was never met and because of that and a report on the bracelets, the London dealer was able to return the samples to his supplier. The supplier, returning to Italy with the bracelets, discovered that his source had fled the country after having defrauded the Italian horological world for a sum close to $2,000,000. Needless to say, he was never paid.

Let us turn to the faking and enhancement of more modern Rolexes. The first trick is the simple removal of the bezel and winder from a steel Datejust and their replacement with gold ones. This is followed by the enhancement of steel bracelets, mostly Jubilee models, by substituting gold links for the center steel links (sometimes with links that were gold color only). This adaptation is fairly easy to spot; a correct Rolex two-tone bracelet will have the gold content stamped on the buckle as a simple number 9, 14, or 18. Today all bracelets are stamped 18. To check whether the watch itself has been enhanced it is necessary to check the model number, principally the last number in watches with a five digit model number. The table below gives the meanings of this number:

Number	Material
0	Stainless Steel
1	Yellow Gold Filled
2	White Gold Filled
3	Stainless Steel with Yellow Gold
4	Stainless Steel with White Gold
5	Gold Shell
6	Platinum
7	14kt Yellow Gold
8	18kt Yellow Gold
9	14kt White Gold

This information is only relevant for watches manufactured after 1977, which was when Rolex first introduced the Quick Set feature. When it was fitted to other models over a period of years, they too were given a five digit model number. It was normal for watches to receive the sapphire crystal at the same time they were changed to Quick Set, so any watches with a four digit model number and a sapphire crystal are likely to have been enhanced.

The most frightening enhancement/fake today is the simple replacement of a steel case with a newly manufactured gold case. These cases are being made for the Datejust, the Submariner, and the Daytona Perpetual. The unscrupulous buy these cases and purchase a steel Datejust, from which they remove the movement, dial and sapphire crystal to fit them in the gold case. The Submariner and Daytona swaps are normally done with two-tone watches, which have the correct color dials, bezels, and winders. This transformation is almost impossible to detect without opening the watch and comparing it with a factory model. However, most people purchasing an almost new watch are not in a position to do this. They are completely unaware of the problem until the watch is returned to Rolex or a Rolex agent for service, usually several years after the purchase.

The author is in a very fortunate position. Over the last six or seven years he has been working with the British Museum on wristwatches and have had access to some of the most sophisticated equipment on this side of a police forensics laboratory. He finds that he is having to use this equipment much more frequently to identify the fakes that come into his hands. The battle escalates like that waged between the West and the Soviet Bloc during the Cold War. The counterfeiters develop a weapon and we develop a countermeasure to it. Then they develop something that evades our countermeasures and we then have to come up with something new. In the end the fakers will always score minor victories, because there are sufficiently uninformed and greedy people out there. But in the end we will win, because there are more of us and our pool of knowledge will always be greater than that of the counterfeiters. Remember, knowledge is our best weapon. Keep your eyes open. If a watch is one that is often faked and the price you are being asked for it seems too good to be true, then it probably is!

370 Fakes: Imitation is the Nastiest Form of Flattery

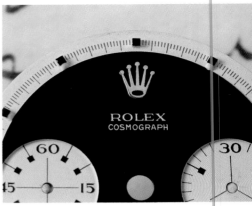

Detail of numerals on rear of fake Rolex Prince (T100) and on correct one (T99). Note the difference in figure styles particularly in the number "3".

CORRECT Rolex Daytona "Paul Newman" dial, showing points to look for; no luminous marker at "12"; tight concentric circles in subsidiary dials and scratch mark on outside of dial where the bezel has made contact.

Fake platinum cased "Rolex" Prince; note the difference in wear between the original dial and the new fake case.

Comparison of case interiors; note the lack of hallmarks and case maker marks in fake platinum case, also note complete lack of wear.

Fakes: Imitation is the Nastiest Form of Flattery 371

Rear of above dial, showing manufacturer's mark and signature "Singer." These should be stamped into the dial, not engraved.

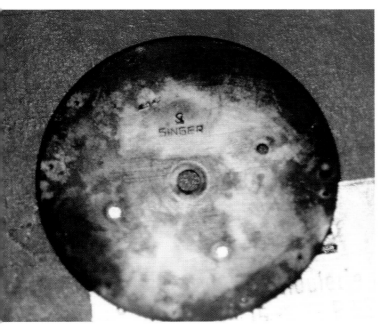

Rear of silver colored dial showing similar markings to the above Daytona.

Fake "grain of rice" bracelet. Note the very amateurish stamping and the mis-spelling of "Swiss." The overall quality of these bracelets, made in the 1990s in imitation of the 1930s Gay Freres originals, is very poor and are most unlikely to fool anyone who knows what they are doing

Rear of a completely fake "President" bracelet, showing lower quality stamping than on factory model.

372 Fakes: Imitation is the Nastiest Form of Flattery

Profile of "hooded" and normal bubblebacks showing the dropped lugs of the "hooded" model (top). If hoods are added to a normal bubbleback the profile will remain the same and so can be easily identified.

Completely fake triple calendar moonphase chronograph, bearing the model number 81806. Rolex never made such a watch; during the 1950s whilst the company was making complicated watches less than six people worked in the department responsible; we have spoken to two of these people and they both deny the company ever made such watches. The quality of the additional parts was also much lower than the original Rolex parts in the watch.

Fakes: Imitation is the Nastiest Form of Flattery 373

Completely fake gold-filled pocket watch, claiming to be from 1907, the effort that has gone into making this watch must be totally unrewarding. The movement is obviously modern, probably from the 1980s.

THE ROLEX FACTORY IN BIENNE
(World's largest production of *OFFICIALLY* Certified Chronometers)

Appendix I

A VISIT TO THE ROLEX FACTORY

This is a reprint of an article in the British magazine Practical Watch and Clockmaker from 1932; I believe it is well worth reading even today.

A factory which has produced two such outstanding watches as the *Rolex Oyster* and the *Rolex Prince* can be expected to have much to show the visitor, but, even so, anyone who is privileged to see the birthplace of these two famous timekeepers is in for some surprises.

Situated in a position which any hotel proprietor would envy, on a slope above Bienne, it is soundly designed and planned. Inside, it is even more attractive, for its equipment is such that it seems as much laboratory as factory. If one is fortunate enough to make a tour under the guidance of the genial Monsieur E. Borer, the presiding technical genius of Rolex, the value of this clever combination of the scientific and practical sides of watchmaking can be appreciated to the full.

A large number of tools and scientific instruments are of Rolex manufacture. In every part of the factory these tools are to be found; the policy of the manufacturers is apparently that they know what they want and make it accordingly. Every stage of production, from the arrival of the raw materials to the final checking of the completed watches is largely handled by such equipment, and the standard of workmanship demanded is so stringent that accuracy has become a commonplace of the Rolex organisation.

For fifty years this tradition has been built up, and the experience gained in countless experiments has been carefully filed and put away for reference in the huge store of drawings and data cards which are kept in the drawing office.

The system adopted for the production of a new model is most ingenious. A complete series of drawings on a large scale are produced, and when setting out the location of each hole, cross lines are used, the outline of each component being constructed around the centres thus obtained. When the drawings are approved, work begins on the press tools. The corresponding centres on the metal from which the tools are made are checked against the cross-line dimensions of the original drawings on an optical machine which is half machine-tool and half microscope; one of the few instruments obtained from outside sources, it has a number of clever modifications which have been carried out in the factory.

Heat treatment of the tools follows. The shop in which this is done is fitted out with electric furnaces and a clever device enabling slightly varying types of steel to be hardened in a manner that obtains the best results. It has been discovered that the right quenching point for a steel which is to be hardened is the point at which its ability to be magnetised temporarily disappears, and so the parts or tools to be treated are subjected to a magnetising field whilst in the furnace. As soon as the steel reaches the temperature at which it offers no practical possibilities of magnetisation, it is ready for hardening. The actual point varies slightly with differing batches of steel, but the magnetic test is an unfailing one.

In the production of press tools some are finally finished to their accurate form by a method which is really using a press tool to make a press tool. The tools are made as near as possible to size, and then a hardened "master-piece", really a replica of the finished part formed to make a tool, is forced through the tool, shaping or "shaving" it to remove any tiny inaccuracies. In the making of some tools, such as those used to press out balance-cocks this system eases the work of the toolmaker considerably and also enables several press tools to be made with exactly the same form and without elaborate and unnecessary work.

Grinding is also used to finish certain tools, and here the work is carried out to an accuracy which must be seen to be believed. When the tool is complete, it is correct to one-hundredth of a millimetre and its heat treatment so perfect that it can stand up to long service with little or no diminution of this accuracy.

The parts of Rolex watches start life in just the same way as do those of other watches - under the press-tool. "Repassing" is the next stage for such parts as cocks, plates and so on, when they are given a second journey through the presses, this time with a special tool of slightly smaller dimensions. It shaves off the edges, trims up, finishes and removes the roughness inseparable from the first process of pressing from strip metal.

Then comes the dotting of centres; it is here that some of the tools so carefully set out under the microscope-tool are met. With the centres accurately set, the work of piercing the holes can proceed, but even after the drills have done their work, a further test ensures that nothing has gone astray.

On leaving the press-work section, the machine shops are the next stop. The chief point of interest here is the illustration of that policy of "one machine per operation" which is in force in the Rolex organisation for many processes. Most of these special tools in the factory are designed along these lines and the resulting accuracy is claimed to justify the slightly increased time involved.

Wheels are cut on either Mikron machines or automatics built by Rolex: the latter are somewhat novel in design but very effective. The cutting on both types is extremely accurate, but nevertheless, microscopic testing of specimens taken from each batch proceeds as a simple routine task. The checking which a wheel undergoes is such that it is practically impossible for a defectively shaped tooth to pass. An enlarged profile is projected on a screen upon is which is laid the master drawing and unless the magnified image and the drawing agree to perfection, production is suspended on the particular machine until adjustments have been made.

Other tests of a similarly stringent kind are applied to the turned items produced by automatics. Pinion blanks, barrel arbours, keyless shafts and other similar components are turned

375

from steel rods on automatic lathes which are cam-controlled. It was pointed out particularly, how much care is taken in the formation of the master cams which control tool movements. Again, the story is one of accurate shaping, heat treatment and subsequent checking against the master drawing.

Pinion cutting is done on another batch of automatic machines. A microscopic examination of a roughly cut pinion, straight off the machine, showed that although several finishing processes had yet to be carried out, the work done in the cutting operation was accurate enough to delight anyone with an eye for mechanical perfection.

But it is in the individual tools for the thousand and one processes peculiar to watch parts that the factory holds its special interest. For instance, a special Rolex microscope is used for balance pivot checking, the accurate curve so necessary for the retention of oil at the correct point being ensured as a result of two examinations. The first check is on the rough pivot, the second on the finished one; the microscope field is so designed that departures from standard are at once visible.

In the shaping of the lever fork, it is not considered good enough to have "just a fork" however accurately it is formed. The contact surfaces are rounded off on a jig-controlled rocking head polisher, and so finished that wastage of energy as the ruby-pin engages with its lever are reduced to a minimum. Slots for the pallets and fork are shaped in the first place on a fully automatic machine controlled by cam-gear strangely reminiscent of striking "locking-plate work." The machine performs its task so quietly that it is not easy to know if it is running or not. Made by Rolex, it embodies enclosed parts to a remarkable extent and the driving motors, which are practically vibrationless in action, are enclosed inside the machine bed.

Like the fork, the pallet slots are tested and examined for accuracy, and after the stones are set, are tested again, this time by the optical projection method against the master drawing.

Roller edge polishing is another process carried out on tools made at the factory. The mirror finish obtained is excellent, and is produced by two polishing discs which work in turn and smooth out any inequalities of surface in a few seconds.

In the regulation of a watch, it is of importance that the index fits the endpiece with accuracy and an even grip. To secure this, a special index turning machine, another clever product of the factory's toolmakers, is used. By accurate slides the tool is governed to turn a perfect seating in the index for the edge of the endpiece, and as the endpiece itself is equally accurately formed, the fit on assembly is the right one.

Other machines which deal with the index include that for producing the curb-pin holes and the insertion of the pins themselves. The pins are turned to an accurate taper and are then forced home by a small press tool; their length is afterwards cut by a method which does not leave any burr to catch the outer coil of the balance spring.

Component assembly is another branch of the factory where the clever brains of the engineers have made it almost impossible for inaccuracies to creep in. An instance is the mounting of the wheels on their pinions. Here a number were checked just as they came off the tools and before any attempt had been made to true up a possible "out of flatness." The result showed that anyone who hoped to find a job as truer in the Rolex works would have a pretty thin time!

Cleaning of parts after manufacture is an important business. No less than seven operations are employed to ensure that the last fragment of adhering oil or metal dust is removed. Each cleaning process is designed to look after a particular form of possible dirt, and when the last has taken place, a blast of hot air dries off the parts and leaves them a gleaming heap.

Escape wheels are dealt with by a special series of machines. On the finishing of the teeth, the actual contact surfaces are given a rounded face to ensure that slight inaccuracies (if any!) in pallet alignment are compensated for and friction properly distributed. Nor is this enough, for in the grinding of the teeth, a dividing head is used to effect a double check against any irregularity of tooth spacing. Again, specimens are checked by optical means to ensure that the result is correct.

Assembly after the accurate control of manufacture, becomes a relatively simple task, but is carried out with extreme care. The progressive system is used the movements being built up stage by stage. A most instructive detail process is to be found in the testing of jewels. Each jewel is tested individually, and as the jewel holes are mounted in "chaton" settings afterwards corrected to the central hole, the result is a pivot hole accurate both for upright and eccentricity. The "chaton" setting is used by Rolex to avoid any pressure being imposed on the jewel itself during the process of pressing home in the plate or cock, and although involving additional work, is claimed by them to justify itself in every way.

A fascinating instrument is the micrometer gauge used to measure endshake. The wheel is mounted and the assembled plate placed in the tool, after which two tiny plungers, operating on the ends of the pivots indicate the endshake by a dial reading. Deviations from the set standard are obvious at once and can be dealt with immediately.

As the watches arrive at the completion of assembly, the story of large production on many machines is reversed. One man does all the oiling, for it is felt that in such an important matter as lubrication, it is best to centralise responsibility at one point. As many as seven different grades of lubricant are used, as it has been found that watches, like other machinery have differing demands in the various parts of their mechanism.

Testing of the movements against accurate master clocks and in the varying positions necessary to ensure timekeeping under all conditions is largely routine. Although adjustment is necessary, as with all newly constructed machinery, the whole aim of the prodigious labour and cost in the earlier process is to reduce the difference between one watch and another and to make the performance constant. The task has been done well and nowhere is this more obvious than in the adjusting department. Examination of the racks of movements under test showed no "stoppers" and but few "off timers" - a great tribute to all concerned and particularly to the escapement assemblers and springers.

Leaving the actual manufacture for the laboratory side proper, one meets with some interesting apparatus. In many other manufacturing organisations, it is emphasised that all types of parts are tested to destruction to ensure that the general standard of quality is being maintained. This is also done at the Rolex laboratories. The teeth of steel keyless wheels are tested by the taking of samples. These are placed on a device in which a wedge is forced down between two adjoining teeth under the pressure of a weight. The weight is moved along a lever until the teeth fracture and the load can be accurately determined. Other similar tests are applied to various parts and their quality ensured.

A particularly interesting branch of research is the microscopic examination of the distribution of oil in, for example, balance pivot holes. Various shapes of pivots can be tried and oil distribution studied under conditions of pressure and endshake. Photographic records are taken for filing purposes.

The master clocks in the laboratory have various tasks. Some are chiefly concerned with the distribution of accurate time around the works, but others deal with research, and one made history when it was used to determine the effect of an unbalanced minute hand on the rate of an astronomical pendulum. The method used was that of a beam of light which gave a variable record showing clearly the disturbance of arc caused by the weight of the hand. Direct contact is maintained between these clocks and the observatory, and timekeeping of an extraordinary standard is the basis of all the work of timing.

I felt that this remarkable place with its endless procession of high-grade movements destined for all parts of the world, was one of the custodians of Swiss watchmaking fame. I turned away with reluctance, for, although much can be seen even in a brief visit, there is much more that is missed or that can only be given a passing glance. To excel the efforts of the technicians in such a factory is a hard task and although it was stressed that continuous attempts are made to improve and advance, it is a fit subject for meditation as to what marvels of precision can be in the minds of those who still look forward and scheme patiently for even finer achievements.

Appendix II
A ROSE BY ANY OTHER NAME.....

Shown below is a list of all names registered by Rolex from their earliest days until 1960, when the registration system changed.

Date	Name	Number	Type
02/Jul/08	Rolex (First registration anywhere)	24001	Corporate
20/Jul/08	Omigra	25904	Brand
19/Oct/09	Elvira	26314	Brand
24/Jan/11	Marconi Lever	28769	Brand
04/Jul/12	Marguerite	31627	Brand
17/Jul/14	King George Lever	36859	Brand
14/Nov/14	Princess Royal Wristlet	36309	Brand
14/Nov/14	Prince of Wales Lever	36310	Brand
14/Nov/14	Crown Jewels Lever	36333	Brand
14/Nov/14	Queen Alexandra Wristlet	36334	Brand
14/Nov/14	The Sailor King Wristlet	36335	Brand
22/Jun/16	Calix Lever	38504	Brand
22/Jun/16	Lexis	38544	Brand
17/Mar/19	Unicorn Lever	43672	Brand
17/Mar/19	LON (League of Nations)	43673	Brand
17/Mar/19	Brex	43674	Brand
10/Apr/20	Falcon	46597	Brand
10/Apr/20	Rolwatco	46598	Brand
24/Sep/20	Lonex	47812	Brand
24/Sep/20	Genex	47813	Brand
24/Sep/20	Rolexis	47831	Brand
24/Sep/20	Irex	47832	Brand
04/Oct/20	King George Lever(Trans. 35859 to HW)	47885	Brand
04/Oct/20	Crown Jewels Lever (Transfers 36333 to HW)	47886	Brand
04/Oct/20	Calix (Transfers 38504 to HW)	47887	Brand
04/Oct/20	Lexis (Transfers 38544 to HW)	47888	Brand
04/Oct/20	Unicorn Lever (Transfers 43672 to HW)	47889	Brand
13/Oct/20	The Eastern Watch	47912	Brand
13/Oct/20	Elvira	47913	Brand
13/Oct/20	Marconi Lever	47914	Brand
13/Oct/20	Coronation Lever	47915	Brand
13/Oct/20	Marguerite	47916	Brand
26/Oct/20	公平洋行 (Translates as Fair Deal Company)	48026	Slogan
03/Nov/20	Prince of Wales Lever (Transfers 36310 to HW)	48058	Brand
02/Jul/21	Hofex	50085	Brand
12/Nov/21	Viceroy	50800	Brand
23/Mar/22	Aqua	51561	Brand
31/Mar/22	The Submarine	51632	Model
31/Mar/22	Diver	51633	Model
09/Nov/22	Aquarium Display	52986	Display
01/Dec/22	Admiralty	53060	Brand
12/May/23	The Buffalo	54080	Model
15/Apr/25	Marconi Veriflat	58784	Model
29/Jul/26	Oyster	62367	Model
28/Feb/27	Huitre	63907	Model
28/Feb/27	Auster	63908	Model
28/Feb/27	Ostica	63909	Model
28/Feb/27	Ostra	63910	Model
28/Feb/27	Snowite	63911	Material
07/Mar/27	Wintex	63998	Model
15/Sep/27	Rolco	65389	Brand
10/Oct/27	Rox	65518	Model
24/Sep/28	President Lincoln	68474	Model
16/Oct/28	Alden	68476	Model
02/Nov/28	Golden Rock	68475	Model
23/Nov/28	The Rolex Prince	68806	Model
21/Jan/29	The Rolex Prince (transferred to Rolex from Wilsdorf)	69165	Model
21/Jan/29	Genex Prince	69311	Model
29/Feb/11	Coronation Lever	28789	Brand
21/Jun/29	Sousmarin	70307	Model
06/Sep/29	All on the tick	70817	Slogan
06/Sep/29	Best by every test	70818	Slogan
21/Oct/29	Ranger	71101	Model
10/Dec/29	Glider	71369	Model
10/Dec/29	La Glissante	71369	Model
06/Jun/30	Image of cherub with a bow and arrow	72577	Corporate
07/Aug/30	Gala prince	73121	Model
07/Aug/30	Prince Ironclad	73122	Model
21/Aug/30	Prince Charming	73288	Model
21/Aug/30	Prince Imperial	73289	Model
21/Aug/30	Prince Elegant	73290	Model
25/Aug/30	Rolex Princess	73231	Model
17/Nov/30	Gliding touch	73781	Model
09/Dec/30	Ultra prima	74108	Model
22/Dec/30	Baldwin Watch	74112	Model
13/Aug/31	Athlete	75691	Model
18/Dec/31	Steelium	76512	Material
21/May/32	Rolesium	77569	Material
14/Jun/32	Rolex perpetual	77724	Model
24/Jun/32	Silgold	77982	Material
24/Jun/32	Wicket	77983	Model
17/Sep/32	Edelglas	78453	Watch glass
11/Oct/32	Steeplechase	78534	Model
07/Mar/33	Minute Reader	79733	Model
30/Mar/33	Zetex (renews 34250)	79904	Corporate
30/Mar/33	Rolex (renews 34251)	79905	Corporate
30/Mar/33	Final (renews 34252)	79906	Corporate
01/Apr/33	Rolesor	79942	Material
22/May/33	Marukoni	80308	Model
20/Sep/33	Damage-Proof	81166	Model
27/Aug/34	RWC Ltd (in cartouche)	83687	Case stamp
21/Nov/34	Coronet with Wilsdorf beneath	84666	Logo
10/Jan/35	Princess Royal Wristlet	85078	Model
10/Jan/35	The Sailor King Wristlet	85079	Model
12/Jan/35	Prince of Wales Lever	85130	Model
20/Dec/35	Scarabee	87363	Model
27/Feb/36	Incognito	87800	Model
27/Feb/36	La Girouette	87807	Model
23/May/36	Speed Model	88400	Model
06/Jun/36	Imperium	88595	Model
13/Jun/36	Schwimmsicher	88596	Model
25/Jul/36	Lilliputian	88753	Model
25/Sep/36	Auto-Suspension	89089	Model & Boxes
03/Nov/36	Rolex (renews 39272) but now extends it to cover everything from cigarettes to paper and explosives!	89513	Corporate
16/Jun/37	Aero Dynamique	90880	Model
28/Jan/44	Geant	106325	Model
28/Jan/44	Oyster Giant	106326	Model
20/Jul/44	Montres Rolex SA Geneve	107846	Corporate
29/Mar/45	Perpetual, the silent self-winder	110116	Slogan
20/Apr/45	Marconi Veriflat	110278	Model

Date	Name	Number	Type
24/May/45	Waferthin	110620	Model
01/Oct/45	Ritedate	112181	Model
23/Mar/46	Rolesor	115061	Material
23/Mar/46	Oyster the silent self-winder	115062	Slogan
17/Apr/46	Rolex (with coronet symbol) For Geneva branch	105046	Corporate
17/April/46	Rolex (with coronet symbol) For Bienne branch	115181	Corporate
29/Jul/46	Oyster (renewing 62367)	116509	Model
23/Oct/46	Rolex (in cartouche with coronet)	117778	Case stamp
28/Nov/46	Rolex (with coronet symbol) Different typeface	118256	Corporate
12/Feb/47	Ostra (renewing 63910)	119147	Model
13/Mar/47	Reflet du gout de geneve	119527	Slogan
30/Jun/47	Rolco	120175	Model
30/Jun/47	Snowite	120176	Material
30/Jun/47	Wintex	120177	Model
30/Jun/47	Ostica	121072	Model
30/Jun/47	Auster	121073	Model
30/Jun/47	Huitre	121074	Model
05/Dec/52	Electroproof	144867	Model
09/Jan/53	Deepsea	144868	Model
09/Jan/53	Electrosafe	144869	Model
26/Jan/53	Explorer	145069	Model
05/Feb/53	Frogman	145231	Model
20/Feb/53	Tuxedo	145369	Model
04/Mar/53	Montres Rolex SA	145594	Case Stamp
07/Apr/53	Golden Dream	146033	Model
09/Apr/53	Rolium	146066	Material??
20/Apr/53	Twinlock	146240	Crown
06/May/53	Multi-Meter	146371	Model
28/May/53	Mono-Meter	146582	Model
29/May/53	Metropolitan	146583	Model
29/May/53	Stratosphere	146584	Model
11/Jun/53	Cosmograph	146700	Model
13/Jul/53	Beau-Brummel	147253	Model
13/Jul/53	Flushfit	147254	Bracelet
13/Jul/53	Titecap	147255	Model/Crown?
29/Jul/53	Reve d'Or / Enseno de Oro / Un Sogno d'Oro	148261	Slogan?
27/Aug/53	Goldarmour	147910	Material
04/Sep/53	Chrono-Time	147904	Model
04/Sep/53	Chrono-Mind	147905	Model
25/Sep/53	Rolon	148101	Model
09/Oct/53	Dual Lock	148262	Crown
09/Oct/53	Dual Safetee Lock	148263	Crown
09/Oct/53	Twin Safetee Lock	148264	Crown
16/Oct/53	Goldkid	148401	Model?
24/Nov/53	Splitpruf	148877	Model/Material?
24/Nov/53	Turn-O-Graph	148878	Model
24/Nov/53	Turn-O-Metre	148879	Model
26/Nov/53	Chameleon	148884	Model
26/Nov/53	Observer	148885	Model
04/Jun/54	What's new in the watch business	152151	Magazine
25/Jun/54	Five Star Series Rolex	151823	Model
25/Jun/54	Four Star Series Rolex	151824	Model
25/Jul/54	One Star Series Rolex	151825	Model
25/Jul/54	Six Star Series Rolex	151826	Model
25/Jul/54	Three Star Series Rolex	151827	Model
25/Jul/54	Two Star Series Rolex	151828	Model
29/Jul/54	Silent Call	152134	Model
31/Jul/54	Diver	152673	Model
31/Jul/54	Deepsea Diver	152674	Model
31/Jul/54	Skin Diver	152675	Model
31/Jul/54	Incomparable-Rolex-The Incomparable	152676	Slogan
11/Aug/54	RWC Ltd in cartouche (renews 83687)	152291	Case stamp
23/Aug/54	Oyster Ship's Anchor	152443	Model
22/Sep/54	Dive-O-Graph	152890	Model
22/Sep/54	Rolex-Engineer	152891	Model
01/Oct/54	Swimpruf	153004	Slogan
01/Nov/54	Lady-Date	153767	Model
09/Nov/54	Snowking	153489	Model
26/Nov/54	Executive	153699	Model
26/Nov/54	Milgauss	153700	Model
26/Nov/54	Tru-Date	153701	Model
26/Nov/54	Veriflat	153702	Model
10/Dec/54	Fathometer	153972	Model
23/Dec/54	Cyclop's Eye	154126	Slogan
10/Jan/55	Deap-See Orchid	154402	Model
10/Jan/55	Physician's Chronometer	154403	Model
09/Mar/55	Orchidee	155146	Model
17/Mar/55	Greenwich Master	155333	Model
10/Apr/55	Orchis	156020	Model
21/Apr/55	GMT-Master	155739	Model
21/Apr/55	Snugfit	155740	Model
21/Apr/55	Twotone	155741	Model
11/Jul/55	Rolex Watch (renews 86820)	157006	Corporate
27/Jul/55	Lilliputian	157730	Model
04/Aug/55	Everflex	157591	Model
04/Aug/55	Propeller	157592	Model
08/Aug/55	La Comtesse	157421	Model
27/Aug/55	Eversure	158054	Model
16/Sep/55	Stratometer	157966	Model
30/Sep/55	Multim In Parvo (MIP)	158137	Slogan
07/Oct/55	Chronoscience	158217	Model
12/Nov/55	Constant-Flo	158668	Model
12/Nov/55	Orchid	158669	Model
12/Nov/55	Sveltine	158670	Model
05/Dec/55	Tropic-Prufe	159455	Model
28/Dec/55	Quartet	159284	Model
19/Jan/56	Benevenuto Cellini	159607	Model
19/Jan/56	ROL-EX-CENTRIC	159608	Model
16/Feb/56	Springless	159885	Model
22/Mar/56	Moistpruf	160274	Model
16/Apr/56	V-Class	160537	Model
05/May/56	Puls-O-Meter	160822	Model
09/May/56	Verifine	160878	Model
09/May/56	Verislim	160879	Model
15/May/56	Leaf-Thin	160921	Model
25/May/56	Spider	161047	Model
29/Aug/56	Fathometer	162244	Equipment
13/Jan/59	Oyster Perpetual (renews 94913)	173597	Model
23/Jan/59	Science King	173778	Model
23/Jan/59	Science Meter	173779	Model
21/Feb/59	Centregraph (renews 95069)	174216	Model
24/Apr/59	Oyster Aviation	175092	Slogan
29/Apr/59	King Midas	175267	Model
29/Apr/59	Microsecond	175268	Model
04/Jun/59	Moneda	175780	Model
17/Aug/59	Statesman (bought from Ollivant & Botsford)	144719	Model
03/Nov/59	Rolex Star Precision	178465	Slogan

Appendix 3
HALLMARKS & OYSTER CASE NUMBERS

There has always been a problem with establishing the date of early Rolex Oyster watches. Rolex has washed its hands over dating early pieces as letters we have received from them makes all too clear.

There is, however, one fixed point we can use: hallmarks. As we know most early Rolex Oyster watches were made for the British market and it was a legal requirement for all precious metal items sold in Britain to be hallmarked by a government department. The history of hallmarks is worth a whole book in itself, but is surely outside the remit of this publication. It is worth noting, though, that the earliest hallmarks in Britain date from the late 12th century making the hallmarking regulations the oldest piece of consumer legislation still in effect.

Under British hallmarking regulations each item was comprised of three parts; a standard mark (9/14/18/22 carat gold), the assay office mark (where it was hallmarked) and the date letter (the year it was assayed). Using the date letter, which gives us the year in which the watch was sent to the assay office[84] and the case number, it is possible to draw up a table of dates and case numbers. However due to the intervention of World War II in 1939, the numbers cease at this point. Being able to date the watches produced between 1926 and 1939 is a considerable improvement on the previous situation we all faced.

It took Rolex almost 30 years (from 1926 to 1953) to use all the numbers between 20,001 and 999,999. At this point the obvious thing to do would have been to add a seventh digit to their case numbering machine(s) and continue into the millions. Rolex, as we have seen from the proceeding chapters, was never a company to follow the obvious path and so it chose to reuse previously issued numbers on the new cases. It, once again, chose not to follow the most logical path and begin at 0001, beginning instead at 100,001, a number previously used in the midst of World War II.

Fortunately they began to reuse these numbers during the period when they were also stamping the date of construction inside the case back. This code consisted of a roman numeral I, II, III or IV representing the four quarters of the year and the last two digits of the year (for example II 54, representing the period April to June, 1954). Using these date codes it is now possible to give definite dates to the previously uncertain period in the mid-1950s.

It should be fairly easy to tell the difference between a watch produced in the 1930s and one from 20 years later, but if for any reason one can not, look inside and check for the date code.

It was not until the late 1950s that Rolex began to use the seventh digit and from this point the numbering sequence became logical and able to be followed with any hope of accuracy.

Please note the number shown below are the hightest we have been able to confirm for a particular year. There is no guarantee that it is the highest and is a possibility of error.

Case Number	Year
20691	1927
24747	1928
28290	1930
29312	1932
29933	1933
30823	1934
35365	1935
37896	1936
40920	1937
43739	1938
71224	1939
99775	1940
106047	1941
143509	1942
230878	1943
269561	1944
302459	1945
387216	1946
529163	1947
628840	1948
710776	1951
937170	I 54
941699	i 1953
952892	1 1954
955466	iv 1953
964789	IV/53
964789	iv 1953
973697	1V 1953
973930	III 1953

---

[84] Because the watches could not be sold without a hallmark, it is safe to assume that they would be sent to the assay office soon after arrival from Switzerland. So the hallmark date would be close to the production date.

Appendix 4

NOTES & QUERIES

This book is the result of over ten years of research on three continents, but it is by no means complete. Despite hours in the bowels of libraries, days spent poring over patent documents and weeks of research in museums, there are still questions left unanswered. Some of them simple and others quite complex. We have chosen to list some of these outstanding questions below in the hope that some of our readers may be able to help, if you can, please contact us through our publishers or e-mail us at UKWATCHES@AOL.COM (JMD) AND WATCHDUDE@AOL.COM (JPH).

1. What happened to Mr. Davis?
2. What was the Oyster Watch Company?
3. What ever happened to Perret & Perregaux?
4. How were they paid, by lump payment, or by units sold?
5. How was Jacques Cousteau involved in the development of the Submariner?
6. What was the relationship between Rolex and Comex?
7. Did Paul Newman ever wear a "Paul Newman"?
8. Were the 1950s lady's jeweled watches really custom made?
9. What did the graduations on the bezel of the original Milgauss represent?
10. Why is the patent for the first Oyster case opener dated over 3 years after the patent for the case itself?
11. When did Rolex first set up the U.S. company and was it their first US company?

Bibliography

Books

Berthoud, Robert. *Repertoire des Titulaires de Brevets Horlogers.*
Switzerland: St. Imer, 1960.
Borer, Emil. *Modern Watch Repairing & Adjusting.* London: N.A.G. Press, 1937.
Chapuis, Alfred & Jaquet, Eugene. *History of thhe Self-Winding Watch.* London: B.T. Batsford, 1956.
Imai, Kesaharu.(Editor.) *Time Spec volume 17.* Tokyo: WPP, 1994.
Kahlert, Muhe & Brunner. *Wristwatches, History of a Century's Development.* Atglen, PA: Schiffer Publishing, 1986.
Landes, David S. *Revolution in Time.* Cambridge: Harvard University Press, 1983.
Pipe, R. W. *The Automatic Watch.* London: Heywood, 1952.
Rolex Watch Co. *The Anatomy of Time, Rolex.* Manchester England: J. Broad, 1955(?)
Rolex Watch Co. *Vade Mecum.* Zurich: Fetz Brothers, 1946.

Periodicals

Horological Journal. 1900 to date. The British Horological Institute.
Journal Suisse D'Horologerie. 1900 to date. The British Horological Institute.
Practical Watch & Clock Maker. 1928 to end. Ceased Publication.

Price Guide

Compiling a price guide is never the easiest thing in the world. Putting it simply, this is a very active and volatile market. If one segment of the market decides it either does or does not want a particular model, any price guide will be out of date the moment it goes to the printer.

Next comes the vexed question of what price do you use? There are at least three different prices for each watch: what someone will give you for it, what you would have to pay to buy it, and what the model watch last brought at auction. The price that others choose depends upon their position in the market. If I wish to buy watches from you it is in my interest to keep the price as low as possible. On the other hand, if I am selling it makes the most sense to keep my prices high. So that there may be no confusion we have chosen to average all three prices. We have to confess this does not make things any simpler, but life was never meant to be simple.

Take this guide as that, just a guide. Please understand that the prices shown relate to the individual watch shown, with that particular dial, in that particular case material, and with that particular bracelet. It could be financially risky to extrapolate from a watch shown to a similar model but with a differing specifications. Because of the impossibility of providing accurate prices, neither the authors or the publishers will be responsible in any way for gain or loss incurred by using this guide.

Please note some watches are so rare that we can only give an estimate of their value. In these situations their guide price is prefixed with E. Prices are in U.S. dollars.

Page	Pos.	Description	Price range
Chapter 1			
15	TL	Silver tonneau	1500/1800
	TR	Silver enamel dial	1000/1200
16		9k Lady's	600/750
17	TL	9k Enamel dial	1000/1200
	BL	9k Demi	1250/1500
	BR	Silver rectangle	1000/1200
18	TL	18k demi-hunter	1750/2000
	TR	9k Cushion	1000/1200
	BL	9k Hunter	1750/2000
19		Silver flexible lugs	1000/1200
20	TL	9k Enamel dial	1000/1200
	TR	Silver demi-hunter	1000/1200
21		GF Cushion	800/1000
22		Silver enamel dial	1200/1500
23	TL	Silver octagon	750/1000
	CR	Silver demi-hunter	2250/2500
24	TL	9k Enamel dial	1500/1750
	BL	9k Enamel dial	1000/1200
25		Silver black dial	1250/1500
26		Nurse's fob	1750/2000
27		Sweep nurse's	1750/2000
28		14k Rebberg	E 1500/1750
29	TL	GF Rectangle	1500/1750
	BL	9k Cushion	1500/1750
30	TL	18k Tonneau	2000/2250
	TR	9k Rectangle	1750/2000
	BL	9k Cushion	1000/1200
31	TL	9k Square	2000/2250
	BR	Silver sweep	1500/1750
32		9k Rectangle	1500/1750
33		Marconi rectangle	1000/1250
34	TL	Sporting model	750/1000
	BL	Hooded steel	3500/4000
35	TL	Silver cushion	800/1000
	BR	2 Tone tonneau	2000/2500
36		Lady's 9k	1250/1500
37		18k Rectangle	2000/2500
38		18 Jewel silver	E 2000/2500

Page	Pos.	Description	Price range
39	TL	9k Cushion	1200/1500
	TR	Lady's exotic dial	E 3000/3500
Chapter 2			
46		9k Hermetic	3500/4000
47	TL	9k Borgel	3500/4000
	BR	Semi-hermetic	3000/3500
48	TL	14k White gold cushion	E 7500/8000
	BR	18k Octagon	5000/5500
49		Black dial	5500/6000
50	TL	9k Enamel cushion	3000/3500
	BL	9k Enamel cushion	3000/3500
51	TL	9k Enamel cushion	3000/3500
	TR	18k Boy's cushion	3000/3500
	BL	18k Boy's octagon	3250/3750
52	TL	Silver black dial	2750/3000
	BL	14k Cushion enamel	4000/4500
53	TR	18k Cushion	4750/5000
	BL	18k Cushion	4750/5000
54		18k Octagon	5500/6000
55	TL	Lady's cushion	1800/2000
	TR	Snowite octagon	1500/1750
56	TL	9k Enamel cushion	3000/3500
	TR	9k Cushion sweep	3250/3750
	BR	9k Cushion	2750/3000
57		9k Cushion sweep	3250/3500
58	TL	Snowite cushion	1250/1500
	BL	Steel boy's size	800/1000
59		Hooded cushion	5000/5500
60	TL	Boy's size	800/1000
	BL	Steel cushion	1000/1250
61	TL	Steel boy's size	1500/1800
	TR	Steel boy's size	800/1000
62	TL	9k Lady's Oyster	1750/2000
	TR	Steel boy's size	1000/1250
	BR	9k Cushion Oyster	3250/3500
63	TL	9k Sweep cushion	3500/3750
	BL	Steel boy's size	1000/1250
64	TL	Steel army	1500/1750

385

Price Guide

Page	Pos.	Description	Price range	Page	Pos.	Description	Price range
64	TR	Steel/gold Viceroy	3500/4000		BL	Steel/pink hooded	5500/6000
	BL	Steel army	1500/1750	96		Ribbed hooded	10.0/11.0K
65	TL	Steel Viceroy	5500/6000	97		Ribbed hooded	10.0/11.0K
	BL	Steel Viceroy	1500/1750	98	TL	Steel/pink hooded	4500/5000
66		Steel Viceroy	2750/3000		TR	Steel/pink Perpetual	3250/3500
67	TL	9k Imperial	3000/3250		B	Steel/gold Perpetual	3250/3500
	BL	Steel cushion	1500/1750	99		Steel/gold hooded	4500/5000
68	TL	Steel Athlete	1750/2000	100	TL	Steel hooded	22.0/25.0K
	TR	GF Pall Mall	1000/1250		BL	Steel/pink hooded	6000/7000
	BL	GF Centregraph	2500/2750		TR	14k hooded	7500/8500
	BR	Steel Everest	3000/3250	101	TL	Steel/pink hooded	6000/7500
69	TL	Steel Royalite	1250/1500		BL	Steel/pink hooded	7500/8000
	BL	2 Tone Oyster	2500/2750	102	TL	Boy's steel Perpetual	3500/4000
	TR	Steel Royalite	1250/1500		TR	Boy's steel Perpetual	4000/4500
70	T	18k Scientific	5500/6000	103	TL	18k Pink boy's Perpetual	12.5/14.0K
	B	Steel Lifesaver	4500/5000		BL	Steel lifesaver	8000/9000
71	T	Steel Oyster Royal	1250/1500		TR	Steel lifesaver	8000/9000
	B	Steel Egyptian	1500/1750	104	TL	18k Round end Perpetual	7500/8500
72		Steel Oyster Royal	3000/3250		BR	Scalloped hooded	7500/8000
73	TL	Steel/Gold Egyptian	3000/3500 ea.	105	TL	Steel giant Perpetual	3000/3500
	BR	14k Viceroy	4000/4500		BR	Steel Perpetual	2000/2250
74	TL	9k Oyster pocket watch	E 3000/3500	106	TL	Tropical	1500/1750
	BL	Silver screw back	1250/1500		TC	Steel/pink hooded	4500/5000
	CR	18k Imperial	4000/4500		C	Steel Perpetual	2250/2500
75	TL	9k Screw back	2000/2500		BC	14k Perpetual	3750/4000
	BL	14k Enamel cushion	4000/4500		TR	Boy's Perpetual	3500/3750
	BR	Steel/gold hooded	3500/4000		BR	Fluted hooded	5500/6000
Chapter 3				107	TL	18k Boy's Perpetual	5000/5500
					CL	Steel cushion Perpetual	4000/4500
81		Harwood	1000/1250		C	18k Perpetual	5000/5500
82	TL	9k Perpetual	E 8000/8500		TR	14k Round end Perpetual	4500/5000
	BL	Steel/gold Perpetual	2500/3000		BR	Steel/gold hooded	4500/5000
83		9k Perpetual	4000/4500	108	TL	18k Rectangle Perpetual	4750/5250
84	TL	9k Perpetual	E 5000/5500		TC	Steel/gold hooded	4500/4750
	TR	Steel black dial	4000/4500		TR	Steel hooded	22.5/25.0K
85	TL	18k Perpetual	5000/5500		CL	Steel Perpetual	4000/4500
	BL	Steel/pink Perpetual	4500/5000		BL	Tropical	1500/1750
86		Steel/gold Perpetual	2750/3000		CR	Flexible hooded	8000/9500
87	TL	Steel Perpetual	2500/2750		BR	14k Perpetual	4250/4500
	TR	14k Perpetual	3500/3750	109	TL	14k N.O.S. Perpetual	5750/6000
	BR	Steel Perpetual	1750/2250		TR	9k Pink Perpetual	4500/5000
88	TL	Steel/gold Perpetual	2500/2750		BL	18k Empire	17.5/20.0K
	TR	Steel Perpetual	1750/2000				
	BC	Steel/gold Perpetual	2000/2250	**Chapter 4**			
	BL	Steel Perpetual	1500/1750				
	BR	Steel/pink Perpetual	2000/2250	115	CR	Silver Brancard	5500/6000
89		18k Perpetual	6000/7000		BL	Steel/gold classic	4750/5250
90	TL	Steel/gold Perpetual	2250/2500	116	TR	9k Classic	5500/5750
	TR	Steel/gold Perpetual	2250/2500		BL	9k Classic	5750/6250
	BL	14k Perpetual	3500/3750	117		Steel step side	10.0/11.5K
	BR	18k Perpetual	4750/5250	118		9k Two color	9000/9500
91	TL	Steel Perpetual	1750/2250	119	CL	18k Brancard	12.0/13.0K
	BR	18k Perpetual	4750/5250		TR	18k Jump railway	12.5/13.5K
92	TR	18k Perpetual	6000/7000		BR	18k Step side	11.0/12.0K
	TL	14k Perpetual	3750/4000	120	TL	18k Railway	12.0/13.0K
	BR	Steel Perpetual	1750/2000		TR	9k Two color	9000/9500
93	TL	Steel Perpetual	4000/4500		BL	Sterling Brancard	5500/6000
	TR	18k Perpetual	4750/5250	121		14k ¼ Century	7000/8000
	BR	Steel Perpetual	1750/2000	122		Striped Brancard	11.5/12.0K
94		Steel/pink Perpetual	5500/6000	123	TL	Engraved Brancard	9000/9500
95	TL	Steel Perpetual	2000/2250		CR	Engraved Brancard	9000/9500
				124		9k Two color	9000/9500

Page	Pos.	Description	Price range	Page	Pos.	Description	Price range
125		Steel Brancard	6500/7000		TR	Steel/pink Firefly	1500/1750
126/7		18k Striped Brancard	11.5/12.0K		BL	Steel Speedking	800/1000
128		14k ¼ Century	7000/8000		BR	Steel square	1500/1750
129	TL	Steel Hunter	4000/4500	172	TL	Steel Royalite	2500/2750
	CR	18k Stepped stripes	12.5/13.0K		BR	Steel rectangle	1750/2000
	BL	14k Curved lugs	8500/9000	173	TL	9k Top Hat	3000/3500
130	TL	18k Flared rectangle	5000/5500		BL	9k Dress watch	1500/1750
	TR	9k Square	3500/4000	174	TL	Steel railway	2500/2750
	BR	9k Rectangle	4000/4500		BL	Steel dress watch	1250/1500
131	TL	Silver brancard	6000/6500	175	TL	9k Rectangle	2750/3000
	TR	18k Railway	12.0/13.0K		BR	9k Rectangle	2250/2750
	BL	18k Gruen Alpina	8000/9000	176	TL	Dennisteel	1000/1250
	BR	Silver Brancard	6000/6500		BL	9k Cushion	1250/1500
132	TL	18k Brancard	11.5/12.5K	177		9k Scientific	7000/7500
	CL	14k Engraved	9500/10.0K	178	TL	Steel/pink hooded	2500/3000
	CR	Two-tone Railway	12.0/13.0K		TR	18k Imperial	5500/6500
	BR	14k ¼ Century	7500/8000		BR	Boy's Perpetual	3000/3500
133	TL	18k Hunter	E 20.0/22.5K	179		Giant Perpetual	3000/3500
	BL	18k Rectangle	6000/7000	180	TL	18k Datejust	5500/6000
	TR	14k White Railway	9500/10.0K		TR	Steel Egyptian	1750/2000
	BR	18k Sweep	13.0/15.0K	181	TL	9k Cushion	1500/1750
Chapter 5					TR	Steel army	1500/1750
					BL	Steel lifesaver	6500/7000
136		Steel Tudor	300/350		BR	9k Viceroy	3000/3500
137		Silver Unicorn	750/1000	182	TL	Athlete Perpetual	4000/4500
138		GF Centregraph	2000/2500		TR	14k Majestic	3500/4000
139	TL	Rolco Rectangle	600/800		BL	Steel/pink Viceroy	3500/4000
	BR	Tudor	500/600		CR	Steel Imperial	2000/2250
140	TL	Marconi	600/800	183	TL	Steel Marconi	1250/1500
	BL	Oyster Lipton	400/600		TC	Steel curved lugs	1800/2000
141	TL	Solar Aqua	350/500		TR	Steel fluted lugs	1800/2000
	BL	Rolex Standard	600/800		BL	9k Round dress	1800/2000
142	TL	Nurse's sweep	1750/2000		BR	14k Hooded	2500/3000
	BL	Nurse's fob	1750/2000	184	TL	Two-tone hooded	3500/4000
143		Wilsdorf PW	3500/4000		TC	Steel Pall Mall	1750/2000
144	TL	Rolex PW	3500/4000		TR	Hooded	6500/7500
	BL	Tudor Submariner	600/750		BL	Steel/pink Precision	1750/2000
145		$20 Coin	3500/4000		BR	Flexible lugs	3000/3500
146	TL	Sterling portfolio	1500/1800	185	TL	Steel/pink Viceroy	4000/4500
	BL	9k Portfolio	2000/2500		TR	Athlete	3500/4000
147		Sterling/enamel portfolio	1750/2000		BL	Two-tone rectangle	3000/3500
148		18k White gold PW	3000/3500		BR	Two-tone hooded	3500/4500
				186		Panerai set	E 7500/8500
Chapter 7				187	TL	9k Cushion	3000/3500
					BR	Athlete Perpetual	7500/8500
158		Viceroy two-tone	5500/6500	188	TL	Two-tone Viceroy	4000/4500
164		Black dial	900/1100		BR	Black dial scientific	E 4500/5000
165		9k Imperial	3750/4250	189	TL	9k Asymmetrical	4000/4500
166	TL	14 Pink hooded	6500/7500				
	BR	Steel Speedking	600/800	**Chapter 8**			
167	TR	Steel Royal	750/1000				
	BR	Steel Speedking	750/1000	192	TL	18k Bombé	4000/4500
168	TL	Steel Royal	750/1000		BR	"German" tropical	2500/3000
	CR	Steel Speedking	750/1000	193		18k Perpetual	3500/4000
	BL	Steel Royal	600/800	194		Steel Perpetual	2000/2500
169	TL	Steel Speedking	600/800	196	TL	Steel Oysterdate	1000/1250
	TR	Steel Speedking	750/1000		BL	18k Pink	2500/2750
	BR	Steel Speedking	2500/3000		TR	18k Tear drop lugs	2500/2750
170	TL	Steel/gold Viceroy	5500/6500	197	TL	Crown dial	1000/1500
	BL	14k Viceroy	6500/7500		BL	Chevron dial	1000/1500
171	TL	Boy's Perpetual	3000/3500	198	TL	Oyster Royal	1000/1500
					BC	18k Pink Dress	3500/4000

Page	Pos.	Description	Price range	Page	Pos.	Description	Price range
199		Steel Perpetual	2000/2500	221	TL	Steel 4048	10.0/11.0K
200	TL	Platinum mid-size	E 25.0/30.0K		BL	Steel/pink 4500	6000/6500
	BR	Everest Perpetual	2500/3000	222	TL	18k 6238	10.0/11.0K
201	TR	9k Dress	1500/2000		BL	Steel 6238	4500/5500
	BL	Soccer watch	E 17.5/20.0K		CR	18k 4500	12.0/14.0K
	BR	Steel bombé	3000/3500	223	TL	18k Paul Newman	26.0/27.5K
202	TL	Tru-Beat	6000/7000		BL	Steel Antimagnetic	4500/5000
	TR	Purple enamel	8000/9000		TR	Steel black dial	4500/5000
	BL	18k Blue enamel	9000/9500	224	TL	Paul Newman	12.5/14.0K
	BR	Steel green enamel	4000/5000		TR	Oyster Paul Newman	13.5/15.0K
203	TL	Steel blue enamel	4500/5500	225	TL	Tiffany Paul Newman	15.0/16.0K
	BL	18k Bombé blue enamel	16.0/17.5K		TR	Steel 6263	6500/7000
	BR	Bubbleback blue enamel	17.5/20.0K		BL	Steel 6263	6500/7000
204	TL	18k Moonphase	32.0/35.0K		BR	Steel 6265	7000/7500
	BL	18k Star moonphase	35.0/37.5K	226	TL	Steel Paul Newman	13.0/14.0K
205	TL	Milgauss	6500/7500		CL	14k Paul Newman	24.0/26.0K
	TR	Turn-O-Graph	3500/4000		TR	Steel 6263	6500/7000
	BL	Textured TOG	3750/4500		BC	Steel black dial	5500/6500
	BR	Steel/gold TOG	4500/5500		BR	18k Pink non-Oyster	8000/8500
206	TL	18k White gold Datejust	6000/7500	227	TL	9k Pink 4537	12.5/15.0K
	BL	18k Moonphase	32.0/35.0K		TC	18k Calendar	30.0/35.0K
	BR	Cloisonne	E 35.0/40.0K		BC	18k Pulsometer	12.5/13.5K
207	TL	18k Star moonphase	35.0/37.5K		CL	18k Pink	9000/10.0K
	TR	18k Black dial	45.0/50.0K		BR	Steel 2508	4000/4500
	C	18k White gold	3000/4000	228	TL	18k 4500 Pulsations	15.0/16.0K
	BR	18k Black dial	45.0/50.0K		CL	18k Flexible lugs	10.0/12.5K
208	TL	Square cloisonne	E 30.0/35.0K		BL	18k Daytona	15.0/17.5K
	TC	18k Moonphase	32.0/35.0K		TC	Steel calendar	16.0/17.0K
	TR	14k Square	3000/3500		C	Steel 3055	3500/4500
	BL	Steel moonphase	27.5/30.0K		TR	18k Black dial 3055	15.0/17.5K
	BR	18k Black dial	45.0/50.0K		CR	Steel Zerograph	17.5/22.5K
209	TL	18K Non-Oyster moonphase	10.0/12.5K		BR	Steel calendar	30.0/35.0K
	BL	18k Square	3000/4000	229	TL	14k Cartier Paul Newman	37.5/40.0K
	TR	14k Bombé	5000/5500		CL	18k Pink	8500/9500
	CR	18k Pink T-bird	3000/3500		BL	18k Pink 3055	17.5/22.5K
	BR	14k Fish	15.0/17.5K		TC	Steel 2918	8000/9500
	BC	18k French	4000/5000		C	18k 3834	11.0/12.5K
210	TL	9k Dress	1500/2000		BC	Steel/pink 4313	6000/7500
	BL	9k Dress	1500/2000		TR	18k Pink 3484	8000/9500
	TR	Black oversize	1500/2000		BR	Steel/pink 4768	8000/9000
	BR	Oysterdate	1250/1500	230	TL	Steel 4768	6500/8000
211		Boxed Oysterdate	1500/2000		BL	Steel square	20.0/25.0K
212	TL	Steel bombé	2500/3000		TC	18k Small pink	17.5/22.5K
	TR	18k Chronometer	2500/3000		BC	18k 3055	17.5/22.5K
	BL	Large Speedking	1000/1250		TR	Steel Antimagnetic	9000/9500
213	TL	9k Dress	1500/2000		BR	Current Daytona	5000/6000
	TR	14k Coca Cola	3250/3750				
	BL	Milgauss	6000/8000	**Chapter 10**			
Chapter 9				235	BL	18k DateJust	4500/5000
					BR	Steel/gold DateJust	1300/1500
217	TR	18k One-button	15.0/17.0K	236	TR	Steel/gold Turn-o-Graph	1500/2000
	BL	Steel square	20.0/25.0K		BL	Steel Date	1500/2000
218	TL	Steel square	20.0/25.0K	237	TL	Steel DateJust	800/1000
	TR	Steel 3184	18.0/22.5K		TR	18k pink purple dial DateJust	6000/7500
	CR	18k 3055	8000/10.0K		BL	Steel DateJust	1000/1200
	BL	Steel 2508	4500/5000		BR	Steel DateJust	1000/1200
	BR	18k 3525	20.0/22.5K	238	TL	18k DateJust	7000/8000
219	TL	Steel/pink 3668	17.5/20.0K		TR	Steel OysterDate	1200/1500
	BL	Steel/gold 4500	5000/5500		BL	Steel/gold DateJust	1800/2000
220	TL	Steel/pink	6000/6500		BR	Steel Air-King-Date	1500/2000
	BL	18k 4062	8000/10.0K	239	TL	14k Bombé	3000/4000
	BR	Steel 3484	6000/7000		TR	Steel/gold T'Bird	3500/4000

Price Guide

Page	Pos.	Description	Price range
Chapter 11			
243		Everest expedition	E 4500/5000
244	TR	Steel 5504	3500/4000
	BR	6350 Honeycomb	3750/4250
245	BL	6150	3750/4000
	BR	6610	3750/4000
246	TL	Military 6350	4000/4500
	BL	5500 Precision	3500/3750
247	TL	6350 Honeycomb	3750/4500
	BR	6350 Flat	3750/4250
248	TL	6298 White dial	3750/4250
	BR	1016 Non-hack	3250/3500
249	TL	6150 Self-winding	3750/4500
	TR	1016 Space Dweller	7000/8000
	BL	1016 Hack	3750/4500
	BR	6350 Honeycomb	3750/4500
250	TL	6350 +Box +papers	4000/4500
	BL	Explorer Date	4500/5000
	TR	1655 Hacking	3250/3750
252	TL	1655 Non-hack	3000/3250
	TR	Steel/gold 6299	3500/3750
	BR	5500 Super Precision	3500/3750
253		16550 Steger Expedition	4000/4500
Chapter 12			
260	BR	Steel 50/165	1250/1500
261	TR	Steel Submariner	3500/4000
	BL	Steel 6536	1500/1750
262	TL	Steel Submariner	1500/2000
	TR	Steel Submariner	1500/2000
	BL	Steel Submariner	3500/4000
	BR	Steel Submariner	1500/2000
263	TL	Steel Submariner	4000/4500
	TR	Steel Submariner	3500/3750
	BR	Steel Submariner	1800/2000
264	TL	Steel Submariner	3000/3500
	BL	Steel Submariner	4000/4500
	BR	Steel Submariner	1800/2000
265	TL	Steel Submariner	1200/1500
266	TR	Steel/gold Submariner	4000/5000
	BL	Steel Submariner	3500/4000
	BR	Steel/gold Submariner	3200/3500
267	BL	Steel Submariner	3500/4000
	CR	Steel Submariner	3000/4000
268	TL	Sea-Dweller prototype	5500/7000
	BL	Comex Submariner	3000/4000
269	TL	Steel Sea-Dweller	1800/2000
	TR	Sea Dweller	2500/2750
	BL	Steel Sea-Dweller	4000/4500
Chapter 13			
274	BL	4272 Kew A Especially Good	7500/8000
	TR	4237 Kew A	6500/7000
275	TL	4257 Kew A	6500/7000
Chapter 14			
278	TR	Steel GMT	2000/2250
	BR	Steel GMT	1500/1800
279	TL	Steel GMT	1400/1600

Page	Pos.	Description	Price range
Chapter 15			
282	TR	18k Day-Date	5000/6000
Chapter 16			
287	TR	Quartz	5000/6000
Chapter 17			
291	TR	9k Egyptian	1750/2000
292	CT	9k Rebberg	600/800
	BL	9k Rebberg	700/900
	BR	15k Lady's	1750/2000
293	BL	Sterling ball watch	2500/3000
294	TL	9k Digital	1500/1800
295	TL	Gold decorative	750/1000
	CT	18k Oyster Perpetual	2500/3000
	TR	Steel/gold hooded	2500/3000
	BL	18k bubbleback	2750/3250
	BR	9k Square deco	1200/1500
296	TL	Platinum dress	1250/1500
	TR	Platinum dress	1000/1200
	CR	Platinum Drummer Boy	8000/10.0k
297	TL	Platinum diamond	7000/8000
298	TL	18k square deco	1000/1200
	TR	18k DateJust	3500/4500
	BR	18k bubbleback	2750/3250
299	TL	18k Top Hat	2000/2500
	BR	Steel/gold Oyster Perpetual	800/1000
300	TL	18k Precision ring	5000/7500
	TR	18k Oyster Perpetual	2000/2500
	CR	18k pink Oyster Perpetual	2000/2500
	BR	18k DateJust	2000/2500
Chapter 21			
332	BR	Silver Tonneau	1500/1800
333	TL	Lady's Perpetual	E 10.0/12.5K
	TR	14k White gold Oyster	E 7500/8500
334		Sweep+subsidiary	4500/5000
335	TL	Silver military	E 5000/6000
	BR	Wilsdorf Oyster	E 2500/3000
336		Steel Bombé 18k rotor	E 12.5/15.0K
337		18k Tudor Auto	4000/4500
338		Franck Muller	E 65.0/75.0K
339	TL	Preziuso	E 65.0/75.0K
	TR	Sterling calendar	E 15.0/20.0K
Chapter 23			
354	TL	Steel DateJust	3000/3500
	TR	Sabot clock	5500/6000
	BL	Steel paper knife	250/300
355	TL	Sabot clock	5000/5500
	TR	Store display	2500/3000
	BR	Fishbowl	5000/6000

Subject Index

Aegler, 10, 12, 77, 79, 111, 112, 152, 289, 315, 328
Aegler, Hermann, 12, 163
Aegler Jean, 10
Aegler S.A. Rolex Watch Co., 12
Aegler, Society Anonyme, Fabrique des Montres Rolex & Gruen Guild, 111
Air-King, 234, 243, 350
Apollo 13, 277, 278
Auto-Rotor, 79, 316, 356

Baumgartner Francis, 41
Bienne, 10, 12, 80, 111, 135, 153, 163, 271
Borer, 151, 152, 159, 285
Borgel, 41
Brancard, 112, 113, 289
Bubbleback, 79, 162, 281, 304, 351, 352, 365

Cartier, Jaques, 41
Chaux de Fonds, 10
chronograph, 138, 193, 215, 352, 367
Chronometer, 77, 112, 151, 162, 163, 192, 195, 216, 241, 242, 243, 255, 281, 285, 315, 317
cloisonné, 195, 332, 352
Comex, 258, 259, 260, 261, 352
Cosmograph, 138, 191, 215, 216, 217
Cuno, Korten, 9

Daily Mail, 43
Datejust, 135, 161, 163, 164, 192, 233, 234, 241, 255, 281, 286, 304, 317, 327, 350, 369
Davis, Alfred James, 10
Day-Date, 192, 233, 281, 281, 282, 286, 305, 317
Daytona, 216, 217, 352, 367
Dennison, 41, 315
Duo-Plan, 112

Egyptian, 160
Everest, 161, 191, 241, 242
Explorer, 151, 192, 241, 242, 243, 244, 245, 255, 290, 305, 317, 328, 350, 352
Extra Prima, 77, 315

Gay Freres, 303, 304, 368
Geneva, 12
Gleitz, Mercedes, 43
GMT Master, 215, 245, 277, 278

Harwood, 78, 79, 331
Hermetic, 42
Huguenin, 281
Hunter, 77, 78, 80, 152, 153, 163, 271, 315, 316

Imperial, 135, 138, 303, 352
Import duty, 12

Jeanneret, 255, 277
Jubilee, 163, 233, 271, 289, 303, 305

Kew, 78, 152, 271, 286, 289, 352

Kodak, 10

Lifesavers, 163

Marconi, 135, 350
Matile, 152, 271
Milgauss, 192, 193, 245, 317, 352

NASA, 151, 277, 278
Neuchatel Observatory, 10

Oyster, 41, 42, 43, 77, 78, 79, 80, 135, 159, 160, 161, 162, 191, 192, 195, 215, 216, 233, 234, 261, 271, 277, 285, 289, 290, 303, 304, 305, 328, 331, 332, 340, 351, 352, 365, 367
Oyster Patent, 42, 345

Paul Newman, 216, 352, 367
Peret, Georges, 42
Perpetual, 43, 77, 78, 79, 80, 81, 112, 135, 151, 160, 161, 162, 164, 192, 233, 234, 289, 290, 303, 316, 350, 351, 352, 369
Perregaux, Paul, 42
Piccard, 255, 256
Practical Watch & Clock Maker, 43, 80
Prima, 77, 315
Prince, 79, 80, 111, 112, 113, 135, 138, 153, 159, 233, 289, 303, 350, 351, 352, 365, 366

Railway, 112, 352
Rebberg, 10, 41, 77, 152, 315 332
Rolco, 135, 350
Rolex Watch Company Ltd, , , 11

Sales promotion, 43
Santos-Dumont, Albert, 11
Sea Dweller, 258, 259, 260, 261, 305
semi-tropical, 42
Space Dweller, 244
Submariner, 138, 191, 192, 193, 244, 255, 256, 258, 259, 260, 261, 277, 290, 305, 332, 350, 352, 369
Super Oyster, 191, 195
Superbalance, 77, 271, 315

Thunderbird, 234
Tiger stripe, 113
Triplock, 191, 255, 258
Tru-beat, 192, 350
Tudor, 135, 138, 244, 258, 286, 331
Turn-O-Graph, 192, 193, 234, 255, 277, 290

Ultra Prima, 77, 315
Unicorn, 135, 350

Wilsdorf, Hans , 9, 41, 113, 135, 151, 159, 163, 315, 327, 328, 329
Wilsdorf & Davis , 9, 10, 11, 41, 111, 151, 289, 304, 315

Zerograph, 215

Index to Model Numbers in Photographs

Case Reference	Page
1	20
4	39
5	339
9	299
20	34
34	32, 37
35	33
73	23, 137
535	31
554	30, 35
678	56
698	48
758	185
765	296
829	133
912	130
951	337
971	115, 123
971A	122, 126
971U	124
1016	249
1045	176
1069	58
1238	291
1370	140
1490	120, 121, 125, 128
1491	118, 120
1500	236
1522	237
1527	131, 133
1601	202, 237, 238
1603	235
1626	236
1655	250, 252
1665	269
1675	279
1680	263
1768	117
1803	282
1862	115, 116
1872	178
1936	180
2081	57
2280	61, 62, 69, 168, 188

Case Reference	Page
2303	217
2317	173
2361	34
2416	62
2420	63
2494	39
2508	218, 227
2518	71
2595	58, 167
2764	84, 87, 88, 108
2849	70
2918	229
2940	87, 91, 93, 95, 100, 108
2942	177
2943	175
3019	62
3055	218, 228, 229, 230
3065	99, 100, 101, 107
3096	59
3116	65, 66, 74, 170, 334
3121	63, 68, 171, 184
3130	83, 108
3131	87, 89, 109, 203
3132	85
3136	140
3139	64, 68, 181
3176	182
3184	218
3260	189
3270	166, 184
3303	295
3348	103
3359	64, 170, 188
3361	133
3372	90, 92, 107, 109
3462	228
3474	68
3478	138
3484	220, 229
3525	218
3529	218, 230
3595	96, 97, 185
3599	100

Case Reference	Page
3612	145
3668	219
3767	103
3877	109
4048	221
4062	220, 229
4220	60, 69, 167, 168, 169
4302	171
4313	229
4392	107
4409	210
4467	180, 235
4486	300
4500	221, 228
4533	208
4537	227
4572	171
4647	60
4768	230, 299
4890	187
4961	107
5003	298
5015	98
5018	336
5026	105, 179
5048	104
5056	166, 169
5500	246, 252
5504	244
5512	265
5513	262, 263, 264, 266, 267, 268
5700	238, 250
6006	178
6036	228
6062	204
6085	201
6090	203, 212
6092	239
6150	245, 246, 249
6200	261
6202	205
6238	222
6241	229

Case Reference	Page
6244	198
6262	224
6263	225
6264	223
6265	225
6266	211
6299	252
6304	230
6309	239
6350	244, 246, 247, 249
6504	295
6508	262
6536	262
6538	261
6541	205, 213
6542	278
6548	200
6556	172
6564	192
6565	194
6567	209
6605	237
6610	245
6619	300
6694	210
6900	300
7247	181
8029	210
8126	108
8171	209
8309	300
8382	202
8437	148
13874	210
16013	238
16520	230
16550	253
16750	278
16803	266
67193	299
69178	298
79090	144
161133	266

Index to Model Numbers in the Text

Case Reference	Page
971	112, 352, 362
1016	242, 352
1019	193, 245, 317, 352
1490	365
1491	352, 365
1625	234
1630	286
1655	352
1662	258
1665	258, 259, 262, 352
1675	277, 278
2240	135, 161
2295	287
2508	352, 367
2540	365
2541	352, 365
3065	161, 162, 352, 365
3121	161
3131	81
3132	81, 352
3361	113, 352, 366
3362	113, 365
3372	81, 162, 304, 317
3462	215
3481	368
3536	352
3591	352
3595	162, 352
3598	162
3599	162, 352
3668	215
3801	352
3937	352, 366
4220	161
4467	233, 304, 352
4487	289, 304
4500	215, 352
4645	195
4768	138
5011	162, 352
5015	162
5018	195
5026	162, 233
5034	215
5056	161, 271
5100	285
5500	243, 352
5504	244, 352
5508	256, 257
5510	244
5512	255, 256
5513	243, 258, 260, 332, 351, 352
5700	244, 351, 352
6034	215
6036	138
6062	191, 352
6084	195
6085	195, 352
6092	195
6150	242, 352
6200	255, 352
6202	194, 255, 277
6204	194, 255
6232	215
6234	215
6238	215, 255
6239	215, 352
6263	215, 216
6266	192
6298	244
6352	242, 352
6466	192, 234
6494	234
6511	281
6536	255, 256
6538	191, 255
6542	277, 352
6556	192
6610	242, 244, 352
6611	281
14270	244
16660	260
16700	278
16710	278
16800	258
17000	286
19018	286

ABRAHAM-LOUIS PERRELET